BROADENING PARTICIPATION IN STEM

DIVERSITY IN HIGHER EDUCATION

Series Editor: Henry T. Frierson

Recent Volumes:

Volume 11: *Beyond Stock Stories and Folktales: African Americans' Paths to Stem Fields* – Edited by Henry T. Frierson and William F. Tate

Volume 12: *Black Female Undergraduates on Campus: Successes and Challenges* – Edited by Crystal Renée Chambers and Rhonda Vonshay Sharpe

Volume 13: *Latino College Presidents: In Their Own Words* – Edited by Rubén O. Martinez and David J. León

Volume 14: *Seeding Success in Indigenous Australian Higher Education* – Edited by Rhonda G. Craven and Janet Mooney

Volume 15: *Maori and Pasifika Higher Education Horizons* – Edited by Fiona Cram, Hazel Phillips, Pale Sauni and Clark Tuagalu

Volume 16: *Black Males and Intercollegiate Athletics: An Exploration of Problems and Solutions* – Edited by Robert A. Bennett III, Samuel R. Hodge, David L. Graham and James L. Moore III

Volume 17: *Infusing Undergraduate Research into Historically Black Colleges and Universities Curricula* – Edited by Jeton McClinton, Mark A. Melton, Caesar R. Jackson and Kimarie Engerman

Volume 18: *The Coercive Community College: Bullying and Its Costly Impact on the Mission to Serve Underrepresented Populations* – Edited by Leah P. Hollis

Volume 19: *The Crisis of Race in Higher Education: A Day of Discovery and Dialogue* – Edited by William F. Tate IV, Nancy Staudt and Ashley Macrander

Volume 20: *Campus Diversity Triumphs: Valleys of Hope* – Edited by Sherwood Thompson

Volume 21: *Underserved Populations at Historically Black Colleges and Universities: The Pathway to Diversity, Equity, and Inclusion* – Edited by Cheron H. Davis, Adriel A. Hilton and Donavan L. Outten

DIVERSITY IN HIGHER EDUCATION VOLUME 22

BROADENING PARTICIPATION IN STEM: EFFECTIVE METHODS, PRACTICES, AND PROGRAMS

EDITED BY

ZAKIYA S. WILSON-KENNEDY
Louisiana State University, USA

GOLDIE S. BYRD
North Carolina A&T State University, USA

EUGENE KENNEDY
Louisiana State University, USA

HENRY T. FRIERSON
University of Florida, USA

emerald PUBLISHING

United Kingdom – North America – Japan
India – Malaysia – China

Emerald Publishing Limited
Howard House, Wagon Lane, Bingley BD16 1WA, UK

First edition 2019

British Library Cataloguing in Publication Data
A catalogue record for this book is available from the British Library

ISBN: 978-1-78756-908-9 (Print)
ISBN: 978-1-78756-907-2 (Online)
ISBN: 978-1-78756-909-6 (Epub)

ISSN: 1479-3644 (Series)

Printed and bound by CPI Group (UK) Ltd, Croydon, CR0 4YY

CONTENTS

ABOUT THE AUTHORS

Sainath Babu, PhD, received his doctorate in Environmental Toxicology from Southern University A & M College (Baton Rouge, Louisiana) and works to execute technical research projects associated with NanoHU. With this center, he has published three book chapters and two conference papers on nanoscience education at HU.

Frank Bayliss, PhD, is Professor of Biology at San Francisco State University. He serves as the current Director of the Student Enrichment Opportunities office that supports the NIH MARC, RISE, BS & MS Bridges, and NSF STC Center grant. He received the Presidential Mentoring Award from President Obama in 2010.

Kelsie M. Bernot, PhD, is an Assistant Professor of Biology, having earned her doctorate in Biochemistry, Cell, and Molecular Biology from Johns Hopkins University. She was awarded the 2018 NC A&T Junior Faculty Teaching Award, and her research focuses on use of high-impact practices to increase the success of historically underrepresented students in STEM.

Teresa Birdwhistell, PhD, is the PI and Program Director for the MARC U*STAR Program at Xavier University of Louisiana. She is a Professor in the Chemistry Department and has served as a research mentor to more than 30 Xavier students. She received her PhD in Inorganic Chemistry from the University of North Carolina at Chapel Hill and her BS in Chemistry from the University of South Carolina.

M. Francis (Fran) Bridges, PhD, Educational Administration, Jackson State Univresity, licensed by the State of Mississippi with a 4A certification in school leadership. As a Retired Educator and community activist she continues to support increasing the representation of minorities in STEM disciplines, culturally responsive teaching and learning, and the role of higher education in fostering a PK-20 seamless education continuum.

Melissa Brocato is Assistant Vice President for Student Success and prior Director of the Center for Academic Success (CAS) at Louisiana State University, an NCLCA National Center of Excellence. She has dedicated her 25-year education career to creating programs and tools that help students learn more effectively and think critically.

Grace E. Byfield, PhD, is Program Manager at the Center for Outreach in Alzheimer's, Aging and Community Health (COAACH), North Carolina A&T State University. She has a PhD in Microbiology, MS in Biology, and BSc in

Botany and has co-authored two other book chapters on URMs in STEM Education in *Diversity in Higher Education.*

Goldie S. Byrd, PhD, is Professor and Director of the Maya Angelou Center for Health Equity at Wake Forest University. She has published over 60 papers and book chapters and in STEM-related journals and has received over US$70 million in STEM-related grants, contracts, and gifts.

Naomi F. Campbell, PhD, is an Associate Professor of Chemistry, Department of Chemistry, Physics & Atmospheric Sciences, Jackson State University, Jackson, MS. Campbell has been active in the POGIL community since 2012. She is a trained POGIL facilitator and has been active in developing and implementing POGIL instruction in the chemistry curriculum.

Eric W. Claville, JD, is the founding director of the Hampton University Pre-Law Institute (HUPLI), graduated over 125 students in pre-law in five years. Graduates were accepted into over 80 law schools with over US$5.5 million in scholarships. With his E4 initiative, he works to increase underrepresented groups in the legal profession.

Michelle O. Fletcher Claville, PhD, is the Principal Investigator for The Nanoscience Project at Hampton University (NanoHU). With over US$4M funding, she has mentored scores of students and faculty from underrepresented groups in STEM fields. Research with her protégés has resulted in 17 peer-reviewed scientific publications and over 50 conference papers.

Dominic P. Clemence-Mkhope, PhD, is a Professor of Mathematics at North Carolina A&T State University, Greensboro, NC. He has been supported by NSF, NASA, and other agencies. His publications include two co-edited conference proceedings: *Mathematical Studies in Nonlinear Wave Propagation* and *Studies on the Dynamics of Human Diseases: Emerging Paradigms and Challenges.* Past honors include Institute for International Education and Victor Rothschild Memorial Fund fellowships.

Roy J. Coomans, PhD, is an Associate Professor, Department of Biology, North Carolina A&T State University, Greensboro, NC, USA. Dr Coomans received a BS in Biology from Eckerd College, St. Petersburg, FL, and a PhD in Biology (Botany) from the University of North Carolina at Chapel Hill.

Tiera S. Coston, PhD, JD, oversees the mentor training program at Xavier. She is a National Research Mentoring Network (NRMN) Trained Facilitator, and supports the teaching and mentoring of faculty through her work in the institution's teaching and learning center. Dr Coston received her PhD in Molecular and Cellular Biology from Tulane University, her JD from Loyola University Law School, and her BS in Biology from Xavier University of Louisiana.

Kathy Cousins-Cooper, PhD, is an Associate Professor at N. C. A&T State University, has spearheaded the implementation of the emporium method at Mathematics Department at N. C. A&T, served as PI for the NSF IUSE grant entitled "Assessing the Impact of the Emporium Model," and worked on several other mathematics educational projects.

Melissa B. Crawford received her bachelor's degree in chemistry, magna cum laude, from Louisiana State University, where she has served as a program manager for the past eleven years. Crawford has mentored more than 350 students, employing holistic approaches to create high-quality STEM ambassadors, who go on to complete graduate degrees.

Ron Darbeau, PhD, is the Dean of the College of STEM and School of Education at University of Arkansas-Fort Smith, is a Fellow of the American Chemical Society (ACS), and serves on the ACS-Committee on Professional Training and the Arkansas contingent of the national *Advancing Coherent and Equitable Systems for Science Education* group.

Maryam Foroozesh, PhD, serves as the Principal Investigator/Program Director of the RISE Program and the Lead PI and Director for the Administrative and Research Enrichment Cores of the BUILD Program at Xavier University of Louisiana. She is an Endowed Professor of Chemistry. To date, she has served as the research mentor to over eighty Xavier students. Dr. Foroozesh received her PhD in Organic Chemistry from Tulane University and her BS in Chemistry from Louisiana State University.

Lisa B. Hibbard, PhD, has taught the general and physical chemistry sequences at Spelman College for over 30 years and is the recipient of several teaching awards. She has served as PI on efforts to introduce innovative pedagogies into courses to improve students' performance and instill in them a love of learning.

Liuli Huang, PhD, is a STEM researcher focused on identifying the patterns in educational learning analytics via interactive visualizations, simulations, and arithmetic models. Her interest is to explore, and research characteristics relevant to STEM learning and practice as to improve students' academic performance.

Kimberly M. Jackson, PhD, works to ensure students have the tools to become effective problem solvers and critical thinkers in a global society. Utilizing best practices, she has created programs to advance the careers of black women scientists. She co-founded Spelman College's Food Studies program and a living-learning community of STEM scholars.

Jennifer Jennings Davis, PhD, is Executive Director of the University of Arkansas Fort Smith's Office of P-20 Collaboration, housing the STEM Education Center and Education Renewal Zone. Both programs are grant funded through the Arkansas Department of Education and focus on providing support to regional schools, administrators, teachers, and students.

A. Alegra Eroy-Reveles, PhD, is an Assistant Professor of Chemistry and Biochemistry at San Francisco State University, and designs and implements interventions to motivate students to persist in STEM majors. As a Latina chemist and SACNAS Leadership Institute fellow, she works to expand the number and diversity of students pursuing careers in science.

Sayo O. Fakayode, PhD, is Department Head, Department of Physical Sciences, University of Arkansas Fort Smith, and uses new pedagogical innovations and inquiry learning to promote critical thinking and to generate student's interest in STEM. Dr Fakayode has received numerous National Science Foundation grant awards to support his fundamental research and STEM education.

Marguerite Giguette, PhD, is the PI and Director for the Institutional Development Core of the BUILD Program at Xavier University of Louisiana. She is a Computer Scientist and serves as Associate Vice President for Academic Affairs. Dr Giguette received her PhD in Computer Science from Tulane University and her MS in Mathematics from Texas A&M University.

Stephanie Givens, MFA, secured funding for the first Upward Bound at Louisiana State University in 2012 and has since grown the program to serve twice as many students. Her innovative practices in partnerships with the LSU campus, local schools, and the community have garnered national attention. She trains other professionals in collaborative best practices.

Georgia Hale, PhD, is the Provost & Vice Chancellor Academic Affairs, University of Arkansas Fort Smith, has authored several journal articles, and is a Co-PI of a grant that includes several higher education institutions in Arkansas and Oklahoma. Dr Hale has severed as a mentor for young women and members of underrepresented groups.

Emorcia V. Hill, PhD, Independent Evaluation Consultant, is the external evaluator for NanoHU. Her work spans the educational continuum and into the workforce. It involves multisites, and employs quantitative and qualitative methodologies. Her evaluation, research, and publications are framed within a multilevel, capacity-building, systems approach, and address diversity, access, inclusion, social networks, scholarship, academic progression, and institutional transformation.

Christine Hohmann, PhD, is a Professor of Biology at Morgan State University and serves as the Director of ASCEND's Research Enrichment Core. Dr Hohmann is an accomplished researcher and has served as the director of several other NIH-sponsored training programs, including NIGMS MARC and NIGMS RISE.

Linda M. Hooper-Bui, PhD, is an Associate Professor of Environmental Sciences and Director of Strategic Initiatives at Louisiana State University. She is interested in mentoring, student persistence, citizen science, resilience of ecosystems to disturbance, and responses of communities to natural and technological disasters.

Eric Hsu, PhD, is Professor of Mathematics at San Francisco State University. He has led numerous STEM education projects, including SF CALL, an NSF INCLUDES project, and four California Math Science Partnerships. He received an NSF CAREER Award for his work on college math teaching.

Kelly Johanson, PhD, is an Associate Professor of Chemistry, and is the Student Training Core Co-Director on the BUILD program at Xavier University of Louisiana. She has mentored over a dozen undergraduate students in her research laboratory including those who began as freshmen and sophomores with no research experience. Dr Johanson received her PhD in Biochemistry from Tulane University, and her BS in Biochemistry from Beloit College.

Shanina Sanders Johnson, PhD, is a proponent of engaging students and providing them with skills to nurture a lifelong pursuit of knowledge. She is also dedicated to mentoring minority women as they pursue STEM degrees. She serves as PI on projects related to authentic learning and culturally relevant curricula.

Farin Kamangar, MD, PhD, is a Professor of Biology at Morgan State University and the Principal Investigator of ASCEND, an NIH-sponsored BUILD program to enhance the diversity of the biomedical research workforce. He has been a teacher and researcher for the past 25 years and has published over 185 peer-reviewed papers.

Margaret Kanipes, PhD, Professor of Chemistry, serves as the Honors Director and Director of the STEM Center of Excellence for Active Learning at NCA&T State University. She has published numerous papers in STEM-related journals and has garnered over seven million dollars in grants as PI/Co-PI.

Seong-Tae Kim, PhD in Statistics, has been conducting research on student performance in college STEM education including a comparison of instructional methods, the impact of socioeconomic status, and student attitudes and behaviors via developing and applying statistical methods. His work has been published and supported by NSF grants.

Heather Lavender, MS, is the Education and Outreach Manager for the Consortium of Innovation for Manufacturing and Materials. Heather is currently finishing a doctoral degree at Louisiana State University with a focus of the science identity and interest of African American girls up to fifth grade.

Nicholas S. Luke, PhD, is an Associate Professor at North Carolina Agricultural and Technical State University. He received his PhD in Computational Applied Mathematics from North Carolina State University. He is a co-author of the book *Mathematical Modeling: Branching Beyond Calculus.*

Saundra McGuire, PhD, is the Director Emerita of the Center for Academic Success and Retired Professor of Chemistry at Louisiana State University She is an internationally renowned expert in the area of learning support and is the author of *Teach Students How to Learn* and *Teach Yourself How to Learn.*

Shiva Mehravaran, MD, MIH, joined ASCEND as the lead evaluator in 2017. She is a dedicated research scientist with over 15 years of experience in the design and implementation of various types of research. She has published over 60 health-related articles and serves as Editor and Peer reviewer for scientific journals.

Paulette Ann Meikle, PhD, is an Associate Dean/Professor of Sociology, College of Communication, Languages, Arts, and Social Sciences, University of Arkansas, Fort Smith. Dr Meikle's research agenda involves integrating aspects of resource management, asset accumulation, women's advancement, gender equality, and community. Dr Meikle has published in several academic journals and with Sage and the Nova Science publishers.

Kathleen Morgan, PhD, serves as the PI and Director of the Student Training Core of the BUILD Program at Xavier University of Louisiana. She is an Endowed Professor of Chemistry and has served as the research mentor to over forty Xavier students. Dr. Morgan received her PhD in Organic Chemistry from Yale University and her AB in Chemistry from Dartmouth College.

Robert H. Newman, PhD, is an Associate Professor of Biology at NC A&T, earned his PhD in Biochemistry and Molecular Biology from Johns Hopkins University. As an ASBMB, ACS, and SEA-PHAGE consortium member, Rob is dedicated to providing authentic research experiences inside and outside the classroom and promoting persistence of URMs in STEM.

Brandon C. Parker is the Program Manager of NanoHU. He received his BS in Chemistry from Southern University and A & M College (Baton Rouge, Louisiana) and MS in Educational leadership from Argosy University (Sarasota, Florida). A PhD candidate in Higher Education Administration, Mr Parker is interested in educational policies associated with interdisciplinary STEM programs.

Michelle Penn-Marshall, PhD, is the Co-principal Investigator/Project Director for the Hampton University Biomedical Research Training Initiative (HUBiRTI) and the Project Coordinator for the Washington-Baltimore Hampton Roads – Louis Stokes Alliances for Minority Participation Program (WBHR-LSAMP). She has trained several students to conduct obesity prevention research in children from rural communities.

Alan R. Peterfreund, PhD, is the Director of SageFox Consulting Group. He has been the lead evaluator for the Student Enrichment Opportunities program since 2002 and was Co-principal Investigator for an NIH-funded research project to study the training and workforce development programs at three high-performing institutions that ran from 2004 to 2010.

Angela W. Peters, PhD, is Vice Provost of Academic Programs and Professor of Chemistry at Claflin University, Orangeburg, SC. She has published previously in Diversity in Higher Education and has received numerous awards, citations, and publications in STEM education. She serves on many advisory boards and councils governing STEM.

Cailisha L. Petty, PhD, is the MAT and Biology Education Coordinator at NC A & T, having earned her doctorate in Teacher Education and Development at the University of NC at Greensboro. She is committed to equity of access in science education for students of color, specifically girls and women, across the spectrum of academic arenas.

Kenneth A. Rath is an educational evaluator at SageFox Consulting Group and has been part of the team evaluating the Student Enrichment Opportunities programs since 2002. His professional work is primarily focused on the evaluation and research of STEM initiatives at the K-12 and higher education levels.

Thomas C. Redd, PhD, is an Associate Professor of Mathematics and an active member of the University General Education Council and the new UNC System Math Pathways project. He has been active with funded initiatives (NSF-BLEND, NSF-iBLEND, NSF-Talent-21, Burroughs Welcome Fund) directed toward broadening participation in STEM disciplines, authoring several articles.

Melissa S. Reeves, PhD, is an Associate Professor of Chemistry at Tuskegee University where she has taught Physical Chemistry since 1995. She is an active member of the POGIL Physical Chemistry Laboratory project and serves on the American Chemical Society Physical Chemistry Exam Committee.

Lahna Roche, is an LPC with 18 years of professional experience in education focusing on student development, particularly metacognition, learning strategies, and skills training. She's worked with diverse student populations including administration in a TRIO program and has published works focusing on metacognition, academic support, learning centers, and college completion.

Checo J. Rorie, PhD, is an Associate Professor, Department of Biology, NC A&T. He received a BS in Biology from Clark Atlanta University and a PhD from UNC-Chapel Hill's Curriculum in Toxicology. He completed a postdoc in Biochemistry from New York University and completed the SPIRE Postdoctoral Program from UNC-Chapel Hill.

Payam Sheikhattari, MD, MPH, is Professor of Public Health at Morgan State University and the Co-principal Investigator of ASCEND. He has substantial experience directing student research centers, and currently directs ASCEND's Student Research Center. Dr Sheikhattari's research focuses on tobacco use in underserved communities.

Gillian B. Silver, MPH, CPH, is the Program Manager of the Morgan State ASCEND Initiative. Her experience includes work at the Association of Schools of Public Health on the W. K. Kellogg Foundation-funded "Affirming a Commitment to Eliminating Health Disparities," the public health workforce crisis, and the undergraduate public health initiative.

Mary A. Smith, PhD, is Professor and Chairperson of Biology at NC A&T, having earned her doctorate in Plant Physiology from Cornell University. Committed to undergraduate STEM education reform and URMs success, she is a GEP consortium member, Partnership for Undergraduate Life Sciences (PULSE) Leadership Fellow and Director of the Maximizing Access to Research Careers MARC program.

Faye E. Spencer-Maor, PhD, is Associate Professor of English, at North Carolina A&T State University in Greensboro, North Carolina, and received a BS in Journalism from Florida A&M University, MA in Journalism from the University of Colorado at Boulder, and PhD in English/Writing Studies from the University of Illinois Champaign-Urbana.

Derrick J. Swinton, PhD, is Chair and Professor of Chemistry at Claflin University, Orangeburg, SC. He has years of experience developing academic programs and initiatives that promote student achievement. He is an adamant believer in creating and implementing learning programs for the advancement of science and broadening STEM participation.

Guoqing Tang, PhD, Professor and Chairman of the Mathematics Department at NCA&T State University, has published over 32 papers in refereed journals and conference proceedings. His research interests lie in the areas of dynamical control systems, scientific computing, and mathematics education. He has received over US$12 million in STEM-related grants.

Gloria Thomas, PhD, Director of the Louisiana State University Center for Academic Success, has had academic training and faculty appointments in chemistry departments at a wide variety of educational institutions. These experiences have provided valuable insights into teaching and learning, diversity and inclusion, higher education trends, and science education.

Verlie A. Tisdale, PhD, is the Dean of the School of Natural Sciences and Mathematics and a Professor of Biology at Claflin University, Orangeburg, SC. She has previously published in Diversity in Higher Education. She serves on college, community, academic, and professional boards, and holds membership in several professional organizations.

Marilyn Tourné, PhD, is an Assistant Professor of Chemistry at Tuskegee University where she has taught General Chemistry since 2012. She has used POGIL instruction in her classrooms since 2014 and will attend her first POGIL Regional Meeting in 2018.

Isiah M. Warner, PhD, is Boyd Professor of LSU System and Vice President of Louisiana State University. His research expertise is in fluorescence spectroscopic analyses, a field in which he is considered a world's expert. He also conducts educational research, focusing on maintaining and enhancing student education in science, technology, engineering, and mathematics (STEM) and encouraging students (particularly underrepresented) to pursue terminal degrees in STEM.

Shannon Watt, PhD, combines her training in science (PhD in Chemistry, Georgia Institute of Technology) with her interests at the intersection of STEM, higher education, and broadening participation. Before serving as Bridges to the Baccalaureate Program Manager at Louisiana State University, she founded diversity-focused programs at the Georgia Institute of Technology and the University of Michigan, taught at Xavier University of Louisiana, and managed research and programs for the Open Chemistry Collaborative in Diversity Equity (OXIDE).

Angela M. White, PhD, is a Teaching Assistant Professor of Biology and Coordinator of Science Initiatives, North Carolina A&T. She earned her PhD in Curriculum and Instruction with a concentration in Educational Psychology from North Carolina State University. She was awarded the 2018 NC A&T Faculty Advising Excellence Award.

C. Dinitra White, PhD, is an Associate Professor, Department of Biology, North Carolina Agricultural and Technical State University. She received a BS in Biology from Johnson C. Smith University and PhD in Microbiology and Immunology from Wayne State University School of Medicine. She is the Director of the NC A&T Pre-Professional Scholars Program.

Clair Wilkins-Green, PhD, is based in the Office of Planning, Institutional Research and Assessment at Xavier University of Louisiana. She is involved in the assessment of all three student training programs discussed in this chapter (BUILD, MARC, and RISE). Dr Wilkins Green received her PhD in Politics from Manchester Metropolitan University and her MA in Political Science from Louisiana State University.

Tyrslai M. Williams, PhD, is the Assistant Director for Strategic Initiatives at LSU. Her primary focus is to provide all students access to higher education in STEM through the conception of intentional programs. Her interests are mentoring, broadening participation, informal science learning, and interpersonal and holistic development of students at all levels.

Zakiya S. Wilson-Kennedy, PhD, is an Associate Professor of Research in Chemistry Education and Assistant Dean for Diversity and Inclusion within the College of Science at Louisiana State University. Her research is published in peer-reviewed journals, such as the *Journal of Science Education and Technology* and the *Journal of Chemical Education*.

Leyte L. Winfield, PhD, has provided advocacy and leadership to enrich and expand the academic environment at Spelman College. Her scholarship is based on a dedication to creating culturally responsive initiatives that promote gender equity in academia and science careers. She has served as the Principal Investigator on several related efforts.

Angela Winstead, PhD, is a STEM educator and active researcher at Morgan State University. Se uses her education and professional experience to encourage the development of student growth and to inspire and empower young scholars to reach their highest potential.

Linus Yu, PhD, is the Department Head of Mathematics at the University of Arkansas Fort Smith, having received his PhD from SUNY at Buffalo, NY in 2008. Under his leadership, the Mathematics faculty have increased publications and presentations by 40% and have received numerous grants.

PREFACE

US students exit undergraduate science, technology, engineering, and mathematics programs at alarming rates (Committee on Prospering in the Global Economy of the 21st Century (U.S.) & Committee on Science Engineering and Public Policy (U.S.), 2007; Committee on Prospering in the Global Economy of the 21st Century (U.S.) & Committee on Science Engineering and Public Policy (U.S.). 2010; National Academy of Sciences, National Academy of Engineering, & Institute of Medicine, 2011). In fact, less than 50 percent of US undergraduates who enter STEM degree programs as aspiring freshmen complete degrees in these areas. This is especially true for minorities, whose departure from STEM degree programs is often twice the rate of Caucasian and Asian students (National Academy of Sciences et al., 2011). As the US population becomes increasingly diverse, the underutilization of US human resources endangers the long-term economic health of the nation.

In 2012, the President's Council of Advisors on Science and Technology (PCAST) issued a clarion call to increase the number and quality of STEM graduates by 1,000,000 (President's Council of Advisors on Science and Technology, 2012). This body of education, industry, and government leaders also advocated for the broad-based adoption of innovative pedagogies to increase student success in STEM degree programs, with an emphasis on increasing the participation of women, minorities, and other UR groups, who they posited would benefit the most from these innovations.

Higher education administrators and educators grapple with how best to transform educational practices in STEM. Noting that diversity is a lever for innovation, this book shares best practices that embody the principles of *inclusive excellence* within STEM. Herein, the dissemination of best practices, adaptation of national models (such as POGIL, peer-led team learning, SCALE-UP, Emporium learning, etc.) for minority populations, and other approaches will contribute to national dialogue on what works. Accordingly, this book provides roadmaps for universities and programing seeking to expand their capacity for advancing student success among groups historically underrepresented in STEM disciplines.

The collective works featured in this book illustrates the development and implementation of high-impact educational practices and programs that have been demonstrated to be effective at broadening the participation of underrepresented groups in the STEM disciplines. For each initiative, the authors describe the origins and structure of the practice or program, the philosophical and theoretical underpinnings, and the institutional context wherein the program has been developed. The authors also summarize evidence of effectiveness and describe implications for local practice. In each chapter, the goal is to provide the reader with an understanding of the innovation and effort sufficient to lead to informed implementation at the local level. Accordingly, the book seeks to provide campus-based faculty, administrators, and diversity professionals with a guide that can be used to develop programs designed to address specific student success and inclusion goals.

REFERENCES

Committee on Prospering in the Global Economy of the 21st Century (U.S.) & Committee on Science Engineering and Public Policy (U.S.). (2007). *Rising above the gathering storm : energizing and employing America for a brighter economic future*. Washington, DC: National Academies Press.

Committee on Prospering in the Global Economy of the 21st Century (U.S.) & Committee on Science Engineering and Public Policy (U.S.). (2010). *Rising above the gathering storm revisited: Rapidly approaching category 5*. Washington, DC: National Academies Press.

National Academy of Sciences, National Academy of Engineering & Institute of Medicine. (2011). *Expanding underrepresented minority participation: America's science and technology talent at the crossroads*. Washington, DC. Retrieved from http://www.nap.edu/catalog/12984.html

President's Council of Advisors on Science and Technology. (2012). Engage to excel: Producing one million additional college graduates with degrees in science, technology, engineering, and mathematics. Retrieved from http://www.whitehouse.gov/sites/default/files/microsites/ostp/pcast-executive-report-final_2-13-12.pdf

PART I

LSU OFFICE OF STRATEGIC INITIATIVES: A GREAT EQUALIZER FOR BROADENING PARTICIPATION IN STEM

Tyrslai M. Williams, Melissa B. Crawford, Linda M. Hooper-Bui, Stephanie Givens, Heather Lavender, Shannon Watt and Isiah M. Warner

ABSTRACT

Louisiana State University (LSU)'s Office of Strategic Initiatives (OSI) is an award-winning office devoted to developing effective, educational approaches that incorporate guidance and exploration, increase students' academic standing, and support measures to improve the institution's diversity, predominantly in science, technology, engineering, and mathematics (STEM) departments. Through the incorporation of three main factors, Mentoring, Education, and Research, *OSI has developed a holistic development model that offers students strategies to overcome those factors that affect their persistence in STEM. OSI houses several programs with a diverse population of students ranging from the high school to doctoral levels. Although varied in student population, these programs unite under the holistic development model to provide support and opportunities to students at each critical educational juncture. OSI's holistic approach has successfully supported over 135 high school, 560 undergraduate, and 100 graduate students. Of the 560 undergraduate students served, 51% were underrepresented minorities and 55% were women. The undergraduate initiatives have garnered 445 bachelor's degrees, with 395 degrees from STEM disciplines, and an impressive overall graduation rate ranging from*

Broadening Participation in STEM: Effective Methods, Practices, and Programs
Diversity in Higher Education, Volume 22, 3–34

ISSN: 1479-3644/doi:10.1108/S1479-364420190000022001

64% to 84%. Through all of the remarkable work performed in OSI, the greatest accomplishment has been the capacity to offer students from mixed backgrounds tools and strategies to thrive at any point in their academic career.

Keywords: Underrepresented; holistic development; mentoring; STEM; research; graduation rate

INTRODUCTION

National Need to Broaden STEM

"The United States takes pride in the vitality of its economy, which forms the foundation of our high quality of life, our national security, and our hope that our children and grandchildren will inherit ever-greater opportunities [...] Without a renewed effort to bolster the foundations of our competitiveness, we can expect to lose our privileged position" (Sciences, Engineering, & Medicine, 2007). This staggering excerpt from the original *Rising Above the Gathering Storm: Energizing and Employing America for a Brighter Economic Future* 2005 report was meant to raise awareness of the imminent threat to America's global competitiveness. The report provided 20 definitive actions designed to safeguard our nation's competitive edge. A follow-up report was prepared five years later, which revisited and updated prior findings. Sadly, the contributors unanimously determined that the outlook of the United States has worsened by considerable measure (Sciences, Engineering, & Medicine, 2010). Two key recommendations include investing in basic scientific research and strengthening our public school system. Federal government research and development (R&D) funding as a fraction of the Gross Domestic Product decreased by 36% from 1964 to 2009 (Federal R&D was 1.92 percent of GDP in 1964 and .74 percent of GDP in 2004). Improvements to our *overall* public school system are minimal. While other countries are rapidly progressing, the US educational system is steadily declining, especially in science and mathematics (Hoffman, 2009). In the 2015 Programme for International Student Assessment (PISA) results, the United States ranked 24th in science and 38th in math out of 71 countries (Desilver, 2017). Every three years, PISA measures the math, science, and reading ability and additional skills among 15-year-olds from developed and developing nations. Additionally, in the 2015 National Assessment of Educational Progress (NAEP), math scores of both fourth- and eighth-graders were lower compared to 2013, and only 40% of fourth-graders and 33% of eighth-graders perform at or above the proficient level in math (Progress, 2015). This gross lack of preparation for science, technology, engineering, and mathematics (STEM) undergraduate degrees directly impacts the rate at which students pursue and persist in these fields, particularly for women and underrepresented minority students (URMs). Overall, women received a higher percentage of bachelor's degrees than men across all racial/ethnic groups (57% vs. 43%, respectively). However, this pattern is not observed in STEM, as only 35% of women versus 65% of men completed STEM degrees (Statistics, 2017). This trend is similar for URMs, who receive STEM degrees at lower rates than their White or Asian counterparts (Statistics, 2017). Thus, the prevailing need for broadening participation in STEM is urgent, as more than half of the children in the US are projected to be part of a minority race/ethnic group by 2020. By 2044, the US Census Bureau expects that more than half of all Americans will belong to a minority group, and by 2060, 20% of the nation's population will be foreign-born. This epidemic, coupled with the projected rapid demographic shift in our nation (Bureau, 2015), has diminished the global competitiveness of the United States and has contributed to our inability to adequately meet our needs through scientific and technological advancement (Sciences et al., 2010). Within higher education, the need for increasing retention has been recognized; however, sustainable efforts that impact institutions overall have been futile.

LSU's Need for Broader Participation

In Louisiana, the state of education is dire. Ranked 49th nationally in terms of education, Louisiana's high school graduation rate of 77.5% is below the national average and only 34% of high school graduates are college ready. As the state's flagship institution, the vision of Louisiana State University (LSU) is to collectively enhance the dissemination of knowledge by increasing student access and success and improving the quality of life for citizens of Louisiana through teaching, research, healthcare delivery, and public service (Administration, 2018).

LSU's mission is to foster first-class learning, the discovery of innovations, and the development of Louisiana's human capital by applying for research and scholarship in advancing intellectual, personal, and professional growth. LSU is one of a few universities nationally to be designated as land-grant, sea-grant, and space-grant, supporting its role of being a leading research-extensive university.

In support of its vision and mission, LSU has made great strides toward more fully representing the demographics of the nation and state with several priorities. Within LSU's Flagship Agenda 2025, *Leading Louisiana. Impacting the World*, there is a commitment to celebrating cultural awareness, catalyzing transformation, fostering collaboration, nurturing ingenuity, and promoting innovation.

As a result, these efforts have led to a steady increase in the diversity of LSU's student population. As represented in Fig. 1, the university's record-setting 2015, 2016, and 2017 spring graduation classes each included the most degrees ever awarded to African American students in any commencement ceremony in the university's history (Planning, 2017). These classes also include record-breaking numbers of Hispanic or Latino students and women. Additionally, the number of students from underrepresented groups who received doctoral degrees increased.

While LSU has made good progress toward recruiting a diverse, qualified cadre of students, the institution does not have a long tradition of recruiting minority students, as the first African American student was not admitted until 1953. Over time, LSU's increasing commitment to diversity and its recruitment

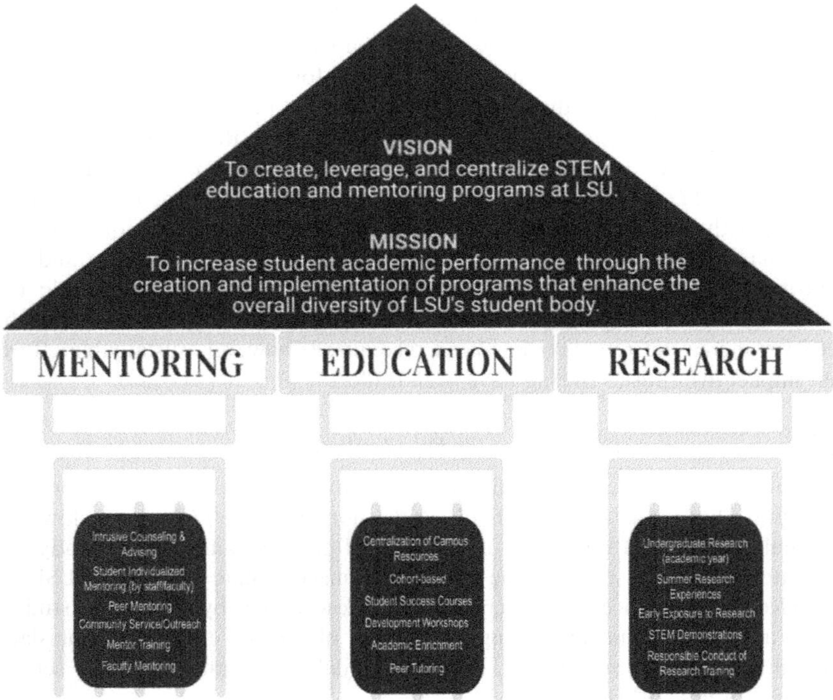

Fig. 1. OSI's Holistic Development Model.

of underrepresented students attracts many African American students. *In fact, the fall 2018 class of incoming freshmen is both the largest and most diverse in our university's history, with more than 40% of the incoming class of STEM majors being a member of a historically underrepresented group.* Many ongoing efforts are also underway to understand the culture and climate at LSU and implement changes to advance inclusivity across the campus and community. While efforts have been developed by various academic and non-academic units, the LSU Office of Strategic Initiatives (OSI) has primarily served as a center focused on identifying barriers and solutions to increase inclusivity for advancing all students, particularly women and URMs in STEM disciplines.

THE LOUISIANA STATE UNIVERSITY OFFICE OF STRATEGIC INITIATIVES

The Vision and Mission of the Office of Strategic Initiatives

The OSI is located on the campus of LSU and houses a variety of diverse programs in STEM for students at the high school, community college, undergraduate, and graduate levels. This unit is led by Dr Isiah M. Warner, Vice President for Strategic Initiatives, and Dr Guoqiang Li, Associate Vice President. In 2001,

OSI was formalized with a vision *to create, leverage, and centralize STEM education and mentoring programs at LSU.* Through its many initiatives, this unit actively supports the educational research and scholarly productivity of faculty and the preparedness and competitiveness of graduate and undergraduate students. Specifically, OSI mission is focused on *developing successful, educational models that integrate mentoring and research, raise students' academic achievement, and support efforts to improve campus diversity, particularly in STEM disciplines.*

The strategies employed by OSI are plentiful; however, the largest strategic impacts include, but are not limited to (1) developing new mentoring activities, (2) providing funding for incoming LSU students, (3) coordinating efforts to pursue and secure education/training grants to support undergraduate students, (4) establishing a close alliance with surrounding universities (e.g., Southern University, Baton Rouge Community College, etc.), (5) supporting K–12 education, teachers, and students, and (6) increasing the number of external awards for students. Over the past 17 years, OSI has surpassed its initial expectations and garnered recognition across the university as an innovative entity designed to support and promote a large number of student-oriented programs at LSU.

OSI's Holistic Development Model

Underrepresented students are often faced with factors that impact their persistence at multiple stages of their academic journey. Factors such as poor self-image, lack of pre-college training, minimal financial support, and the absence of social integration directly impact their determination to succeed academically, particularly at the undergraduate level (Braxton, Hirschy, & McClendon, 2004; Hossler, Ziskin, Gross, Kim, & Cekic, 2009; Locks, Hurtado, Bowman, & Oseguera, 2008; Seidman, 2005). One strategy that has been leveraged by many is the use of cohort experiences and/or learning communities as educational models. These models allow students from similar fields to work collaboratively, engage in peer mentoring, build social and academic networks, and expand team-working skills to ignite self-efficacy and determination. As a result, increased retention, graduation, and success rates have been noted (Lei, Gorelick, Short, Smallwood, & Wright-Porter, 2011). As such, OSI works to engage students in its programs through cohorts by using a holistic development model.

OSI's holistic development model is rooted in the Social Cognitive Career Theory (SCCT; Lent, Brown, & Hackett, 2002), which has been used considerably in research as a method to examine career development in STEM disciplines (da Silva Cardoso et al., 2013; Fouad & Santana, 2017; Hardin & Longhurst, 2016; Herrera & Hurtado, 2011; Kier, Blanchard, Osborne, & Albert, 2014; Lent et al., 2001, 2013). Based on Bandura's Social Cognitive Theory (Bandura, 2001), SCCT has been used to clarify cognitive and motivational activities that impact psychosocial behaviors relevant to STEM education and careers, including academic and research performance. Self-efficacy is the grounding principle within the SCCT model and suggests that an individual's interest (e.g., in a research career) is directly influenced by their confidence in their ability to perform relevant tasks (i.e., student self-efficacy) and in their convictions about the feasibility of doing so (i.e., student expected outcomes).

Looking closely at the SCCT model, the intersection of interests, performance, and outcome expectancy influence perceived career effectiveness and the motivation to persist (Bakken, Byars-Winston, & Wang, 2006; Lent & Brown, 2013). In addition to self-efficacy, other variables, including science-identity, are reliable indicators of persistence in a science career (Byars-Winston, Estrada, Howard, Davis, & Zalapa, 2010; Estrada, Woodcock, Hernandez, & Schultz, 2011). Through our development model, students from underrepresented groups have engaged in high-impact experiential learning experiences that *include holistic mentoring, early exposure to research, and academic enrichment to ensure the success of all students* (Fig. 1).

OSI's Three Pillars of Success

Though multiple factors contribute to the success of underrepresented students at critical junctures, OSI has determined that three important factors, *Mentoring, Education*, and *Research*, support the targeted increase in academic success. These factors serve as the pillars of the holistic development model employed by OSI in developing and sustaining programming. When the three pillars are combined, they provide the vital holistic experience URMs need to overcome those factors that affect their persistence in STEM.

Mentoring Pillar

The mentoring pillar of the development model is the most widely used pillar of the model but also the most crucial measure for student success. The mentoring relationship between a mentor and protégé is an interaction that supplies support and insight on various topics that can make or break the success of the protégé. Students who are mentored at the undergraduate level favor increased grade point averages (GPAs) and higher retention rates, graduate within shorter time frames, and enter graduate programs at a higher rate (Salinitri, 2005; Wilson et al., 2012). OSI acknowledges these effects and employs mentoring in an intrusive, peer, faculty/staff, and tiered manner (Blake-Beard, Bayne, Crosby, & Muller, 2011; Campbell & Campbell, 2007).

On a campus as large as LSU's, student contact is very limited and often requires a more direct or *intrusive approach* in supporting and mentoring students. Mentoring through an intrusive approach requires your program leadership team to acquire an active effort in pursuing and maintaining communication and involvement of students. This mentoring approach allows the mentors to go above and beyond in engaging students in program activities and campus resources. Using this method, mentors are required to initiate contact with students, continually extend support, and oftentimes inquire about setbacks and potential difficulties the students may encounter. This level of mentoring requires every interaction to be strategic and intentional with students in order to catalyze an increase in students' academic performance (Clark, 2006).

An equally important and often overlooked style of mentoring that is critical to the success of students is *peer mentoring*. In particular, those students who have not become acclimated to the university setting benefit from an additional support system such as peer mentoring. Studies show that peer mentoring has a positive effect on the retention and persistence of all students, especially those who are underrepresented (Astin & Astin, 1992; Gregerman, Lerner, von Hippel, Jonides, & Nagda, 1998; Hathaway, Nagda, & Gregerman, 2002; Walters, 1997). In recognizing the value of peer mentors, OSI uses this type of mentoring in an effort to complement faculty support in guiding students toward academic success. Peer mentors create an inclusive, personal learning environment for students that is essential for professional and personal development. In adapting and sustaining successful college student behaviors, peer mentors are there to also instill a supportive relationship. This dynamic, reciprocal relationship develops over time and benefits both the mentor and protégé (Allen, Russell, & Maetzke, 1997). Over the years, OSI has found that the peer mentor relationship reinforces group connectivity and provides positive leadership opportunities for both mentor and protégé (Crawford, Wilson-Kennedy, Thomas, Gilman, & Warner, 2018).

Faculty and staff mentoring combined provides a strong foundation and support system for student academic advancement. When faculty and OSI staff engage in a mentoring relationship with students, the objective is to heighten their experiences and academic success. When paired with any of the styles mentioned above, the mentoring relationship becomes a *tiered mentoring approach* (Wilson et al., 2012).

OSI developed this holistic mentoring model to address the main factors that impact STEM degree completion. Through this model, OSI has acknowledged that one type of mentoring style may not work best for all students, and often, a culmination of mentoring approaches is necessary for underrepresented students. *At the core of this model is the philosophy that any student, who has access to necessary support structures, is motivated to embrace new learning paradigms and willing to embark on new experiences, can become successful.*

At the heart of the mentoring pillar is a focus on accelerating students' cognitive growth and encouraging engagement in mentoring incoming students. This mentoring approach heightens the efforts of faculty, permeates the institutional culture, and prepares the next generation of innovative leaders in the STEM workforce.

Education Pillar

The educational pillar focuses on the advancement of students. It largely speaks to the efforts that OSI employs to enrich the academic standing of students through programming, seminars, and workshops. This strategy is designed to support the typical classroom learning environment. OSI's education pillar heavily involves the gathering of campus resources including but not limited to supplemental instruction (SI) sessions, LSU's Center for Academic Success (CAS) workshops, and campus tutoring in order to centralize the best resources for student success. OSI collaborates with CAS to help students identify their personal learning style, use metacognitive learning strategies, and develop successful habits and behaviors (Wilson et al., 2012).

OSI offers its educational components to students in multiple methods: workshops, seminars, small group activities, one-on-one sessions, and *Success Courses*. Of these practices, the most widely used strategy was a series of *undergraduate success courses*. These courses were designed to educate students in a flexible, adaptive method that would benefit them not only academically but also professionally. Workshops and seminars offered were leveraged to provide the ultimate academic enrichment experience for students by soliciting campus departments such as CAS, Campus Life, Career Services, and Student Health Center along with the OSI staff engaging presentations to provide a holistic learning experience for its students.

Research Pillar

The last pillar of OSI's Holistic Development Model is the research pillar. Undergraduate research experiences have been shown to have a strong correlation to the academic enrichment of undergraduate students (Lopatto, 2004, 2007; Seymour, Hunter, Laursen, & DeAntoni, 2004). Those who do engage in research, benefit through improved persistence, enhanced motivation to learn, increased self-efficacy, and a heightened sense of independence. Additionally, literature has shown that students entering into research in earlier years (freshman or sophomore) are more likely to have an enhanced undergraduate educational experience and pursue a graduate degree (Crawford, Wilson-Kennedy, Thomas, Gloria, Gilman, & Warner, 2018; Stanford, Rocheleau, Smith, & Mohan, 2017). While this finding can be linked to *all* students (STEM and non-STEM), *underrepresented students*, who are exposed to active learning such as undergraduate research earlier on in their collegiate career, tend to have increased academic success. Accounting for these benefits, OSI has developed and promoted student-focused seminars and workshops that inform students of the benefits of performing research, how to properly secure research, and how to handle their research ethically. OSI has developed partnerships with faculty interested in providing avenues for students to conduct research during the academic year and summer including intramural, extramural, and international opportunities.

OSI's Initiatives

Over the years, OSI has housed eight programs at a steady state. Currently, the office houses four educational programs that range from the high school through the doctoral levels (Fig. 2). These students are diverse in terms of ethnicity, gender, and socioeconomic status. They also differ in affinity groups such as STEM disciplines, sexual preference, and religious standing. This "mixed" cohort experience acquired by each program offers students the opportunity to develop an inclusive community equipped with opportunities for collaboration across the

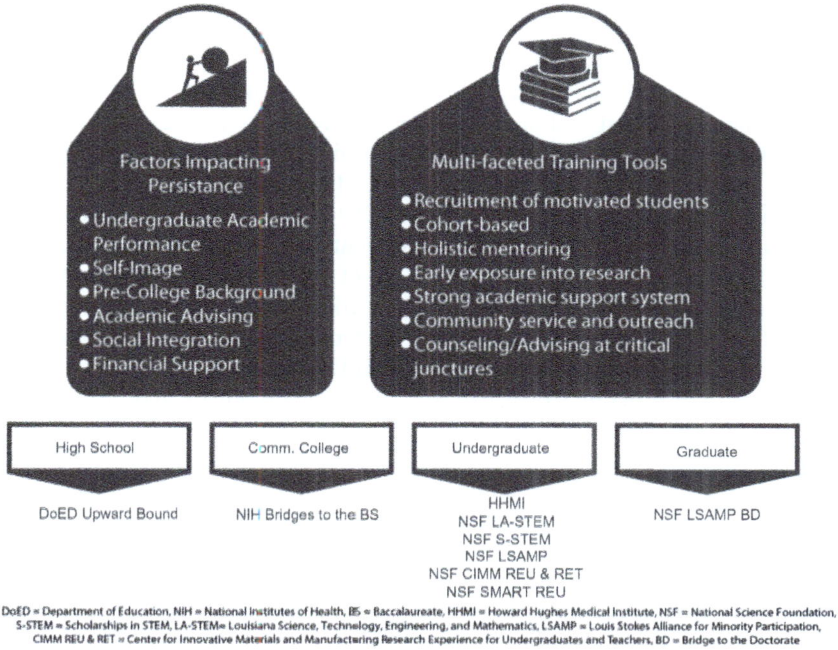

Fig. 2. OSI Educational Programs and Multi-faceted Training Tools.

various programs, soft skill development and expansion, and long-lasting friendships/mentorships.

While OSI has a unique offering of programs that serve a diverse population of students, these programs all integrate under the Holistic Development Model to provide students with varying levels of comprehensive experiences built upon the three pillars of success. Herein, each program will discuss how the pillars of OSI's holistic model are employed.

HIGH SCHOOL INITIATIVE

Upward Bound Program – College Access Granted

Introduction

Upward Bound is a federal Department of Education TRIO college access program that has been operational since 1964. Although this program is broadly focused, there are STEM opportunities, particularly during the summer institute. The population served by LSU's Upward Bound program is eligible high school participants who are potential first-generation college students, low-income individuals, or individuals who are at high risk for academic failure. The goal of Upward Bound is to end the cycle of poverty through educational opportunities and access. In 2012, LSU Upward Bound (LSU-UB) began serving students at one local high school; in 2017, the program was refunded and granted an additional award to serve a separate high school. The initiative to broaden participation and college access to more high schools around the Baton Rouge area is an imperative mission as persons from minority groups make up more than 50% of the population in East Baton Rouge Parish. Residents in the capital city of Baton Rouge live in poverty at a rate 6.6% higher than the national average (Quick Facts: East Baton Rouge Parish, Louisiana; UNITED STATES [Median Household Income, (2013–2017)], (2013–2017)). While it is well established that economically disadvantaged students do not graduate at the same rate as the national average, with only 73.3% of low-income students graduating from high school in 2013 nationally, when compared to 67.7% in Louisiana (DePaoli, Balfanz, & Bridgeland, 2016), the work and need for programs that exude a focus on eliminating this gap is dire in Louisiana.

The LSU-UB program's objective is to holistically motivate and provide low-income and first-generation students with the comprehensive skill set needed to obtain a college degree. "First generation" is an underrepresented group, and the term brings a common denominator of barriers and challenges for students and their families when it comes to educational access and opportunity. Many students have educational, social, and internal struggles that must be overcome, and Upward Bound strives to assist these students through their journey to success.

Mentoring

How does LSU-UB engage and bolster low-income and first-generation students who do not recognize their potential or believe that college is in their future? LSU's collaborative endeavors to tackle this question brought about immersive learning experiences for exposure and engagement. These experiences include *intensive, deliberate mentoring for students who struggle with barriers* such as academics, achievement vs affiliation, imposter syndrome, and understanding the need for higher education. Creating seamless partnerships with LSU faculty and their graduate students more closely connects the participant to the real academy and establishes informal and formal mentorship.

LSU-UB mentors typically encompass a 1:10 mentor to student ratio during the summer semesters. These mentors regularly interact with the students in a near-peer mentor relationship. This relationship is important in establishing a connection between the students and the program. LSU-UB mentors dine with UB students in the university cafeterias, participate in cultural and educational events on campus, share their academic success stories, and chaperone off-campus events. Mentors provide encouragement (academic, cultural, and social), reinforcement, and motivation through teaching workshops, tutoring, and spending time with students individually or in small group settings.

Education

LSU-UB was formed by the Office of Academic Affairs with the guidance of leading science professors. The key to making LSU-UB's services a success has been the close relationship with other LSU academic units. When designing the curriculum for the participants, STEM faculties were involved in creating pre-collegiate preparatory services for their potential future students. Participants receive intensive, engaging academic instruction and tutoring on the LSU campus and hands-on experience in core academic areas during the school year and summer. During the summer sessions, students are privy to information that they will learn during the next school year. This initiative is strategic in leveling the academic setting for underrepresented students and giving students the opportunity to familiarize themselves with those topics that may be more difficult to grasp during the academic year.

The LSU-UB model works twofold with regard to enriching experiences. The faculties are also able to enhance and even learn skills pertinent when working with underrepresented students before entering college. This first-hand professional development experience will, therefore, better equip them to work with underrepresented students at the college level.

Research

LSU-UB students are introduced to research at an extremely early point in their educational careers. These students who are typically in the 10th through 12th grades are granted the opportunity to perform experiments in a laboratory setting. During the course of the summer programming, students who enroll in the chemistry laboratory course as an elective engage in experiments in top-rated laboratories with tenured faculty, attend lectures by nationally recognized researchers, and participate in small group science demonstrations and facility tours.

One of the most recent components of UB is its partnership with "Teaching through Technologies (T3) Alliance" program to provide LSU's UB the needed support to implement a curriculum detailing the scientific background and hands-on use of these three technologies. Students will also receive instruction in STEM communication, leadership, and growth mind-set and will participate in a team or individual service project. During this project, students will use their newly acquired technical skills to address and solve a problem in their local community. The vision of T3 is to provide a replicable, adaptable, and scalable curriculum and a comprehensive support mechanism that can be widely adopted to increase the STEM interest and engagement of high school students from underrepresented minority groups.

Outcomes

The LSU-UB program has served 136 students with 62% female and 38% male students populating the program. The program has successfully supported 87% African American and 2% Hispanic underrepresented students. Through the Upward Bound program, more than 37 students identified as low-income, first-generation, and high-risk students, and more than 67 students identified as low-income and first-generation students. Overall, LSU-UB has garnered a very successful track record with towering results. Furthermore, the greatest outcome

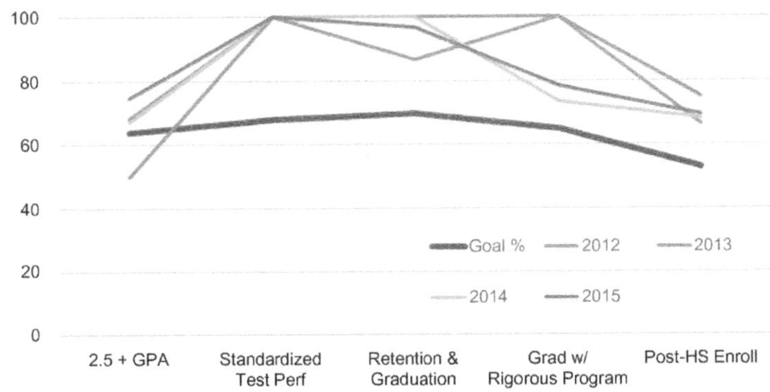

Fig. 3. LSU's Upward Bound Goals and Outcomes.

remains with the underrepresented students who are succeeding at noticeably higher rates through their participation in the LSU-UB program.

LSU-UB established ambitious, yet attainable goals, given the baseline data for the local school and district of 32%. With the current framework of services, the program has far surpassed these goals with the exception of the GPA (57.7%) in the first year of the award (2012) (Fig. 3); though the outcome is much higher than the before mentioned local and district baseline. Overall, LSU-UB is increasing and strengthening the pipeline for underrepresented college-bound students to achieve academic success through a holistic approach.

COMMUNITY COLLEGE INITIATIVE

NIH Bridges to the Baccalaureate—Bridging to Success

Introduction

The NIH-sponsored Bridges to the Baccalaureate (Bridges) program serves students in the Science, Technology, Engineering, and Mathematics (STEM) Division of Baton Rouge Community College (BRCC), an institution with a substantial URM population. Whereas a number of grant-sponsored activities are designed to benefit all BRCC STEM Division students, others are specifically tailored for US citizens enrolled full-time at BRCC in biomedical or behavioral sciences from varying backgrounds. These groups include underrepresented minorities (racial/ethnic), first-generation college students, individuals with disabilities, and financially disadvantaged (Pell-eligible) students. Many participants meet more than one of these criteria.

BRCC students are much more likely to be "non-traditional" students with real-life obligations (e.g., elder- or child-care) than they are to be the "traditional" first-time, full-time, recent high school students that typically encompass four-year institutions. This does not mean that BRCC STEM students are any less motivated than their LSU counterparts. However, their situations do give rise to a number of considerations: (1) The oft-assumed two-year, full-time path to an AS degree and transfer may not be realistic for students who have significant personal or family obligations, (2) Students who have jobs often are not in a position to "step away" to try research for a summer without risking those jobs and, consequently, their livelihoods, and (3) Policies and practices based on "traditional" students often have unintended consequences for "non-traditional" students.

The goal of the Bridges program is to increase the number of underrepresented students who complete AS degrees in STEM at BRCC and then transfer to and complete BS degrees in biomedical or behavioral sciences programs at LSU. In order to increase the number of underrepresented students achieving this goal, the Bridges program must also acknowledge the unique issues of "non-traditional" students and adapt the three pillars of the OSI Holistic Development Model to support the students in a way that meets their needs.

A key element in preparing BRCC biomedical and behavioral students for successful transfer to LSU is the Bridges Research Scholars (Bridges Scholars) program, for which up to 12 BRCC students per cohort are selected through an application, recommendation, and interview process. Newly selected Bridges Scholars complete a six-session orientation during the spring semester before their first summer of research as a Bridges Scholar, in which they learn more about the process and culture of research, select a faculty mentor and research project, receive advice from more senior Bridges Scholars, and commit to fulfilling the expectations of the Bridges program and their research groups. Bridges Scholars who maintain academic and program standards participate in activities, engage in intensive mentoring with Bridges staff, and receive research wages.

Mentoring

Proactive mentoring is woven throughout the Bridges Scholars' experience. Bridges Scholars' activities are largely coordinated through one Bridges team member at BRCC and one at LSU – both with advanced science degrees and teaching experience. This allows the program staff to become familiar with each Bridges Scholar's situation, strengths, weaknesses, goals, and needs; it also allows each Bridges Scholar to become comfortable confiding in and seeking out the two team members. Bridges Scholars receive regular e-mail communications to encourage the Scholars to maintain the good habits they have developed. At a minimum, Bridges Scholars are required to meet with team members at the beginning, middle, and end of each semester to discuss each student's academic and personal progress, goals, challenges, and resource needs. Bridges team members offer individual mentoring and, as needed, connect Bridges Scholars with appropriate resources on their respective campuses and within their communities. In addition, near-peer mentoring within and across cohorts helps to increase the confidence and self-efficacy of Bridges Scholars as they embark on unfamiliar territory.

Education

During the academic year, Bridges Scholars focus primarily on their academic courses and part-time research. They participate in relevant workshops sponsored by various LSU units (e.g., Communication across the Curriculum (CxC), Science Café, Olinde Career Center, the LSU Discover Undergraduate Research Program). This approach to academic-year professional development accommodates Bridges Scholars' specific interests, needs, and schedules while allowing them to become comfortable interacting with LSU staff in these units. Bridges Scholars also gain practice in presenting their research in a variety of venues – in a poster session at BRCC, at the regional LSU Undergraduate Research Conference and Discover Day events, and at national conferences. First-year Bridges Scholars attend and present their research at the Annual Biomedical Research Conference for Minority Students (ABRCMS) and may later participate in conferences within their disciplines. Because ABRCMS serves as their first conference – and for many, their first airline flight – Bridges Scholars participate in an intensive "Conferences 101" preparatory workshop that covers a broad range of topics such as airport security, conference attire, and business cards. Bridges Scholars have successfully competed with their counterparts from four-year institutions, earning conference presentation awards on the local, regional, and national levels.

During their second summer, Bridges Scholars focus on building their presentation skills and preparing for post-Bridges opportunities. In collaboration with CxC, second-year Bridges Scholars receive intensive training in designing and giving oral scientific presentations. These efforts culminate in an end-of-summer symposium, where each Bridges Scholar presents a 10-minute talk to the community of Bridges staff and research mentors. In conjunction with the LSU Olinde Career Center, Bridges Scholars also learn about the process of applying for internships, external research experiences, and jobs; this includes identifying and applying for appropriate positions, crafting and adapting a CV or résumé, navigating a career fair, and preparing for an interview. Bridges Scholars complete this training just before the LSU fall Career Expo, which allows them to position themselves successfully for the opportunities they will encounter after completing the Bridges program.

Research

Bridges students engage in undergraduate research up to 40 hours/week during summer and semester breaks, no more than 15 hours/week during semesters for a maximum of two calendar years.

In addition to completing at least 10 weeks of research at LSU during each summer of participation, all Bridges Scholars continue their learning outside the research lab. Through the LSU Summer Undergraduate Research Program (SURP), Bridges Scholars are trained in laboratory safety, conducting ethical research, maintaining appropriate research records, and developing poster presentations for the LSU Summer Undergraduate Research Forum (SURF). They also learn about a variety of research fields and career paths by interacting with SURP students and personnel from other disciplines at weekly seminars, events like Astronomy Night, and site visits such as the Tulane Primate Research Center. Bridges Scholars also embark on the two-year, Bridges-specific curriculum described above, which complements the discipline-specific research training they receive in the lab and through SURP.

Outcomes

In general, the timeline of the Bridges program has corresponded to increases in BRCC STEM enrollment, BRCC STEM AS degree completion, and BRCC-to-LSU STEM transfer. Thirty-one students from a variety of STEM majors have participated in the Bridges Research Scholars program and have conducted research in 10 academic units across the LSU campus. While it is relatively early in the process (five years), outcomes to date include the following:

- 16 students transferring to four-year institutions (13 to LSU);
- four students earning BS degrees;
- one student enrolled in a PharmD program;
- seven students remaining enrolled at BRCC; and
- eight students entering the workforce or military (most due to catastrophic personal or family losses in the Baton Rouge flood of August 2016).

The 2015 and 2016 cohorts of Bridges Scholars (21 Scholars) have earned a total of ten regional and three national conference presentation awards. Bridges Scholars demonstrate that, given the opportunity, community college students can compete and succeed on the same playing field as their counterparts from four-year institutions. Bridges Scholars and course-based undergraduate research experience participants have shared their experiences across the BRCC STEM Division and helped to build momentum in student motivation. BRCC students are excited to emulate their classmates' success. Overall, the Bridges program offers a progressive yet effective pathway for those "non-traditional" students to pursue undergraduate degrees with the same achievement as their counterparts.

UNDERGRADUATE INITIATIVES

HHMI Professors Program – The Beginning of OSI

Introduction

The Howard Hughes Medical Institute (HHMI) Professors Program was OSI's pioneering initiative established after Professor Isiah Warner was named one of 20 HHMI Professors in 2002. The US$1 million funding received supported this grant designed to investigate and solve the ongoing problem of poor performance by STEM undergraduates who demonstrated high-ability, interest, and work ethic throughout high school. *The theoretical framework of the OSI Model was developed through the HHMI Professors Program Initiative, combining mentoring, education, and research.* When provided with effective learning strategies, encouragement, and guidance, along with financial support, the HHMI Professors Program demonstrated that students experiencing challenges could overcome those obstacles and excel. Furthermore, they have persisted as undergraduates and doctoral students.

The HHMI Professors Program primarily targeted undergraduates with a GPA range of 2.5–3.0, a group of students that are often overlooked as potential scientists and engineers. When given the opportunity, these students have advanced individually and often empower and mentor their peers. The program also supported community college students through an additional initiative aimed at assisting students with transitioning to LSU to complete STEM bachelor's degrees (HHMI BRCC). This component provided skills training, personal development, and undergraduate research opportunities for BRCC students. Thirty-four students were supported through this effort. Another notable segment of transfer students included 18 participants who were involved with the undergraduate programs and were not in the HHMI BRCC initiative. In total, these two groups of students represent 35% of the total participants served.

Mentoring

The mentoring aspect of this foundational program is based on a hierarchical mentoring ladder that fosters collaboration and multi-directional mentoring between individuals at varying levels of education including faculty, graduate students, undergraduate students, community college, and K–12 students and teachers (Fig. 4).

Within the HHMI Professors Program, mentoring by faculty, staff, and peers was instrumental in unlocking the unlimited potential of its participants by providing individualized guidance and support. In addition to the tiered mentoring received from faculty and research advisors, students were required to meet individually with program staff four times per semester (Wilson et al., 2012). They all engaged in peer mentoring activities and some served as mentors to younger students. The increased accountability reinforced lessons learned through program workshops, individual meetings, and so on and cultivated a culture of excellence. When faced with obstacles, students with strong and effective mentoring relationships were more likely to overcome these challenges with minimal effect on academic performance. Through peer interactions with the undergraduate participants, the community college students received additional support with transitioning to LSU.

Education

Historically, the approach has been to teach students effective strategies for learning science by applying what they have learned in the classroom to mentoring and research. Through the creation of a series of four levels of Mentoring, Education, and Research *Success Course* designed for first through fourth year students, undergraduates are enriched by learning basic metacognitive techniques and

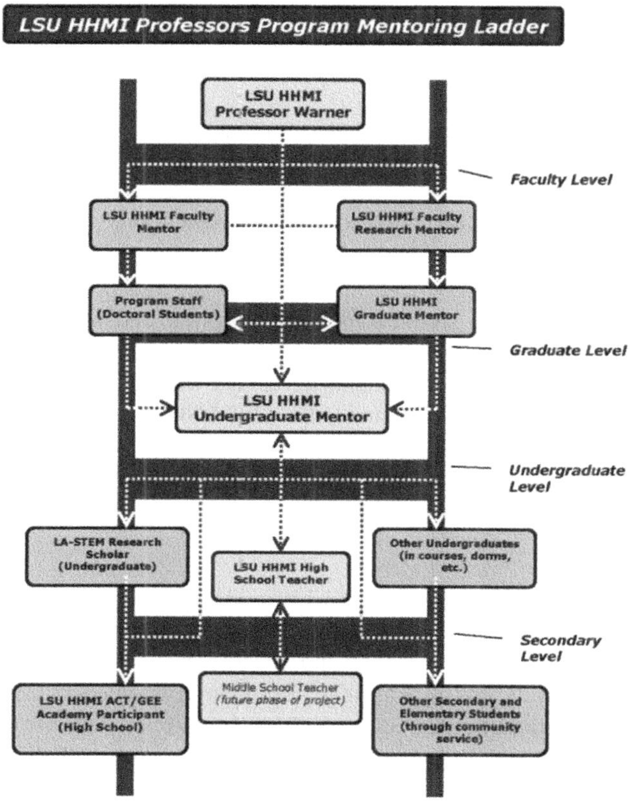

Fig. 4. HHMI Professors Program Hierarchical Mentoring Ladder.

learning strategies, preparing for research, developing a growth mindset, overcoming imposter syndrome, and working with a peer mentor. In turn, these students go on to teach those learned strategies and serve as mentors to peers and high school students. Metacognition describes the ability to think about thinking, be consciously aware of oneself as a problem solver, and monitor one's mental processing (Samuelson, 1982). Often students struggle within STEM because they lack these abilities and in many cases are ill-prepared as incoming first-year undergraduates. When coupled with individualized mentoring and undergraduate research, OSI's *Success Courses*, holistically accommodate students at various levels of development and aid in increasing the confidence and favorable outcomes of a group that was initially lower performing. Furthermore, the HHMI Professors Program participants engaged in *Success Courses* and activities with students in other OSI programs including higher achieving scholars from the Louisiana Science, Technology, Engineering, and Mathematics (LA-STEM) program (discussed in the next section). The synergy between the two programs strengthened the effectiveness of both in terms of broadening participation.

Students needing additional help were required to utilize campus resources such as the LSU CAS, the LSU Health Center, and other available support services.

Research

Undergraduate research was a vital component of the HHMI Professors Program, particularly because the students entered the program with academic challenges. Research provided a vital connection between classes and real-world application. In preparation for research, participants gathered information about research in their field and compiled a list of professors whose research interested them. After receiving guidance from staff and peer mentors, participants met with faculty to secure a research position. Students received training in areas such as research ethics, lab safety, and best practices through the *Success Courses.* Additionally, students developed writing and presentation skills by presenting during various courses, the OSI LSU Undergraduate Research Conference, as well as other local and national conferences. Participants in HHMI BRCC were provided opportunities to conduct undergraduate research at LSU as community college students, which increased STEM persistence.

Outcomes

The HHMI Professors Program served 149 students through its community college and undergraduate initiatives. The six-year graduation rate in STEM for all students served is 65%. For women and Black participants, the rates are 64% and 61%, respectively. Sixty percent of all underrepresented minority students completed undergraduate degrees in STEM. In comparison with all majority participants (White and Asian), with a rate of 71%, women and all minority students are within 7% and 11%, respectively, reducing the achievement gap. Of the 34 HHMI BRCC participants, 25 (74%) transferred to LSU as participants in our undergraduate program. Fourteen of these transfer students completed bachelor's degree in STEM, while one transfer student transitioned into the veterinary medicine program and continued to complete a doctoral degree in this field. Altogether these transfer students represent 60% of the HHMI BRCC population. Due to the large representation of HHMI BRCC Scholars in the overall program, graduation rates reported include the transfer of student participants. These outcomes are quite remarkable, as these students are traditionally considered as struggling students, destined to depart from their STEM degree programs or withdraw from the university. To achieve this success at a level greater than the national average (40%) and other comparative institutions (see OSI outcomes section) is a great testament to the strength and flexibility of the OSI model.

HHMI Professors Program participants have also pursued graduate degrees. Twenty-one participants have advanced degrees, including nine STEM doctoral degrees, six master's degrees, five medical degrees, and one doctorate of veterinary medicine. Another two graduates are currently pursuing doctoral degrees. As with the graduation rates, their acquisition of advanced degrees proves that any student can succeed, despite pre-college background and academic performance as incoming students when provided with holistic support through mentoring, education, and research.

LA-STEM Research Scholars Program – The Model Program

Introduction

Established in 2003 with funding from the National Science Foundation, the Louisiana Board of Regents, LSU, and Research Corporation, The Louisiana Science, Technology, Engineering, and Mathematics (LA-STEM) Research Scholars Program was created to expand student access and increase persistence in STEM. Inspired by the University of Maryland Baltimore County (UMBC) Meyerhoff Scholars Program, LA-STEM focused on developing leaders and researchers, cultivating a culture of inclusion in STEM, and supporting student achievement (Crawford et al., 2018). LA-STEM admitted high-ability students (GPAs of 3.5 or greater, ACT composite score minimum of 24) who demonstrated the potential to succeed in STEM. Students were selected after distinguishing themselves as scholars and leaders who were committed to promoting diversity in the sciences. LA-STEM Scholars exemplified a strong dedication to mentoring, enthusiasm for diverse cultural experiences, as well as a passion for serving their community. The LA-STEM Program served a total of 249 students over 12 years. Forty-four percent of participants were underrepresented and 53% were women.

Through the creation of a close-knit and supportive community of faculty, staff, and students, LA-STEM Scholars received holistic comprehensive training and mentoring to ensure successful completion of their undergraduate degrees and matriculation into doctoral programs. They were also required to maintain the highest academic standards, conduct undergraduate research, engage in peer mentoring, and volunteer for outreach initiatives.

LA-STEM Scholars received unique advantages due to its highly comprehensive structure including the following:

- cohort-based structure;
- generous financial support;
- automatic admission into the LSU Ogden Honors College;
- integration into a supportive scholarly community among students, staff, and faculty;
- early exposure to learning strategies to increase undergraduate success;
- participation in an eight-week Summer Bridge Program for incoming students;
- yearlong guidance and holistic support;
- peer and faculty mentoring relationships;
- academic advising and professional counseling, particularly during critical junctures;
- research opportunities during the academic year;
- summer research experiences at institutions nationwide and internationally;
- peer tutoring;

- increased knowledge of the opportunities available in one's major field;
- comprehensive graduate school preparation;
- professional skills enrichment;
- leadership skills development; and
- networking opportunities

Mentoring

Similarly to the HHMI Professors Program, participants in the LA-STEM Program engaged in quality tiered mentoring relationships with staff, faculty, and peers. These individualized interactions provided opportunities for students to establish a development plan, communicate and resolve any challenges, and build confidence as researchers and scholars through strong mentoring relationships. Students were required to meet with program staff a *minimum of two times per semester*. Those with academic challenges had additional monthly meetings and were required to receive additive support from other campus resources. LA-STEM Scholars also engaged in mentoring relationships with their research advisors and graduate students within their group. A number of these students held leadership roles on campus and served as mentors to their peers and younger students through outreach initiatives. Often these students were empowered to share the knowledge gained with others, which demonstrates the exponential impact of intrusive and effective mentoring. Students have noted the increased success of their peers and family in terms of persistence and degree completion through the graduate level as a result of the transference of information received through being involved with the LA-STEM program.

Education

A key element of the LA-STEM Research Scholars Program was its comprehensive eight-week Summer Bridge Program, which has been largely cited by students as the primary component that fostered a strong sense of community between LA-STEM and HHMI Professors Program participants. Through engaging team-building activities, intensive workshops on the college transition and expectations, and academic courses, students acquired vital skills in a smaller setting. Although participants were largely incoming first-year students, older students in both programs also participated and contributed as informal mentors. Following the Bridge Program, scholars continued to receive academic enrichment, research training, professional development, and preparation for graduate school through the *Success Courses*.

Research

LA-STEM Scholars gained research experiences very early, often securing positions during Summer Bridge. These interactions enriched their learning community and allowed for the increase of self-efficacy and identity as scientists. As a result of early integration into research, LA-STEM Scholars performed better academically through applying concepts learned in academic courses to their research. Engagement in undergraduate research throughout the academic year (*for at least 10 hours per week*) and for a minimum of two summers was a program requirement. This increased the competitiveness of our students for extramural summer research programs, international research experiences, local and national awards, and graduate program admission. Additionally, LA-STEM scholars were often assigned independent projects and garnered peer-reviewed publications.

Outcomes

To date, 161 LA-STEM Scholars successfully completed the program as *LA-STEM Graduates*. The LA-STEM Program currently boasts an overall six-year graduation rate in STEM of 87%. The rate of 87% for women and 81% for all underrepresented minorities is within 4% and 10%, respectively, of all majority participants (White & Asian: 91%). It is worth mentioning that these rates include *all* scholars served, regardless of the duration of time spent in the program. These rates are two to three times more than the national average. To date, 78% of students who successfully completed the program as LA-STEM Graduates have earned or are pursuing post-baccalaureate degrees. Forty-three percent of the LA-STEM Graduates finished with a minimum cumulative GPA of 3.7. Lastly, LA-STEM Graduates have garnered considerable national award/fellowship recognition, including the following:

- Barry M. Goldwater Scholarships (8);
- Gates Cambridge Scholarship (first in LSU's history);
- Morris K. Udall Scholarships (first & second in LSU's history);
- National Science Foundation Graduate Research Fellowships (18);
- Fulbright Scholarships (2);
- HHMI Gilliam Fellowship (1);
- Harry S. Truman Fellowship (1); and
- National Defense Science & Engineering Graduate Fellowship (1).

When looking closely at the effects of the LA-STEM program, it is important to highlight that 60% of *all* students served have completed or are pursuing advanced degrees. This demonstrates the far-reaching impact of intervention and support from minimal to comprehensive levels, particularly in universities with limited resources for effective programming. The Summer Bridge and first-year experiences have proven to be great contributors to increasing persistence and matriculation into advanced degree programs. The LA-STEM program is an exemplary model that demonstrates the *astronomical impact* that proven comprehensive holistic support can have on broadening participation.

S-STEM – Small Investment Yields Large Gains

Introduction

With funding from the National Science Foundation, the S-STEM Scholars Program provided scholarships to participants within the College of Science, who were academically talented science and mathematics students from economically disadvantaged backgrounds. The program maintained a steady state of 30 participants per year and provided support to a total of 112 students (61% women and 49% underrepresented) from 2007 to 2015. Scholars received primary guidance aimed at ensuring academic success in the STEM disciplines resulting in undergraduate degree completion. In addition, the S-STEM program gave its scholars an opportunity to participate in a broad range of academic enhancement and outreach activities. In the S-STEM program, requirements were kept to a minimum, as many of the participants had part-time employment to offset educational expenses.

Mentoring

In an effort to leverage the program benefits, S-STEM Scholars received individualized mentoring with program staff in addition to informal mentoring experienced through group activities. Similarly to other programs, those students needing additional help were connected with other support offices that provided SI, tutoring, and counseling. Scholars were also mentored by faculty and peers across other OSI undergraduate programs.

Education

S-STEM Scholars were required to participate in monthly group meetings, where they interacted with peers from OSI's LS-LAMP Program of similar size with comparable goals. In these meetings, scholars received training on a broad range of topics to help them navigate and succeed as undergraduates and prepare for careers in STEM. Topic examples included goal setting, time management, and developing powerful personal statements. These meetings were pertinent for S-STEM Scholars, particularly those who balanced academics and part-time employment.

Research

While research was not required for participation, it was encouraged. Those S-STEM Scholars who were involved in other OSI programs such as LS-LAMP, HHMI, or LA-STEM did engage in undergraduate research and benefited from the support of their faculty advisor, postdocs, and graduate students.

Outcomes

Although among the least comprehensive of the OSI programs, the S-STEM outcomes were comparable to that of the LA-STEM program. The overall STEM graduation rate for all S-STEM scholars served is 84%. Women and underrepresented minority students graduated at comparable rates (84% and 83%), respectively. Both groups had rates within 2–3% of all majority students (White and Asian: 86%), and women graduated at rates equivalent to men. While the primary emphasis was focused on their experiences at LSU, additional initiatives were employed to prepare students for matriculation into graduate programs. Forty-nine percent of S-STEM Scholars served either have completed or are pursuing advanced degrees, including 11% at the doctoral level. Of the S-STEM scholars who have completed or are pursuing advanced degrees, 33%

are women and 26% are URMs. These remarkable outcomes, particularly given the limited program resources, confirm that even the smallest amount of *strategic and effective* interventions can result in considerable impact.

LS-LAMP Program-cultivating Pathways

Introduction

The Louis Stokes Louisiana Alliance for Minority Participation (LS-LAMP) Program is a statewide initiative funded by the National Science Foundation and is aimed at increasing the number of minority students completing baccalaureate STEM degrees. The program also focuses on creating avenues for these students to pursue terminal degrees and careers in STEM. The Louisiana Alliance is one of more than 40 alliances nationwide. To date, the LSU LS-LAMP program has served 99 participants (100% underrepresented and 66% women). These students have conducted undergraduate research, received individualized mentoring, and benefited from academic enrichment and professional development.

Mentoring

As with the other programs, LS-LAMP participants are required to meet individually with program staff a minimum of twice per semester (four times for those with challenges). They also build mentoring relationships with their respective research advisors and peers including those from other OSI programs such as those in S-STEM. LS-LAMP students have also engaged with graduate-level students through undergraduate research involvement and interaction with Bridge to the Doctorate Fellows.

Education

In conjunction with students in the S-STEM Scholars Program, LS-LAMP participants received professional development and training through monthly group meetings. Examples of topics addressed include applying for summer research programs, giving powerful presentations, and research ethics. They are also exposed to identifying extramural undergraduate research opportunities and preparedness, the graduate school process, and career exploration.

Research

LS-LAMP Scholars are required to conduct research for *at least 10 hours weekly*. These scholars are provided with resources, guidance, and training on selecting compatible research advisors. Participants present their work at local and national conferences. As members of the Louis Stokes Alliance for Minority Participation, LS-LAMP Scholars engage with fellow scholars from across the country at the national LS-AMP Conference. Additional benefits of the program include the ability to present at conferences exclusive to LS-AMP and direct access to the Louis Stokes Midwest Center of Excellence resources for summer research opportunities.

Outcomes

To date, four LS-LAMP students have completed PhDs in STEM, with another 11 students currently pursuing doctoral degrees (10 in STEM). Fifteen scholars have completed master's degrees, of which three are referenced above for their current pursuit of PhD degrees. Four scholars are currently pursuing master's degrees. Two scholars have earned and two are pursuing medical degrees, and one has earned a law degree. The benefits of the LS-AMP statewide and national affiliations have enriched the level of programming and opportunities available to students in the pursuit of doctoral degrees.

SMART REU – Advancing SMART Opportunities

Introduction

The Smart Polymer REU serves rising undergraduate seniors and juniors with an emphasis on women, underrepresented minorities, and students of community colleges. Half of the participants are directly from Historically Black Colleges and Universities (HBCUs) or other minority-serving institutions. Although not through overt selection, many of the participants are first-generation college students. Smart Polymer is an interdisciplinary project across LSU's science and engineering departments. The research focuses on a theme of smart polymer composite materials and structures that have integrated multi-functionality, with capabilities including sensing, actuating, healing, and adapting in addition to classical tasks, such as load-bearing.

Mentoring

While the students who participate in the Smart Polymer REU have only a brief experience (10-week summer research program) within the OSI Development Model, when compared to the other programs, they still receive adequate mentoring from various avenues. Smart REU students are surrounded by multiple mentors (faculty, postdocs, graduate students, peers, and OSI staff), who encourage and further motivate these students through OSI's holistic approach. REU students are not only developed scientifically in this program, they are also advanced professionally. Students receive weekly SMART seminars where they have the opportunity to discuss research and avenues for pursuing a doctoral degree in STEM. They also gain insight into the skills and strategies needed to evolve as young and budding scientists.

Education

In addition to gaining valuable research skills, participants receive enrichment designed to further increase their academic success. Students are offered faculty seminars and participant presentations as avenues for presenting their ongoing research and project barriers, followed by feedback from peers and faculty mentors. Such activities enhance student confidence and groom them for future public presentations on a larger scale. Students also engage in ethical workshops that present a wide overview of ethics training modules of research and profession.

The SMART Polymer REU is very invested in increasing the number of students pursuing doctoral degrees. The program also offers its students graduate education workshops and panels to assist students with a better understanding of the process of applying for, selecting, and pursuing graduate degrees.

Additionally, students have an opportunity to broaden cultural awareness within their REU cohort through the *SMART Cultural Exchange Night*. An off-campus event such as a cultural exchange night promotes inclusivity due to the informal structure of the setting, where students and faculty mentors interact through candid dialogue about family or cultural traditions and practices, thus deepening understandings of similarities and differences among diverse groups.

Research

The goals of the project are to provide exceptional research experiences for diverse undergraduates at the interface of materials and structures and prepare undergraduates for advanced studies and research careers while establishing a network among each other and faculty mentors. The program also motivates and promotes inclusivity through awareness and cultural interactions between students and faculty with various backgrounds.

The majority of participants are from smaller institutions and typically do not have access to cutting-edge research components; however, through the SMART Polymer REU, participants are surrounded with the state-of-the-art equipment, advanced software, and researchers that are at the top of their fields.

Outcomes

The primary desired outcome of this new program is to increase the number of students who enroll in graduate programs. The first-year cohorts of SMART Polymer REU are currently juniors and graduating seniors. Of this first cohort, three were women participants (37%) and three were underrepresented minorities (37%). To date, only two of the eight students from this cohort have completed undergraduate studies and are preparing to pursue graduate degrees.

Bridge to the Doctorate Program – Priming the Doctoral Pipeline

Introduction

Louisiana Broadening Resources for Increasing Diversity in Graduate Education (LA-BRIDGE) serves students from underrepresented groups to pursue PhD degrees in STEM disciplines. Students who received bachelor degrees from an LS-AMP institution are chosen for the Louisiana State University Bridge to the Doctorate (LSU BD) Fellowship.

The LSU BD Initiative is based on strategic institutional approaches to building a diverse workforce at the doctoral level. The vision of LSU BD is to offer a national model for the production of underrepresented scientists and engineers with doctoral degrees in STEM. LSU has served as a bridging institution within the LS-LAMP community for eight cohorts of BD Fellows. The goals of LSU BD are not only to recruit highly qualified Fellows from underrepresented groups but also to support and retain the fellows through their doctoral track. Additionally, LSU BD strives to prepare all Fellows for a career in research at a professional level upon leaving the institution. These efforts of supporting and retaining fellows are performed through programmatic structures that enable the success of BD Fellows, especially as of recent with cohort 8 initiatives focusing on preparing Fellows for the NSF Graduate Research Fellowship Program (GRFP).

Mentoring

Intrusive mentoring is a critical tool when assisting students in their transition from an undergraduate program to a graduate program. Students who transition from primarily underrepresented institution require a more energetic approach to mentoring. Thus, they are engaged in one-on-one mentoring meetings with the LSU BD leadership team to determine the students' current status as an incoming BD Fellow and create a plan based on the Fellows' departmental, personal, and professional goals and obligations.

LSU BD Fellows are also offered assistance with setting and maintaining these goals throughout their term. This assistance is captured using Individual Development Plans with the assistance of the LSU BD leadership team. Fellows create their initial IDPs within the first few months of entering the program to create clear goals and strategies for achieving those goals. LSU BD leadership team consults the home department on developing a timeline most efficient for each Fellow as well as offers information on departmental expectations and milestones. The initial goals determined by the Fellow and leadership team are required at the end of each semester during one-on-one mentor meetings to reflect on the progress of each goal. During these meetings, Fellows provide progress updates and set new goals within the IDP profile. Fellows are highly encouraged to continue using the IDPs established with their chosen graduate advisors at the completion of the BD fellowship in an effort to strengthen the mentor relationship across the STEM departments at LSU.

Education

Oftentimes students from HBCUs typically are unaware of the rigors and expectations of graduate school, leading to an ill-prepared start when compared to their counterparts. As a result, LSU BD has recognized these issues and created a *Bridges Orientation* to ease the transition of students and minimize the devastation associated with the first year of graduate school. This two-day Bridge Orientation is designed to prepare incoming Fellows prior to the beginning of the fall semester and cover a multitude of topics that center around self-growth including *metacognition, time management, balancing multiple priorities, self-efficacy, imposter syndrome*, and *managing expectations*. Fellows are also educated on topics that assist in strengthening peer and faculty relationships such as *emotional intelligence, strategic planning, interpersonal skills*, and *faculty–student relationships*.

During the academic year, the LSU BD program holds monthly meetings and workshops to support those initial topics covered during the Bridge Orientation but further advance the fellows through academic, research, and professional development. LSU BD also focuses on nurturing the development of each Fellows' "soft skills" or "intrapersonal skills." Interpersonal development workshops and training topics include but are not limited to *mentoring, effective communication, branding, managing social media presence*, and *networking are offered.*

Research

The LSU BD program supports 21 BD-eligible departments for Fellows to perform research within. These departments house a large number of graduate faculty who are available to serve as faculty research advisors for BD Fellows. LSU BD proudly supports each Fellow throughout the process of selecting a research mentor. This is a very critical component of the Fellows' graduate journey, as typical STEM doctoral programs are roughly six years long and students spend on average 40–50 hours in a lab setting. This lengthy process requires the faculty–student dynamic to be a strong match from the beginning.

Fellows are offered assistance in the further development of research skills through the workshops and monthly meetings. Fellows will cover topics that are briefly reviewed in graduate programs such as *managing citations, research ethics, communicating your research*, and *applying for awards*. Topics as such are important attributes of research that are often overlooked or poorly explained for students, but with access to proper training and resources, students can experience a positive effect on their overall graduate studies.

Outcomes

LSU BD programs have successfully supported 102 fellows over the eight cohorts, of which all Fellows were representative of underrepresented groups and received a STEM BS or master's degree from an LS-AMP institution. Compellingly, the vast majority of LSU BD Fellows were African American (88%), while Hispanic/Latino (6%) and Native American (2%) Fellows were also recruited. Of these 102 LSU BD Fellows, 62% have been women. To date, 34 doctoral and 15 master's degrees have been granted to LSU BD Fellows, while 26 are currently matriculating. LSU BD Fellows have been able to garner admirable positions including serving in faculty, teaching, industry, postdoctoral, and other STEM positions.

OSI's Outcomes

Since 2001, OSI has been a national model for broadening participation, primarily in STEM, through comprehensive and holistic student development. Overall, OSI has served more than 135 high school, 560 undergraduate, and 100 graduate students and has had great success with creating an inclusive environment for all students. Although some of OSI's individual programs consisted of only underrepresented minority students (due to funding agency requirements), 51% of all scholars served through our undergraduate efforts were URMs and 55% were women. The overall distribution of women by race can be found in Fig. 5.

Of the undergraduates served through non-summer programs (547 students), 445 (81%) have received bachelor's degrees, of which 395 (72%) were in STEM

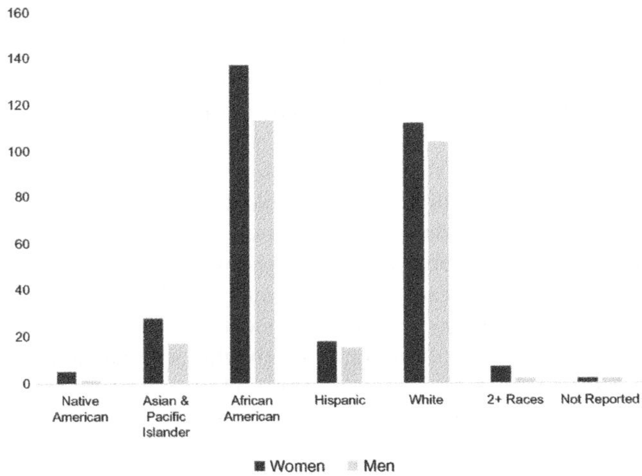

Fig. 5. OSI Undergraduate and Community College Students Served by Race and Gender.

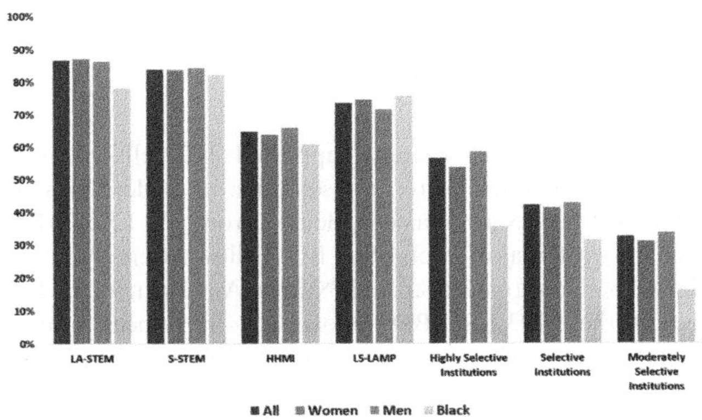

Fig. 6. OSI's Six-year STEM Graduation Rate Comparison for All Undergraduate Initiatives.

disciplines and 33 (6%) are currently pursuing BS degrees. A substantial number of those STEM degree recipients were women (55%) and URMs (48%), proving the model to be exemplary for broadening participation and increasing persistence in STEM.

When examining the undergraduate initiatives (HHMI, LA-STEM, LS-LAMP, and S-STEM), the overall graduation rates in STEM range from 64% to 84%. As seen in Fig. 6, in comparison with moderately selective, selective, and highly selective institutions as reported by the Center for the Institutional Data Exchange and Analysis at the University of Oklahoma, whose overall rates range from 32% to 57%, all four of the undergraduate OSI programs are superior (Whalen, 2017). *Remarkably, even scholars in the HHMI Professors Program, who were initially overlooked due to their academic challenges, graduated at an overall rate of 7% higher than the highly selective institutions.* These outcomes further demonstrate the power of the OSI model in terms of its ability to *effectively support students at all achievement levels.*

At the doctoral level, 33% of OSI's BD Fellows have received STEM PhDs and another 15% completed master's degrees. As a result of OSI's primary focus centered on matriculation into graduate programs, a considerable number of undergraduates have also received advanced degrees. When examining these students, 141 of the undergraduate participants have completed advanced degrees, including 30 STEM PhDs and 32 STEM master's. Forty-two have completed degrees in health-related fields (MD, DDS, and DVM). Another 81 are currently pursuing advanced degrees, including 48 STEM PhDs, 2 MD/PhDs, 7 STEM master's, and 18 in health-related fields.

As a result of OSI's efforts, in June 2018, the office was awarded the Presidential Award for Excellence in Science, Mathematics, and Engineering Mentoring (PAESMEM) as a 2014 organizational awardee (Presidential Awards for Excellence in Science, 2018). PAESMEM is the highest honor granted to mentors who work to bolster STEM talent. The National Science Foundation (NSF) administers the honorary awards program on behalf of the White House Office of Science and Technology Policy (OSTP). OSTP selects both individuals and organizations to receive the honor. This year, 27 individuals and 14 organizations were recognized (Presidential Awards for Excellence in Science, 2012–2018). While OSI's remarkable outcomes have yielded a plethora of awards for individual participants and national recognition for LSU and OSI, *our greatest achievement has been our ability to level the playing field for students from diverse backgrounds, through providing a supportive community in which any student can thrive.*

OSI's Lessons Learned

Over the past 17 years, OSI has developed and housed several programs, however, not without overcoming a few challenges and shortcomings. Throughout the years, several lessons were learned and deemed important to continue providing strong mentoring programs that increase students' success at the high school, undergraduate, and graduate levels. One of the most valuable lessons learned is the importance of not only creating student development activities but also creating a *strong sustainability plan from the beginning*. This sustainability plan should be one that aligns its initiatives with those of its respective university but is also independent of full institutional support. Ideally, a healthy balance between the following key elements would aid in securing the longevity of effective programming:

- early advocacy by senior administrators;
- increased buy-in of faculty;
- better exposure and dissemination of best practices;
- secured donor support; and
- scalable programs.

Early advocacy of senior administrators is an undeniable effect that would guide the opinions of other administrators in securing the longevity of an office such as OSI. Encouraging an increased number of faculty involvement in programmatic activities would not only create more mentors for students but also offer an increased *"buy-in" of faculty* who then will become advocates for the work performed. Ensuring the accomplishments and impacts of the office are widely *disseminated locally and nationally* would also strengthen the foundation for sustainability by generating knowledge for others to use. Pursuing *donor support* at the inception of the program would provide financial assistance for effective programs who are no longer eligible to receive federal support. Lastly, when developing programs that have demonstrated extreme success with smaller cohorts, *it is critical to identify scalable best practices* to meet the needs of an institution.

CONCLUSION

Through this trifecta of mentoring, research, and education, participants in the LSU-OSI programs are engaged in activities that support their academic and professional development. Within OSI, most students learn to effectively steer their academic paths and become leaders who teach others what they have learned. In all, OSI students are successfully navigating through their high school, undergraduate, and graduate studies in notable ways. Because of OSI's holistic approach, these students are breaking barriers, rewriting their futures, and embarking on incredible paths to higher education. While there are several hallmarks of success for OSI programs individually, the office has developed a model that addresses the issues faced by underrepresented students and women when pursuing degrees. OSI will continue to evolve its *holistic approach* while further pushing the boundaries for broadening participation for all students.

REFERENCES

Administration, L. S. U. (2018). *Strategic plan*. Retrieved from https://www.lsu.edu/administration/ir/strategic-plan.php

Allen, T. D., Russell, J. E., & Maetzke, S. B. (1997). Formal peer mentoring: Factors related to protégés' satisfaction and willingness to mentor others. *Group & Organization Management*, *22*(4), 488–507.

Astin, A. W., & Astin, H. S. (1992). *Undergraduate science education: The impact of different college environments on the educational pipeline in the sciences. Final report*. Authoring Institution: California University, Los Angeles, Higher Education Research Institution. Sponsoring Agency: National Science Foundation, Washington, D.C.

Bakken, L. L., Byars-Winston, A., & Wang, M.-F. (2006). Viewing clinical research career development through the lens of social cognitive career theory. *Advances in Health Sciences Education*, *11*(1), 91–110.

Bandura, A. (2001). Social cognitive theory: An agentic perspective. *Annual Review of Psychology*, *52*(1), 1–26.

Blake-Beard, S., Bayne, M. L., Crosby, F. J., & Muller, C. B. (2011). Matching by race and gender in mentoring relationships: Keeping our eyes on the prize. *Journal of Social Issues*, *67*(3), 622–643.

Braxton, J. M., Hirschy, A. S., & McClendon, S. A. (2004). *Understanding and reducing college student departure*. (Vol. 3, Issue 3). San Francisco, CA: Jossey-Bass.

Bureau, U. S. C. (2015). *New census bureau report analyzes U.S. population projections*. Retrieved from https://www.census.gov/newsroom/press-releases/2015/cb15-tps16.html

Byars-Winston, A., Estrada, Y., Howard, C., Davis, D., & Zalapa, J. (2010). Influence of social cognitive and ethnic variables on academic goals of underrepresented students in science and engineering: A multiple-groups analysis. *Journal of Counseling Psychology*, *57*(2), 205.

Campbell, T. A., & Campbell, D. E. (2007). Outcomes of mentoring at-risk college students: Gender and ethnic matching effects. *Mentoring & Tutoring: Partnership in Learning*, *15*(2), 135–148. doi:10.1080/13611260601086287

Clark, M. R. (2006). Succeeding in the city: Challenges and best practices on urban commuter campuses. *About Campus*, *11*(3), 2–8.

Crawford, M. B., Wilson-Kennedy, Z. S., Thomas, G. A., Gilman, S. D., & Warner, I. M. (2018). LA-STEM research scholars program: A model for broadening diversity in STEM education. *Technology & Innovation*, *19*(3), 577–592.

da Silva Cardoso, E., Dutta, A., Chiu, C.-Y., Johnson, E. T., Kundu, M., & Chan, F. (2013). Social-cognitive predictors of STEM career interests and goal persistence in college students with disabilities from racial and ethnic minority backgrounds. *Rehabilitation Research, Policy, and Education*, *27*(4), 271.

DePaoli, J. L., Balfanz, R., & Bridgeland, J. (2016). *Building a grad nation: Progress and challenge in raising high school graduation rates. Annual Update 2016*. Authoring Institution: Civic Enterprises; Johns Hopkins University, Everyone Graduates Center; Alliance for Excellent Education; America's Promise Alliance. Sponsor: AT&T; State Farm; Target Corporation.

Desilver, D. (2017). *US students' academic achievement still lags that of their peers in many other countries*. Retrieved May 19, 2017. http://www.pewresearch.org/fact-tank/2017/02/15/u-s-students-internationally-math-science/

Estrada, M., Woodcock, A., Hernandez, P. R., & Schultz, P. (2011). Toward a model of social influence that explains minority student integration into the scientific community. *Journal of Educational Psychology*, *103*(1), 206.

Kennedy, J. V. (2012). The sources and uses of US science funding. *The New Atlantis*, *36*, 3–22.

Fouad, N. A., & Santana, M. C. (2017). SCCT and underrepresented populations in STEM fields: Moving the needle. *Journal of Career Assessment*, *25*(1), 24–39.

Gregerman, S. R., Lerner, J. S., von Hippel, W., Jonides, J., & Nagda, B. A. (1998). Undergraduate student-faculty research partnerships affect student retention. *The Review of Higher Education*, *22*(1), 55–72.

Hardin, E. E., & Longhurst, M. O. (2016). Understanding the gender gap: Social cognitive changes during an introductory stem course. *Journal of Counseling Psychology*, *63*(2), 233.

Hathaway, R. S., Nagda, B. A., & Gregerman, S. R. (2002). The relationship of undergraduate research participation to graduate and professional education pursuit: An empirical study. *Journal of College Student Development*, *43*(5), 614–631.

Herrera, F. A., & Hurtado, S. (2011). *Maintaining initial interests: Developing science, technology, engineering, and mathematics (STEM) career aspirations among underrepresented racial minority students*. Paper presented at the Association for Educational Research annual meeting, New Orleans, LA.

Hoffman, L. (2009). *Numbers and types of public elementary and secondary schools from the common core of data: School year 2007-08. First Look. NCES 2010-305*. National Center for Education Statistics.

Hossler, D., Ziskin, M., Gross, J. P. K. Kim, S., & Cekic, O. (2009). Student aid and its role in encouraging persistence. In J. C. Smart (Ed.), *Higher education: Handbook of theory and research* (Vol. 24, pp. 389–425). Netherlands: Springer.

Kier, M. W., Blanchard, M. R., Osborne, J. W., & Albert, J. L. (2014). The development of the STEM career interest survey (STEM-CIS). *Research in Science Education*, *44*(3), 461–481.

Lei, S., Gorelick, D., Short, K., Smallwood, L., & Wright-Porter, K. (2011). Academic cohorts: Benefits and drawbacks of being a member of a community of learners. *Education*, *131*(3), 497–504.

Lent, R. W., & Brown, S. D. (2013). Social cognitive model of career self-management: Toward a unifying view of adaptive career behavior across the life span. *Journal of Counseling Psychology*, *60*(4), 557.

Lent, R. W., Brown, S. D., Brenner, B., Chopra, S. B., Davis, T., Talleyrand, R., & Suthakaran, V. (2001). The role of contextual supports and barriers in the choice of math/science educational options: A test of social cognitive hypotheses. *Journal of Counseling Psychology*, *48*(4), 474.

Lent, R. W., Brown, S. D., & Hackett, G. (2002). Social cognitive career theory. *Career Choice and Development*, *4*, 255–311.

Lent, R. W., Miller, M. J., Smith, P. E., Watford, B. A., Lim, R. H., Hui, K. ... Williams, K. (2013). Social cognitive predictors of adjustment to engineering majors across gender and race/ethnicity. *Journal of Vocational Behavior*, *83*(1), 22–30.

Locks, A. M., Hurtado, S., Bowman, N. A., & Oseguera, L. (2008). Extending notions of campus climate and diversity to students' transition to college. *Review of Higher Education*, *31*(3), 257–285.

Lopatto, D. (2004). Survey of undergraduate research experiences (SURE): First findings. *Cell Biology Education*, *3*(4), 270–277.

Lopatto, D. (2007). Undergraduate research experiences support science career decisions and active learning. *CBE—Life Sciences Education*, *6*(4), 297–306.

Planning, L. O. O. B. A. (2017). *LSU's record-setting 2015, 2016, and 2017 spring graduation records. Baton Rouge, Louisiana: Higher Education Research Institution, Louisiana State University.*

Presidential Awards for Excellence in Science, M. E. M. (2012–2018). *Nominations.* Retrieved from http://paesmem.net/nominations

Presidential Awards for Excellence in Science, M. E. M. (2018). *Exceptional science, mathematics and engineering mentors announced.* Retrieved from http://paesmem.net/new-awardees#oa

Progress, N. A. o. E. (2015). *2015 Mathematics & reading assessments.* Retrieved from https://www.nationsreportcard.gov/reading_math_2015/#?grade=4

Quick Facts: East Baton Rouge Parish, Louisiana; UNITED STATES [Median Household Income, (2013_2017)]. (2013–2017). Retrieved from https://www.census.gov/quickfacts/fact/table/eastbatonrougeparishlouisiana,US/PST045218

Salinitri, G. (2005). The effects of formal mentoring on the retention rates for first-year, low achieving students. *Canadian Journal of Education/Revue canadienne de l'education*, *28*(4), 853–873.

Samuelson, J. A. (1982). Metacognition: Thinking about thinking. *The Review of Education/ Pedagogy/Cultural Studies*, *8*(2), 133–141.

Sciences, N. A. o., Engineering, N. A. o., & Medicine, I. o. (2007). *Rising Above the gathering storm: Energizing and employing America for a brighter economic future.* Washington, DC: The National Academies Press.

Sciences, N. A. o., Engineering, N. A. o., & Medicine, I. o. (2010). *Rising above the gathering storm, revisited: Rapidly approaching category 5.* Washington, DC: The National Academies Press.

Seidman, A. (Ed.). (2005). *College student retention: Formula for student success.* Washington DC.

Seymour, E., Hunter, A. B., Laursen, S. L., & DeAntoni, T. (2004). Establishing the benefits of research experiences for undergraduates in the sciences: First findings from a three-year study. *Science Education*, *88*(4), 493–534.

Stanford, J. S., Rocheleau, S. E., Smith, K. P., & Mohan, J. (2017). Early undergraduate research experiences lead to similar learning gains for STEM and Non-STEM undergraduates. *Studies in Higher Education*, *42*(1), 115–129.

Statistics, N. C. f. E. (2017). Indicator 24: STEM degrees. *Status and Trends in the Education of Racial and Ethnic Groups.* Retrieved from https://nces.ed.gov/programs/raceindicators/indicator_reg.asp

Walters, N. B. (1997). *Retaining aspiring scholars: Recruitment and retention of students of color in graduate and professional science degree programs.* ASHE Annual Meeting Paper.

Whalen, S. K. (Ed.). (2017). *2016-17 CSRDE STEM retention report. Fall 2006 through Fall 2015 cohorts of first-time, full-time STEM students in 186 colleges and universities.* Norman, OK: The University of Oklahoma; Center for Institutional Data Exchange and Analysis.

Wilson, Z. S., Holmes, L., Sylvain, M. R., Batiste, L., Johnson, M., McGuire, S. Y. ... Warner, I. M. (2012). Hierarchical mentoring: A transformative strategy for improving diversity and retention in undergraduate STEM disciplines. *Journal of Science Education and Technology*, *21*(1), 148–156.

EMPOWERING UNDERGRADUATE STUDENTS TO LEAD RESEARCH: THE ASCEND PROGRAM AT MORGAN STATE UNIVERSITY

Farin Kamangar, Gillian B. Silver, Christine Hohmann, Shiva Mehravaran and Payam Sheikhattari

ABSTRACT

The focus of this chapter is to describe the methods and results of ASCEND, an innovative program that empowers undergraduate students to lead research projects. ASCEND, which stands for "A Student-Centered Entrepreneurship Development Training Model to Increase Diversity in the Biomedical Research Workforce," is funded by the National Institutes of Health and is being implemented at Morgan State University, a historically black university in Baltimore, Maryland. The results are thus far very promising and show that placing undergraduate students in leading research positions and surrounding them with like-minded peers enhances their sense of science identity, leadership, peer support, and research capabilities. It is hoped that students who participate in ASCEND will pursue graduate training and become future successful biomedical researchers.

Keywords: Biomedical research workforce; entrepreneurial mindset; underrepresented minorities; undergraduate students; science identity; peer support

---

Broadening Participation In STEM: Effective Methods, Practices, And Programs
Diversity in Higher Education, Volume 22, 35–53

ISSN: 1479-3644/doi:10.1108/S1479-364420190000022002

BACKGROUND AND MOTIVATION FOR THE PROGRAM

ASCEND, which stands for "**A** Student-**C**entered **En**trepreneurship **D**evelopment Training Model to Increase Diversity in the Biomedical Research Workforce," is funded by the National Institutes of Health (NIH) and is being implemented at Morgan State University (MSU), a historically black university in Baltimore, Maryland (Kamangar et al., 2017). ASCEND is part of a wider effort, the Diversity Program Consortium (DPC) (Hurtado, White-Lewis, & Norris, 2017), to help address some of the disparities in our biomedical research workforce. African Americans, Hispanics, Native Americans, and Hawaiians/Pacific Islanders make up less than 10% of the faculty research positions in the United States (Valantine & Collins, 2015), while they constitute over 30% of the US population. A major leak in the educational pipeline occurs between undergraduate and graduate degrees. The groups mentioned above collectively earned 22% of US bachelor's degrees in science and engineering in 2014, but they earned only 13% of the doctoral degrees in these fields (National Science Foundation, National Center for Science and Engineering Statistics, 2017). Therefore, in 2014, the NIH announced the **Bu**ilding **I**nfrastructure **L**eading to **D**iversity (BUILD) Initiative (National Institutes of Health, n.d.), as part of the DPC, primarily to encourage and empower undergraduate students to pursue graduate school in the field of biomedical sciences. MSU's ASCEND program is one of the 10 programs selected nationally to receive a BUILD award. BUILD awards, including ASCEND, allow for faculty and institutional development, in addition to student development.

Novelty was a key feature that NIH required for all BUILD programs. Before BUILD, NIH had funded various diversity-related programs, such as Maximizing Access to Research Careers (MARC), for decades. These programs typically provide intensive research mentoring to students in faculty-initiated research, fund students to complete their undergraduate degree, provide role models, facilitate taking the Graduate Record Exam (GRE) and other tests necessary to be admitted to graduate school, and provide travel funds to national research meetings, among other things. While these programs have been very successful in training some distinguished researchers, there are still several challenges to achieving a substantial and sustainable increase in diversity (Krulwich, Saran, & McGee, 2012). Therefore, NIH postulated that additional elements are needed to enhance these training programs and asked for novel programs that addressed other components.

ASCEND uses many of the useful elements of previous programs (e.g., appropriate mentoring and financial assistance), yet has some distinguishing novel features, such as empowering students to take leading positions in research, competing for research funding, and enjoying substantial peer support. We hypothesized that ASCEND's intervention, described below, would enhance students' sense of science identity (Hanauer, Graham, & Hatfull, 2016; Hurtado, Cabrera, Lin, Arellano, & Espinosa, 2009), research competency (Hurtado et al., 2009), self-efficacy (Brown, Lent, Ryan, & McPartland, 1996; Byars-Winston, Estrada, Howard, Davis, & Zalapa, 2010; Chemers,

Zurbriggen, Syed, Goza, & Bearman, 2011; Hanauer et al., 2016; Phillips & Russell, 1994), and peer support (Dennis, Phinney, & Chuateco, 2005).

INSTITUTIONAL CONTEXT

MSU is a historically black university founded in 1867, which celebrated its sesquicentennial in 2017. While the university has always served all races and ethnic groups, the large majority of students are still African American and from underserved backgrounds; close to 70% are recipients of Pell Grants (a federal grant for undergraduate students with financial need). A large percentage of students are first-generation college students who may not have strong support networks and are at high risk of dropping out. Therefore, providing excellent training and a supportive environment remains an important mission of MSU.

Over its 150-year history, the institution has grown from a small biblical institute, with only a handful of students, to an institution with approximately 7,700 undergraduate and graduate students. Currently, the university offers tens of bachelors, masters, and doctoral degrees. Parallel to its growth in numbers and academic disciplines, the university has made major strides to enhance its status as a doctoral-granting research university. In 1988, the state designated the campus as Maryland's Public Urban University (Morgan State University, n.d.). In May 2017, MSU was officially designated (by state law) as Maryland's Preeminent Public Urban Research University. MSU currently has a Carnegie classification of Doctoral University – Moderate Research (R3) and plans to obtain a classification of Doctoral University – Higher Research (R2).

The training and missions of MSU, as well as student body composition, made it an excellent environment to study the effects of a novel intervention to increase diversity in the research workforce.

CONCEPT AND RATIONALE FOR THE ASCEND'S TRAINING MODEL

The concept of ASCEND – that is, empowering a person without substantial research experience to lead research projects – has its roots in primary health care research. The Almaty (Alma-Ata) Declaration of 1978, a major milestone in the field of public health, identified primary health care as the key to reaching the goal of "Health for All" (World Health Organization, UNICEF, 1978). Conducting health research in remote areas of developing countries was challenging, as the number of local experts was insufficient, and outside experts were not intimately familiar with the health issues of those areas. Therefore, the World Health Organization (WHO) offered health administrators and health workers in these areas short-term (one- to three-week) workshops on how to select a research topic, how to write a proposal, and how to collect data. The workshops were followed by several months of data collection and then another workshop on statistical analysis and interpretation of data. The underlying assumption was that the local health administrators and health workers would know of and choose to work on problems that are prevalent and important in their region. Given that they were the owners of the research, they persevered and were often able to collect the data, even though they were not professional researchers.

In the early 1990s, Iranian medical universities adopted this WHO model and organized one-week workshops for medical students on how to select a research topic, how to write a research proposal, and how to conduct a research study. These universities also established student research centers (SRCs) as a place for peer support and the exchange of ideas, as well as periodic workshops on a variety of topics such as basic research, life skills, and so on (Rezaee-Zavareh, Khosravi, & Alavian, 2016). While formal evaluations have not been conducted, there is indirect evidence that SRCs have been successful. After 25 years, these centers are still funded and have expanded from one university in 1993 (Tehran University of Medical Sciences) to over 20 universities across the country. Many currently successful Iranian biomedical researchers are former members of these SRCs. The number of Iranian publications indexed in Scopus rose from 1,174 in 1998 to 51,112 in 2016; this 43-fold increase may be partly attributed to student training programs that started in the early 1990s.

The ASCEND program is adopting and improving upon the research training model described above. We hypothesize that using this model for undergraduate research training in historically black colleges and universities (HBCU) will increase the students' long-term interest in health research, partly by increasing their research motivation, research competency, sense of self-efficacy, and sense of science identity (Byars-Winston et al., 2010; Campbell et al., 2014; Carlone & Johnson, 2007; Estrada et al., 2016; Hanauer et al., 2016; Trujillo & Tanner, 2014). Empowering students to take leading roles in research creates an internal locus of control, which is positively correlated with job satisfaction and performance (Judge & Bono, 2001) as well as self-efficacy (Sherer et al., 1982). Being involved in the entire research process should also contribute to self-efficacy, which has been associated with commitment to research (Brown et al., 1996; Chemers et al., 2011; Phillips & Russell, 1994).

ASCEND's model is consistent with scientific findings that autonomy, a sense of purpose (meaning), mastery, a sense of progress, a sense of membership (belonging), and little fear are important contributors to happiness at work and the urge to continue. In her book *Evolve!*, Rosabeth Moss Kanter, a professor at Harvard Business School and Chair and Director of the Harvard Advanced Leadership Initiative, suggests three primary sources of motivation and initiative in high-innovation companies: meaning, membership, and mastery (Kanter, 2001); all of these elements have been included in the ASCEND model. Choosing the research topic, either individually or via a democratic group process, may lead to a topic that is meaningful to students. Membership and peer support are emphasized in all parts of ASCEND. The ASCEND Scholars work toward mastery as they experience the entire research process, from selecting a topic to presenting findings to colleagues.

Harvard Business School professor Teresa Amabile and psychologist Steven Kramer reviewed diaries from nearly 12,000 work days and found that the happiest and most productive days were those marked by a sense of progress

(Amabile & Kramer, 2010). A meta-analysis of data from 420,599 people in over 63 countries suggested that autonomy is a very important predictor of personal happiness and well-being, more so than wealth (Fischer & Boer, 2011). Further, in his book *Drive*, Daniel Pink argues that the motivation trifecta is autonomy, mastery, and a sense of purpose (Pink, 2011). These findings from work satisfaction in business are echoed in academia. A national survey shows that faculty at public universities are most satisfied (above all other factors) because their work provides flexibility for determining course content (90.6%) and autonomy and independence (85.8%) (M. Eagan et al., 2014). We posit that these factors will also apply to the undergraduates we are training to become self-directed researchers (Candy, 1991).

ASCEND'S STUDENT-CENTERED RESEARCH TRAINING INITIATIVES

ASCEND is implementing three student training initiatives: the SRC, an annual Summer Research Institute (SRI), and a two-year scholarship program (ASCEND Scholars Program). All three initiatives are intended to enhance peer support, sense of science identity, research competency, and sense of self-efficacy. All research training and project topics are health-related and within the mission of NIH. Fig. 1 illustrates the ASCEND training program's logic model. The results thus far show that the SRC is highly sustainable and scalable, but the other components have also shown remarkable success.

The SRC: The SRC is an MSU-registered student organization created to attract, train, and enable its student members to conduct health research. The SRC has a dedicated renovated suite (the renovations were supported by ASCEND) that includes a student lounge and two meeting spaces with the state-of-the-art technology. The SRC is open to all full-time undergraduate students with a grade point average of 2.75 or higher.

The center is designed to be *of* the students, *for* the students, and *run by* students (with the support of a staff coordinator and a faculty advisor). The SRC

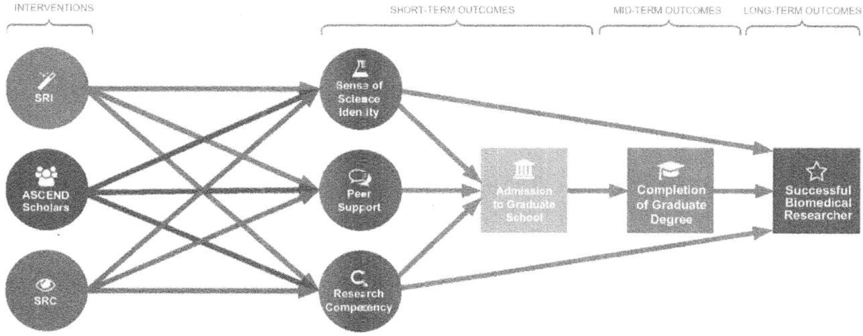

Fig. 1. Logic Model for ASCEND Student Programs.

was founded in 2016, with 37 founding members, of whom 9 were selected by the faculty advisor to serve as executive members. The founding members wrote the constitution and bylaws of the SRC and started various activities of their choosing, including forming research clubs, organizing journals clubs, and providing community service. Over time, student ownership and leadership grew. In 2017, the executive board was selected by SRC members themselves rather than the faculty advisor, and the new board devised more activities and attracted more students. The membership grew from 37 in February 2016 to 133 in February 2018, which is an initial indicator of success. Members are students of various disciplines (biology, psychology, public health, architecture, political science), and interdisciplinary research and collaborations are highly encouraged.

Two salient SRC activities include the Health Research Concepts Competition (HRCC) and forming research clubs. To participate in the HRCC, student applicants are required to submit a 600-word abstract on an innovative health research concept, along with a letter of support from a faculty mentor. For each round of competition, the student with the winning concept is awarded a US$1,000 prize. Concepts that are found to be meritorious advance to the proposal submission phase and are eligible for up to US$5,000 in funding support. As of February 2018, there has been one round of competition, and a second round has been announced. The research clubs provide an opportunity for students to have discussions and activities that focus on a specific area of health research. So far, two research clubs have been formed, the Design-Health Club and the Students Mentored in Applied Research Techniques (SMART) Club, and they each have faculty advisement.

To encourage students' increased involvement in activities and the decisions that affect them, the 2017–2018 Executive Board implemented the SRC Membership Incentive Program. This program consists of four levels of membership: bronze, silver, gold, and diamond (Appendix 1). This point system (Appendix 2) includes several dimensions: leadership, research, academic excellence, and community service. Achieving higher levels offers members benefits that will help with their academic and career success. Appendix 3 outlines the benefits that the SRC offers.

The SRC is designed to empower students to develop leadership skills, receive peer support from like-minded students, own their research, and identify as scientists. Initial evaluation results are shown later in this chapter.

The SRI. The SRI, another ASCEND initiative, is a research training camp held annually for 25–30 students, usually rising sophomores or juniors, from multiple disciplines (e.g., biology, psychology, chemistry, engineering). Students attending this eight-week workshop-style program familiarize themselves with the concept of health, the branches of science that contribute to health knowledge, the scientific method, and how scientists think and work. By the end of the program, the students write and receive feedback on a preliminary research proposal. The students learn to work in interdisciplinary teams, and the teams are guided by a (also interdisciplinary) team of five faculty members and five near-peer mentors. This interdisciplinary model mimics the real research world

(National Academies of Sciences, Engineering, and Medicine, 2005) and also fosters the development of communities of practice across MSU's campus (Ladeji-Osias et al., 2015). In accordance with ASCEND's principle of placing students in leading positions, the participants select their own research topics and develop their own proposals. We hypothesize that proposing research topics, discussing the feasibility of each topic, and writing proposals contribute to greater competency and self-efficacy in research.

The ASCEND Scholars Program: The goals of this two-year training program are to enhance students' research competency and self-efficacy, enhance their science identity, and provide a supportive environment of peers, near-peers (graduate students), and faculty mentors. Each year, 20 of the SRI participants who meet the selection criteria are accepted into the ASCEND Scholars Program. During the program, ASCEND Scholars – with the support of mentors – polish the research proposals that they started during the SRI, submit Institutional Review Board (IRB) or Institutional Animal Care and Use Committee (IACUC) applications, and then submit the proposals to ASCEND leadership to be considered for internal funding; if approved, they then conduct their research projects. To mimic real research conditions faced at universities – and to strengthen the sense of entrepreneurship – not all proposals are funded; only those that are considered competitive at an undergraduate research level are funded. The remaining 5–10 SRI students who are not selected to be ASCEND Scholars can choose to become an ASCEND Associate or participate in another MSU research training program such as RISE if they wish to receive research training; on the other hand, they may find that they are not interested in continuing down the research path. ASCEND Associates do not receive tuition support or a stipend from ASCEND but can participate in many of the ASCEND Scholars Program's activities such as the research project and workshops.

The Scholars also undertake a host of other activities to assist them with their undergraduate studies and to prepare them for graduate school. Examples of these activities include the following: working with faculty mentors to prepare individual development plans, attending weekly interdisciplinary seminars, traveling to (and some presenting at) scientific meetings such as the Annual Biomedical Research Conference for Minority Students (ABRCMS), and participating in writing seminars, GRE preparation classes (for juniors), mock graduate school interviews, résumé writing and building, and critical-thinking workshops. Also, each semester, the Scholars are required to register for and attend an ASCEND-specific course. The course has a student-facilitated seminar format and integrates participants' research experiences with the science to facilitate the understanding of major health care/prevention theories, principles, practices, and procedures. This course is important for its content but also for the dedicated time it provides students to work with their groups on their research projects, and the opportunity faculty and near-peer mentors have to monitor and support the Scholars.

ASCEND'S DISTINGUISHING FEATURES

ASCEND implements several components that enhance it beyond traditional programs. One distinguishing feature of ASCEND is its emphasis on students taking the lead in research. For example, during the SRI, each group chooses one of the research topics put forth by their team members and develops a proposal under the guidance of faculty and near-peer mentors. Examples of students undertaking leadership roles and initiating health research concepts through the SRC were provided earlier.

A second important feature of ASCEND is that students experience the entire research process from soup to nuts: they come up with a topic, write a proposal, apply for ethics approval and funding, conduct the research, analyze the data, and disseminate the results. This process should enhance research competency and self-efficacy and thus help students feel substantially more at ease when they enter graduate school, where they are expected to take the lead and complete their research (Chang, Eagan, Lin, & Hurtado, 2011).

A third important feature of ASCEND is its heavy emphasis on teamwork and peer support. This is manifested in the SRC (an environment in which students can learn from each other), the ASCEND SRI (group work on proposals and group presentations), and the ASCEND Scholars Program (cohorts of 20 students). Peer support and peer-led teaching are especially important for minority students who may have fewer scientific role models in their families and social networks (Ghee, Keels, Collins, Neal-Spence, & Baker, 2016; Salto, Riggs, Delgado De Leon, Casiano, & De Leon, 2014; Snyder, Sloane, Dunk, & Wiles, 2016; Toven-Lindsey, Levis-Fitzgerald, Barber, & Hasson, 2015; Vaughan, Sanders, Crossley, O'Neill, & Wass, 2015).

The fourth feature is competition. Only some of the proposals submitted by the SRC members are considered meritorious and will be funded. Likewise, during the SRI, each group of trainees votes to choose a topic from among those suggested by peers; only a subgroup of all proposals developed through the SRI is funded.

A fifth feature is encouraging interdisciplinary science. The SRC members, SRI students, and ASCEND Scholars are from highly diverse fields, from biology to psychology and engineering, supported by interdisciplinary seminars and conferences.

Some of the features mentioned above have been used individually in previous training programs to varying degrees. Collectively, they make ASCEND unique. Most other undergraduate research training programs use what we call the apprenticeship training model: students are placed in a lab where they conduct and master a specific research technique (e.g., running gels), gaining more mastery and independence over time. While the apprenticeship model has been helpful in training many distinguished scientists over the years, one potential shortcoming is that the students may not understand how their efforts contribute to the overall project goals, and some may abandon the idea of pursuing a career in research. Another potential disadvantage is that this model is very resource-intensive (Harrison & Grant, 2015), typically requiring one-on-one mentoring, with training outcomes often being critically dependent on the quality of the mentoring environment (M. J. Eagan et al., 2013; Weinberg & Lankau, 2011). Typically, there is less peer support and less competition than what ASCEND offers.

In contrast to apprenticeship, we call what ASCEND offers an "entrepreneurship" model for training undergraduate students in biomedical research. In the context of ASCEND, the term "entrepreneurship" has a slightly different meaning than in its business context. In the business context, entrepreneurs are those who have novel ideas, own their idea, take a risk, and develop a product that is intended to be marketed. In the context of ASCEND, students will be entrepreneurial researchers. They will develop their individual research idea (rather than a faculty member's research idea), own the idea, and develop a proposal, and the product will be the students' research and the scientific knowledge gained.

EVALUATION OF THE SUCCESS OF THE ASCEND TRAINING MODEL

ASCEND's student training efforts are designed to make students successful biomedical researchers. In the short-term, we expect that the activities will enhance students' sense of science identity, research competency, and peer support, which lead to higher admission rates to graduate school. Fig. 1 illustrates ASCEND's intended short-term, mid-term, and long-term outcomes. At this point, only data on short-term outcomes (i.e., sense of science identity, peer support, research competency, and admission to graduate school) are available or being collected for the SRC, SRI, and ASCEND Scholars Program. In the following section, we provide some of the quantitative and qualitative evaluation results for the SRC as well as initial results from the quantitative evaluation of the SRI.

Quantitative evaluation of the SRC: A questionnaire with 27 items was prepared using items from several previously administered surveys to evaluate the effectiveness of the SRC in enhancing four domains: (1) peer support for research and science, (2) sense of science identity, (3) research self-efficacy, and (4) social and academic self-efficacy. Peer support questions asked about the number of friends with whom they can share scientific work. The items in domains 2–5 were on a Likert scale. The respondents were asked to provide two responses to each item; one current and one reflective about the time before becoming an SRC member. The questionnaire was completed by 46 SRC members. Comparing now to before SRC membership, all domains showed statistically significant improvements, with large effect sizes. Responses to each domain showed very good psychometric properties, with high Cronbach's αs, ranging from 0.70 to 0.91 for each item. The number of friends with whom study participants had scientific and research interactions increased from a median of 2.9–7.5 (Wilcoxon rank sum p-value < 0.001). On a scale of 1–5, the sense of science identity increased from a mean of 3.1 to 4.4 (paired t-test p-value < 0.001), the mean research self-efficacy score increased from 2.5 to 4.1

(paired t-test p-value < 0.001), and the mean social and academic self-efficacy increased from 3.4 to 4.3 (paired t-test p-value < 0.001). Longer duration of membership was associated with larger increases in each of these items. After joining the SRC, these four items (sense of peer support, science identity, research self-efficacy, and social and academic self-efficacy) showed very strong positive correlations with each other.

Qualitative assessment of the SRC: The results shown above are consistent with the findings of a qualitative study that our team has conducted. SRC members who participated in this study were of the belief that they owe the SRC for their positive changes, accomplishments, and aspirations. They agreed that the SRC has a welcoming environment where they feel a sense of belonging and form strong bonds with like-minded students; engages them in activities that improves their social, academic, leadership, and research skills; helps them set up a clear direction and inspires them to aim higher in their career paths; provides the opportunity for networking with students and faculty at other HBCUs and minority institutions through conferences and workshops; and offers workshops that help them prepare for successful graduate school applications. When asked about challenges and needs, most SRC students have stated that they are extremely satisfied with their SRC experience, and they would not change anything about it.

In summary, both quantitative and qualitative results collected thus far suggest that SRC is effective in increasing students' sense of science identity, peer support, research self-efficacy, and social and academic self-efficacy.

Quantitative assessment of the SRI: Students' self-perception of their research knowledge and skills was tested using an eight-question survey. At the end of the summer, the students were asked to score themselves for their knowledge after the SRI and also reflect back to the beginning of the summer and score where they were back then. The results showed a substantial increase in perceived research knowledge and capabilities. For example, in SRI 2016, on a 10-point scale, the mean self-perceived score of knowledge was 8.2 after the SRI, compared to 6.0 when they reflected back to the beginning (paired test p-value $<$ 0.001). The Cronbach's α was 0.80 for this instrument.

ENGAGING STAKEHOLDERS

To enhance the chances for success of this program, the ASCEND research team made substantial efforts to engage stakeholders, among whom are MSU students, faculty members, and institutional leaders as well as NIH officials. Student engagement was made possible through their own leadership, as described above. Many faculty at MSU are highly committed to the success of their students and were therefore engaged with ASCEND for that reason. Further, faculty engagement was made possible by providing funds for enhancing their research (e.g., by providing funds to purchase equipment and travel) or enhancing student-centered pedagogy. The ASCEND research team gave several presentations to university officials (including to the Board of Regents and the MSU President and his cabinet) about the impact of ASCEND on the institution. Given that the goals of ASCEND (e.g., enhancing student success, enhancing the research status of the university) are highly aligned with those of the university, the university administration is highly supportive of the program. NIH officials are heavily engaged in working with the ASCEND program and monitoring its progress and therefore established the BUILD initiative using the cooperative agreement mechanism, which necessitates a high level of interaction.

SUSTAINING THE ASCEND INTERVENTIONS BEYOND THE NIH FUNDING PERIOD

ASCEND's activities are currently funded by the NIH BUILD grant awards. It is imperative to find ways to sustain the effective interventions beyond the funding period. Some potential sustainability strategies are (1) to integrate the effective interventions within the routine university operations and (2) to seek alternative sources of funding. An example of the former is transforming the SRI curriculum and experience into an approved university course. Alternative sources of funding can be federal awards, state funds, or philanthropic donations. The fact that ASCEND's goals are aligned with MSU's strategic goals helps in securing such funds. Included among MSU's five main strategic goals are enhancing student success and enhancing MSU's status as a doctoral research university. To obtain such funds, the intervention should be demonstrated to be highly effective and not exorbitantly expensive.

A good example is the SRC, which has shown to be very effective, as indicated by increasing student membership and favorable quantitative and qualitative evaluations. Maintaining the SRC does not require substantial resources; a part-time faculty advisor and a staff member satisfy the initial needs. A relatively small amount of funding will also be needed to support student-initiated projects and benefits for high-ranking members.

Student research activities and the pursuit of graduate school happen only in a favorable, nurturing milieu. The fact that MSU is planning to enhance its status as a research institution is a major asset. Using the NIH BUILD award funds, the ASCEND program has strived to improve the faculty's research competencies, strengthen MSU's research infrastructure, enhance the culture of research within MSU, and strengthen MSU's collaborations with research-intensive institutions. These activities, which are summarized below, enhance the chances of obtaining further grants and availability of faculty with research expertise to work with students and thus sustaining student research activities.

Faculty development. ASCEND offers grant writing workshops, pilot research project grants, community-based participatory research grants, funds for travel to conferences, course redesign awards, statistical support, and mentoring workshops. Grant writing workshops are intended to familiarize junior faculty with the NIH grant writing process. ASCEND faculty and staff organize several of these workshops each year, on topics such as the principles of grant writing, writing specific aims, preparing budgets, preparing NIH-format biosketches, and revising proposals based on reviewer comments.

ASCEND also provides funds for course redesign awards, the purpose of which is to increase students' interest in biomedical research by encouraging faculty to make their undergraduate health-related courses more interdisciplinary, student-centered, and research-training focused and thus more appealing to students. Some students can be turned off by "boring" science classes early in their college years and thus leave STEM majors before they get a chance to participate in biomedical research (Gasiewski, Eagan, Garcia, Hurtado, & Chang, 2012). Research over the past decades have shown that students learn and retain information more effectively if they are actively engaged in constructing their own knowledge (Miller, Pfund, Pribbenow, & Handelsman, 2008). ASCEND funds two to six awards for such curricular changes every year.

Infrastructure development: ASCEND has renovated three physical spaces in the science complex: two Active Learning Centers and one space for the SRC (described previously). The Active Learning Centers – classrooms with modular furniture, whiteboards, and other student-centered learning features – are intended to enhance group work and further engagement of students in learning activities. ASCEND funds are also being used to purchase furniture and technology to make additional classrooms better for student-centered learning.

To increase MSU's capacity to conduct biomedical research, specifically by providing laboratory training and services for faculty and students, ASCEND has established a Core Laboratory. ASCEND has purchased equipment and is supporting technical staff for the Core Laboratory, which will offer preparatory technology, separations technology, analytics technology, histologic analytics technology, and augmentation of biomedical teaching, such as 3-D human anatomy and virtual microscopy.

Research collaborations. ASCEND's local research collaborators include faculty at Johns Hopkins University and the University of Maryland. These research partners review pilot grant proposals submitted to ASCEND, give talks at SRI and SRC seminars, host field trips for SRI students, judge proposals that result from the SRI, host ASCEND Scholars in their labs to work on population-based research studies, and provide the Scholars with guidance on successfully applying to graduate programs. In addition, they may partner with, or mentor, some of our junior faculty.

ASCEND also plays an important role in cultivating the culture of research at MSU, by, for example, increasing the number of MSU faculty who are knowledgeable about NIH (and similar) research grant processes, creating a community of practice through which faculty share new pedagogical techniques and research skills, and enhancing collaborations with research-intensive institutions. ASCEND has taken several measures to enhance the faculty's research skills, improve the research infrastructure, and provide additional resources, which are described above. Some years from now, if successful outcomes are achieved, the institution will have faculty who are more familiar with research and will be able to secure more research grants, thus having resources to train students.

CHALLENGES AND SOLUTIONS

Creating a training program that puts the students in leading research positions has been a wonderful experience but has been challenging at times and required substantial rethinking or restrategizing.

While we did have an overall idea of what the SRC was going to be, we let the students determine the direction of the growth and development of this center. This practice, while clearly important and leading to stellar results, required long hours of programming, intent listening, synthesis, and implementation. The faculty mentor and staff had several retreats with student leadership to provide an environment that allows for creativity and debate.

Students often come up with research topics that are too broad, implausibly difficult to conduct, or not feasible with the available expertise and equipment at MSU. The key is to find a balance between leading the students in the right direction and allowing freedom for their creativity and leadership. It is also important to search for services that accommodate the students as much as possible. For example, when ASCEND's core laboratory or faculty labs cannot accommodate the study, we may need to find research laboratories at one of ASCEND's research partner institutions. It is important to consider incentives that encourage such collaborators to allow the use of their labs.

Another challenge is the multidisciplinary nature of the program. Students and faculty come from multiple disciplines (e.g., public health, biology, chemistry, and psychology). While all of these disciplines are relevant to health, they each use some of their own terminology and research methods, which may make training and group work difficult. Our solution to this problem was that the instructors and near-peer mentors spend some time together before the SRI to discuss issues and harmonize language and terms across disciplines. In addition, the SRI's curriculum includes important cross-cutting concepts, such as the meaning of health, general research methodology, and proposal writing.

Another challenge has been finding the right mentor, one that has similar research interests and expertise to those of the mentees. The students come up with a variety of research topics, and appropriate mentors are usually – but not always – available among MSU or research partner faculty. A solution is to ask faculty who are interested in working with students to lecture during the SRI. Our experiences over the past two years show that students' proposed topics are sometimes inspired by what they hear from their faculty and near-peer mentors. While the students feel that they own the topic, they are also proposing a research topic for which a good research mentor is available. Another solution is what the SRC has started as the HRCC program. Each HRCC project is endorsed by a faculty member, therefore, guaranteeing that the right mentor is available.

Further, time management has been a challenge for students. ASCEND Scholars have to spend substantial time polishing their research proposals, conducting their research projects, and attending a variety of seminars and workshops. These activities could potentially take time away from their other studies and may impact their GPA, which may negatively affect the success of their graduate school applications. Solutions to date include consolidating some of the activities into a weekly for-credit course, adjusting some of the program requirements as well as supporting students through life-skill classes. Another solution has been engaging students in the SRC, where they conduct projects at their own pace.

In all cases, program faculty and staff are responsive to issues as they arise to help the students achieve their goals and experience success at MSU.

LESSONS LEARNED AND RECOMMENDATIONS

Some components of the ASCEND program, particularly the SRC and SRI, have shown great promise in enhancing research skills, peer support, sense of leadership, and sense of science identity among students. However, ASCEND is still a work in progress. While the main focus will remain on student leadership and student-centered activities, we will continue to conduct evaluations and fine-tune the interventions. We will also try to test the most successful aspects in other settings and environments, primarily to assess replicability and scalability. We will share the results and experiences gained with other institutions.

FUNDING

The work reported in this publication was supported by the NIH Common Fund and Office of Scientific Workforce Diversity (USA). The publication of this article was supported by NIGMS under three linked awards: UL1GM118973, RL5GM118972, and TL4GM118974. The work is solely the responsibility of the authors and does not necessarily represent the official view of the NIH.

ACKNOWLEDGMENTS

The writing team acknowledges the efforts of all the members of the ASCEND team. In particular, we wish to thank Drs Jocelyn Turner-Musa, Cleo Hughes-Darden, Gloria Hoffman, and Michael Koban for their continued work with the project. We also greatly appreciate the literature review provided by Dr Jummai Apata.

REFERENCES

Amabile, T. M., & Kramer, S. J. (2010). *What really motivates workers.* http://www.hbs.edu/faculty/Pages/item.aspx?num=37331

Brown, S. D., Lent, R. W., Ryan, N. E., & McPartland, E. B. (1996). Self-efficacy as an intervening mechanism between research training environments and scholarly productivity: A theoretical and methodological extension. *The Counseling Psychologist*, *24*(3), 535–544. doi:10.1177/0011000096243012

Byars-Winston, A., Estrada, Y., Howard, C., Davis, D., & Zalapa, J. (2010). Influence of social cognitive and ethnic variables on academic goals of underrepresented students in science and engineering: A multiple-groups analysis. *Journal of Counseling Psychology*, *57*(2), 205–218. doi:10.1037/a0018608

Campbell, A. G., Skvirsky, R., Wortis, H., Thomas, S., Kawachi, I., & Hohmann, C. (2014). NEST 2014: Views from the trainees-talking about what matters in efforts to diversify the STEM workforce. *CBE Life Sciences Education*, *13*(4), 587–592. doi:10.1187/cbe.14–04-0068

Candy, P. (1991). *Self-direction for lifelong learning. A comprehensive guide to theory and practice*. San Francisco, CA: Jossey-Bass.

Carlone, H. B., & Johnson, A. (2007). Understanding the science experiences of successful women of color: Science identity as an analytic lens. *Journal of Research in Science Teaching, 44*(8), 1187–1218. doi:10.1002/tea.20237

Chang, M. J., Eagan, M. K., Lin, M. H., & Hurtado, S. (2011). Considering the impact of racial stigmas and science identity: Persistence among biomedical and behavioral science aspirants. *The Journal of Higher Education, 82*(5), 564–596. doi:10.1353/jhe.2011.0030

Chemers, M. M., Zurbriggen, E. L., Syed, M., Goza, B. K., & Bearman, S. (2011). The role of efficacy and identity in science career commitment among underrepresented minority students. *Journal of Social Issues, 67*(3), 469–491. doi:10.1111/j.1540-4560.2011.01710.x

Dennis, J., Phinney, J., & Chuateco, L. (2005). The role of motivation, parental support, and peer support in the academic success of ethnic minority first-generation college students. *Journal of College Student Development, 46*(3), 223–236.

Eagan, M. J., Hurtado, S., Chang, M. J., Garcia, G. A., Herrera, F., & Garibay, J. (2013). Making a difference in science education: The impact of undergraduate research programs. *American Educational Research Journal, 50*(4), 683–713. doi:10.3102/0002831213482038

Eagan, M., Stolzenberg, E., Lozano, J., Aragon, M., Suchard, M., & Hurtado, S. (2014). *Undergraduate teaching faculty: The 2013-2014 HERI faculty survey*. Los Angeles, CA: UCLA.

Estrada, M., Burnett, M., Campbell, A. G., Campbell, P. B., Denetclaw, W. F., Gutiérrez, C. G., ... Zavala, M. (2016). Improving underrepresented minority student persistence in STEM. *CBE Life Sciences Education, 15*(3). doi:10.1187/cbe.16-01-0038

Fischer, R., & Boer, D. (2011). What is more important for national well-being: Money or autonomy? A meta-analysis of well-being, burnout, and anxiety across 63 societies. *Journal of Personality and Social Psychology, 101*(1), 164.

Gasiewski, J. A., Eagan, M. K., Garcia, G. A., Hurtado, S., & Chang, M. J. (2012). From gatekeeping to engagement: A multicontextual, mixed method study of student academic engagement in introductory STEM courses. *Research in Higher Education, 53*(2), 229–261. doi:10.1007/s11162-011-9247-y

Ghee, M., Keels, M., Collins, D., Neal-Spence, C., & Baker, E. (2016). Fine-tuning summer research programs to promote underrepresented students' persistence in the STEM pathway. *CBE Life Sciences Education, 15*(3), 28. doi:10.1187/cbe.16-01-0046

Hanauer, D. I., Graham, M. J., & Hatfull, G. F. (2016). A measure of college student persistence in the sciences (PITS). *CBE Life Sciences Education, 15*(4), 54. doi:10.1187/cbe.15-09-0185

Harrison, S., & Grant, C. (2015). Exploring of new models of research pedagogy: Time to let go of master-apprentice style supervision? *Teaching in Higher Education, 20*(5), 556–566. doi:10.1080/13562517.2015.1036732

Hurtado, S., Cabrera, N., Lin, M., Arellano, L., & Espinosa, L. (2009). Diversifying science: Underrepresented student experiences in structured research programs. *Research in Higher Education, Research in Higher Education, 50*(2), 189–214. doi:10.1007/s11162-008-9114-7

Hurtado, S., White-Lewis, D., & Norris, K. (2017). Advancing inclusive science and systemic change: The convergence of national aims and institutional goals in implementing and assessing biomedical science training. *BMC Proceedings, 11*(12), 17. doi:10.1186/s12919-017-0086-5

Judge, T. A., & Bono, J. E. (2001). Relationship of core self-evaluations traits – Self-esteem, generalized self-efficacy, locus of control, and emotional stability – With job satisfaction and job performance: A meta-analysis. *The Journal of Applied Psychology, 86*(1), 80–92.

Kamangar, F., Silver, G., Hohmann, C., Hughes-Darden, C., Turner-Musa, J., Haines, R. T., ... Sheikhattari, P. (2017). An entrepreneurial training model to enhance undergraduate training in biomedical research. *BMC Proceedings, 11*(12), 18. doi:10.1186/s12919-017-0091-8

Kanter, R. M. (2001). *Evolve: Succeeding in the digital culture of tomorrow*. Boston, MA: Harvard Business Review Press.

Krulwich, T. A., Saran, S., & McGee, R. (2012). Developing talent to increase diversity in biomedical sciences workforce. *Mount Sinai Journal of Medicine, 79*(4), 497. doi:10.1002/msj.21330

Ladeji-Osias, J., Hohmann, C., Hargett, S., Brown, L., Hughes-Darden, C., & Reece, M. (2015). The impact of undergraduate research in STEM at Morgan State University on the production of doctoral degrees in engineering and the sciences. In J. McClinton, M. A. Melton, C. R. Jackson, K. Engerman (Eds.), *Infusing undergraduate research into Historically Black Colleges and Universities Curricula (Diversity in Higher Education, Volume 17)* (pp. 143–175), BioMed Central.

Miller, S., Pfund, C., Pribbenow, C., & Handelsman, J. (2008). Scientific teaching in practice. *Science (New York, N.Y.)*, *322*(5906), 1329–1330. doi:10.1126/science.1166032

Morgan State University (n.d.). Brief History of Morgan State University. Retrieved from http://www.morgan.edu/about/history.html

National Academies of Sciences, Engineering, and Medicine. (2005). *Facilitating interdisciplinary research.* Retrieved from http://www.nap.edu/catalog/11153/facilitating-interdisciplinary-research

National Institutes of Health.(n.d.). Building Infrastructure Leading to Diversity (BUILD) Initiative. Retrieved from https://www.nigms.nih.gov/training/dpc/Pages/build.aspx.

National Science Foundation, National Center for Science and Engineering Statistics. (2017). *Women, minorities, and persons with disabilities in science and engineering: 2017* (Special Report NSF 17-310). Arlington, VA. Retrieved Tables 5-6 and 7-7, from https://www.nsf.gov/statistics/2017/nsf17310/data.cfm

Phillips, J. C., & Russell, R. K. (1994). Research self-efficacy, the research training environment, and research productivity among graduate students in counseling psychology. *The Counseling Psychologist*, *22*(4), 628–641. doi:10.1177/0011000094224008

Pink, D. (2011). *Drive: The surprising truth about what motivates us.* New York, NY: Riverhead Books.

Rezaee-Zavareh, M. S., Khosravi, M. H., & Alavian, S. M. (2016). Student research committee as a model for development of medical research projects. *Archives of Iranian Medicine*, *19*(11), 820. doi:0161911/AIM.0014

Salto, L. M., Riggs, M. L., Delgado De Leon, D., Casiano, C. A., & De Leon, M. (2014). Underrepresented minority high school and college students report STEM-pipeline sustaining gains after participating in the Loma Linda University Summer Health Disparities Research Program. *PloS One*, *9*(9), e108497. doi:10.1371/journal.pone.0108497

Sherer, M., Maddux, J. E., Mercandante, B., Prentice-Dunn, S., Jacobs, B., & Rogers, R. W. (1982). The self-efficacy scale: Construction and validation. *Psychological Reports*, *51*(2), 663–671. doi:10.2466/pr0.1982.51.2.663

Snyder, J. J., Sloane, J. D., Dunk, R. D. P., & Wiles, J. R. (2016). Peer-led team learning helps minority students succeed. *PLoS Biology*, *14*(3), e1002398. doi:10.1371/journal.pbio.1002398

Toven-Lindsey, B., Levis-Fitzgerald, M., Barber, P. H., & Hasson, T. (2015). Increasing persistence in undergraduate science majors: A model for institutional support of underrepresented students. *CBE Life Sciences Education*, *14*(2), 12. doi:10.1187/cbe.14-05-0082

Trujillo, G., & Tanner, K. D. (2014). Considering the role of affect in learning: Monitoring students' self-efficacy, sense of belonging, and science identity. *CBE Life Sciences Education*, *13*(1), 6–15. doi:10.1187/cbe.13-12-0241

Valantine, H. A., & Collins, F. S. (2015). National Institutes of Health addresses the science of diversity. *Proceedings of the National Academy of Sciences of the United States of America*, *112*(40), 12240–12242. doi:10.1073/pnas.1515612112

Vaughan, S., Sanders, T., Crossley, N., O'Neill, P., & Wass, V. (2015). Bridging the gap: The roles of social capital and ethnicity in medical student achievement. *Medical Education*, *49*(1), 114–123. doi:10.1111/medu.12597

Weinberg, F. J., & Lankau, M. J. (2011). Formal mentoring programs: A mentor-centric and longitudinal analysis. *Journal of Management*, *37*(6), 1527–1557. doi:10.1177/0149206309349310

World Health Organization, UNICEF. (1978). *Primary health care: Report of the International Conference on Primary Health Care, Alma-Ata, USSR.* Retrieved from http://www.who.int/iris/handle/10665/39228

APPENDIX 1: SRC INCENTIVE MEMBERSHIP-LEVEL REQUIREMENTS

Level	Requirements
Bronze	• Submit required documents (current transcript, current cv/resume, semester course schedule, signed photo release forms, and completed research interest(s) form) within 48 hours of initial membership acceptance • Attend two SRC-sponsored events per semester • Attend two SRC-sponsored community service events per semester • Attend one general body meeting per semester • *Cumulative GPA of 2.75*
Silver	• All bronze-level requirements • *Cumulative GPA of 3.0* • *60 accumulated points*
Gold	• All bronze-level requirements • *Cumulative GPA of 3.2* • *90 accumulated points*
Diamond	• All bronze-level requirements • *Cumulative GPA of 3.5* • *120 accumulated points*

APPENDIX 2: SRC MEMBERSHIP INCENTIVE PROGRAM POINT SYSTEM

Contribution	Points Awarded
Befriend/Like/Follow ASCEND and the SRC on social media	1 pt. per network
Visiting the SRC	1 pt. per day
Attending an additional SRC-sponsored event	2 pts.
Attending an additional SRC-sponsored community service event	2 pts. per hour
Bringing a friend to an SRC event	2 pts. per friend
Recruiting new member	5 pts. per member
Attending an additional SRC general body meeting	3 pts.
Attending an SRC executive board meeting	3 pts.
Attending an SRC committee meeting	3 pts.
Attending an MSU office of student activities' student organization meeting	4 pts.
Representing the SRC during an MSU office of student activities' event	5 pts.
Joining an SRC committee	5 pts.
Serving as a notetaker for a committee meeting	5 pts.
Submitting a research concept to the SRC	7 pts.
Submitting a research proposal	8 pts.
Submitting a research internship/job application	8 pts.
Submitting a graduate school application	8 pts.
Receiving a research award/recognition	8 pts.
Planning and hosting an SRC event	10 pts.
Participating in a research competition	10 pts.
Participating in a research conference	10 pts.
Presenting research	10 pts.
Presenting a poster at a conference	15 pts.
Hosting a research conference	15 pts.
Receiving a research internship/job offer	15 pts.
Being accepted to graduate school	15 pts.
Receiving a scholarship offer	15 pts.
Presenting an oral presentation at a conference	20 pts.
Being published	20 pts.
Serving as an SRC executive board member	20 pts.

APPENDIX 3. SRC MEMBERSHIP INCENTIVE PROGRAM BENEFITS

Benefit	Membership Levels			
	Bronze	Silver	Gold	Diamond
Access to the SRC	●	●	●	●
Use of SRC computers	●	●	●	●
Free Printing	●	●	●	●
Access to the SRC kitchen	●	●	●	●
Access to the SRC restroom	●	●	●	●
Use of SRC books	●	●	●	●
Access to SRC conference rooms	●	●	●	●
Invitation to general training and workshops	●	●	●	●
Research opportunities	●	●	●	●
SRC T-shirt		●	●	●
SRC paraphernalia		●	●	●
SRC pin		●	●	●
SRC graduate stole		●	●	●
Faculty mentor match		●	●	●
Materials for research projects		●	●	●
Opportunity to create an ad hoc committee		●	●	●
Attendance at SRC retreat(s)			●	●
Recognition at SRC awards ceremony			●	●
Recommendation letters			●	●
Event planning opportunities			●	●
Extended bearcard access during finals week			●	●
Extended bearcard access during the weekend				●
Paid conference registration fee				●
Paid graduate school application fee				●
Post-graduate testing prep financial assistance				●
Post-graduate testing fee				●

ADVANCING STEM BY TRANSFORMING PEDAGOGY AND INSTITUTIONAL TEACHING AND LEARNING: THE CREATION OF A STEM CENTER OF EXCELLENCE FOR ACTIVE LEARNING

Margaret I. Kanipes, Guoqing Tang, Faye E. Spencer-Maor, Zakiya S. Wilson-Kennedy and Goldie S. Byrd

ABSTRACT

This chapter highlights the creation of a STEM Center of Excellence for Active Learning (SCEAL) at North Carolina Agricultural and Technical State University. The overarching goal of the STEM Center is to transform pedagogy and institutional teaching and learning in order to significantly increase the production of high-achieving students who will pursue careers and increase diversity in the STEM workforce. Some of the STEM Center's efforts to reach its goals included supporting active learning classroom and course redesign efforts along with providing professional development workshops and opportunities to garner funding to cultivate student success projects through the development of an Innovation Ventures Fund. Outcomes from this Center have led to several publications and external grant funding awards

Broadening Participation In STEM: Effective Methods, Practices, And Programs
Diversity in Higher Education, Volume 22, 55–72

ISSN: 1479-3644/doi:10.1108/S1479-364420190000022003

to continue implementation, assessment, and refinement of active learning innovations and interventions for STEM student success for years to come.

Keywords: Curricular reform; faculty development; active learning; student success; faculty community of practice; pedagogical innovations

North Carolina Agricultural and Technical State University (NC A&T) has been recognized nationally as a leader in the production of STEM baccalaureate degrees awarded to African Americans (NSF, 2010). However, the national need for underrepresented STEM graduates persists, as the number of US-born graduates in STEM fields will fall far short of the trained professionals needed to replace the large number of projected retirees over the next 20 years (National Academy Press, 2007; NSF, 2003). According to a recently published National Science Foundation's Science Resource Statistics report (NSF, 2010), minorities in general and African Americans in particular continue to be significantly underrepresented undergraduate and graduate degree holders in STEM, with underrepresented minorities accounting for roughly 16%, 11%, and 6%, respectively, of the total bachelor's, master's, and doctoral degrees awarded in STEM in the US from 1995 through 2006. African Americans account for roughly 8%, 6%, and 3%, respectively, of those awarded degrees during the same period of time. Additionally, the six-year graduation rate of African American and Hispanic STEM undergraduates is roughly 24%, compared to about 40% of White students (Higher Education Research Institute, 2010; National Science Foundation, 2017; The College Board, 2005). Paradoxically, recruiting, retaining, and graduating minority students in STEM remain a challenge despite expanding career opportunities in both private and public sectors as well as in academia for those with training and skills in STEM disciplines. At NC A&T, STEM undergraduate retention and graduation rates have been quite low over the past few years, hindering timely production of baccalaureate degree holders competitive for STEM graduate study or workforce. Underachievement among minority student populations majoring in STEM contributes to the disproportionately low numbers of African American students graduating from STEM disciplines (National Academy Press, 2007).

One of the major causes of underachievement of minority STEM students is their under-preparedness in prerequisites for, and underperformance in, some entry-level math and science courses, such as College Algebra and Trigonometry for Scientists and Engineers I and II, Pre-calculus, Calculus I, Concepts of Biology I and II, General Chemistry VI and VII, College Physics I and II, and General Physics I and II. Since these courses often serve as prerequisites for a wide range of higher-level courses in math and other subjects, they are considered "gateway" courses for a vast range of majors, especially those in high-demand STEM fields. Success in these gateway math and science courses is crucial in retaining and graduating STEM majors, as mathematics success is considered as a key diagnostic for a student to complete a STEM degree (Adelman, 1999; 2006). Improving student learning and performance in these key gateway courses has been a challenge to higher education institutions across the country (NSSE Annual Results, 2011), and it poses an even greater challenge to minority-serving institutions like NC A&T.

WHY IS A STEM CENTER OF EXCELLENCE FOR ACTIVE LEARNING NEEDED?

With these facts and data in mind, a unique convergence of minds, experiences, and desires converged in the former College of Arts and Sciences (CAS) at NC A&T to create a STEM Center of Excellence for Active Learning (SCEAL). Five chairs from the Departments of Biology, Math, Physics, Chemistry, and English, and the Dean of the CAS found themselves working together to review and discuss the data and trends cited earlier and consequently put together a plan to address and change these trends. Information and data gathered, along with information obtained through meetings with campus partners from the university administration, the state, and others, prompted a strategic plan to target student success in the first two years of college as a priority. The chairs and campus faculty from the units mentioned above were challenged with developing innovative ideas to address these issues and ensure the success of all students at the university. As leaders of the only unit at the university that encounters 100% of all students at the university, it was imperative that the former CAS unit played an important role in leading efforts to address challenges around student success. A student-centered active learning environment for 12 key gatekeeper mathematics and science courses was planned. This student-centered active learning environment would increase students "time on task," build confidence, enhance mastery, and improve performance, while also enabling faculty to engage students in an innovative learning paradigm to stimulate self-motivated inquiry-based learning. As noted in a BIO2010 report, mathematics and science instruction needs innovations that engage students more actively in the learning process.

The effort to plan to create and implement SCEAL was built upon some existing efforts by the faculty in the former CAS in addressing student success issues. The Mathematics Department initially participated in a number of efforts such as the 2010 Wabash National Study to address student performance in college algebra and calculus, a UNC performance funding program in piloting a small-scale Math Emporium Project for non-STEM college algebra courses, a CAS Innovation Ventures Fund grant on student success which funded a pilot project on implementing the SCALE-UP model in some Calculus I classes, and site visits to other universities and community colleges such as Guilford Technical Community College, Virginia Tech, Alcorn State (Math Emporium Model), NC State University, Clemson, and the University Maryland Baltimore County (UMBC) (SCALE-UP Model). These preliminary small-scale efforts helped conceptualize and lay a foundation for large-scale implementation of the two research-based student-centered active learning models.

At the same time, 2011 saw the publication of the controversial book *Academically Adrift: Limited Learning on College Campuses* by Josipa Roksa and Richard Arum (2011). This book shared the results of the study of 2,003 undergraduates from over 24 universities including North Carolina A&T State University that showed that college students do not learn much in college. More specifically, the study reported, "In the first two years of college, students on average improve only 0.18 standard deviations, this translated into a 7-percentile point gain [...] [...]" (2011, p. 35). In addition, the study indicated that nearly half, 45%, of college students make no gains in learning during their first two years of college. In historically black colleges and universities (HBCUs) campuses, like NC A&T, administrators and faculty saw problems with retention between the first and second years of matriculation and wanted to address the issues exposed by *Academically Adrift* and the specificity of learning at a minority-serving institution which was not specifically addressed in that study.

Roksa and Arum (2011) also reported that "Students may not be spending a lot of time studying, but that may reflect the finding that they (students) are not being asked to do much by their professors" (p. 35). They suggested that almost 50% of the students sampled reported that they had been required to do very little writing in the courses they had (Roksa and Arum, 2011, p. 36). These findings, along with the years of teaching experiences of the CAS team indicated that in order to make a difference for students and teachers, new strategies had to be created and implemented to address both student and teacher needs – something to encourage and reward innovation and change. The team operation on the notion that although NC A&T is a STEM-focused institution, all disciplines are vital and contribute to student success. Therefore, the idea of a STEM Center of Excellence in Teaching and Learning was born. The team believed, just as Roksa and Arum (2011), "[...] when faculty have high expectations, students learn more" (p. 36). The newly developed STEM Center would provide a space for faculty to develop innovations in teaching with high expectations for their students and provide a way for faculty to share their innovations and activities with other faculty members. There are many efforts to improve student learning, but the bottom line is that the culture of higher education must change to address student learning. The creation of a STEM Center is just one way to attempt to positively affect student learning and teaching excellence in North Carolina.

DEVELOPMENT OF A STEM CENTER OF EXCELLENCE FOR ACTIVE LEARNING

As a result of the work and research of this team, a proposal for the creation of a STEM Center of Excellence for Active Learning was written. In 2013, NC A&T received a five-year, US$1,760,000 gift from the North Carolina GlaxoSmithKline (GSK) Foundation to develop a STEM Center. The overarching goal of the SCEAL is to transform pedagogy and institutional teaching and learning at NC A&T to significantly increase the production of high-achieving students who will pursue careers and increase diversity in the STEM workforce. The objectives of the Center are to (1) build a strategic pipeline for high-achieving STEM students, (2) build and implement mathematics and science models for student-centered active learning, and (3) build a community of practice for student-centered learning. SCEAL'S long-term expected outcomes are to recruit and advance significantly larger numbers of students in STEM, assure STEM student success and high graduation rates, increase the number of STEM graduates, and increase student motivation to pursue graduate and professional degrees as well as contribute significantly to the global STEM workforce. Summarizing the primary objectives of the STEM Center led to the following SCEAL framework with three key initiatives: Faculty Development, Student Success, and Inquiry and Active Learning (Fig. 1). Assessment and dissemination coupled with external partnership development are also foundational activities that support SCEAL's framework. This book chapter focuses on NC A&T's efforts to drive STEM student success through the development and implementation of a SCEAL. SCEAL framework initiatives on Faculty Development as well as Inquiry and Active Learning will be emphasized in this book chapter in more detail.

FACULTY DEVELOPMENT: KEY TO STEM STUDENT SUCCESS

Academic faculty is key innovation stakeholders in US colleges and universities. This is especially true in the area of student success. Numerous studies have directly linked faculty development to improvements in undergraduate education. However, very limited research has focused on the unique development

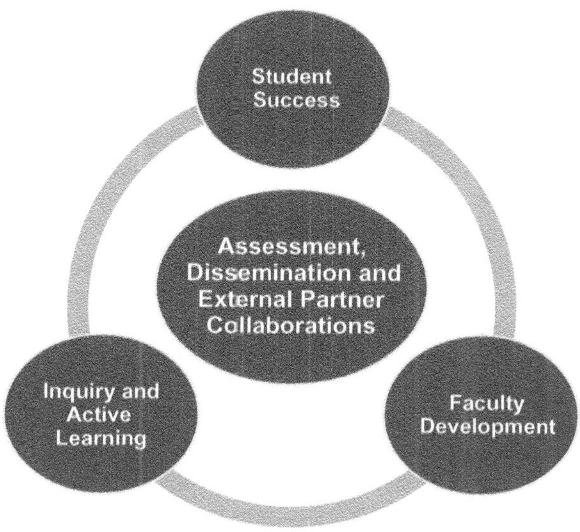

Fig. 1. STEM Center of Excellence for Active Learning Framework.

needs of faculty in HBCUs. Moreover, very little is understood about how the development of faculty through communities of practices at HBCUs can lead to innovations that advance student success.

The National Academies, the Presidential Council of Advisors on Science and Technology (PCAST), and others have noted the critical role that evidence-based teaching methodologies play in supporting the retention and persistence of all students, particularly ethnic and racial minorities, first-generation, and low-income students, who are typically underrepresented in STEM (Anderson et al., 2011; Handelsman et al., 2004; National Research Council, 2012; President's Council of Advisors on Science and Technology, 2012; Wieman, 2007, 2012). Faculty learning communities can play an important role in providing support for faculty members engaged in pedagogical innovation (Austin, 2011). These communities "[...] provide opportunities for faculty members to interact with others as they explore new assumptions and try out new approaches to teaching and learning. Learning communities enable faculty members [...] to explore and experiment with new approaches to teaching in an environment that simultaneously provides challenge and support" (Austin, 2011). These network communities also offer structures through which universities can incentivize systemic pedagogical change.

Given that studies have directly linked faculty development to improvements in undergraduate education (Eble & McKeachie, 1985; Rouseff-Baker, 2002), leaders within the former CAS hypothesized that faculty development within the context of faculty collaborations could be used to build a culture of innovation around student success and curricular reform. Key campus constituencies at the University under study identified its CAS as the primary incubator for testing its hypothesis because this unit, comprising both STEM and non-STEM departments, offered the majority of STEM gateway and general education courses within the university. The CAS, as the primary unit for STEM gateway and general education, became the chief mechanism for building an innovation culture around student success and developed the SCEAL to centralize these efforts. Herein, a set of interventions were designed and included an endowment for faculty and student success that would support faculty development and an internal award competition (the Innovation Ventures Fund) that would support pedagogical innovations and student success initiatives through faculty collaborations. Further, access to support structures was provided through both an academy of teaching and learning and a center of excellence for active learning in STEM. The dean for the CAS promoted scholarly teaching as a method for advancing through the academy to the rank of full professor, advocated for and secured extramural funds dedicated to faculty development, and negotiated with the University's business and finance office to build renovated classrooms that were needed to adopt and adapt new models of education. Collectively, these efforts were designed to promote faculty collaboration around student success initiatives with a focus on pedagogical reform.

FROM TRADITIONAL CLASSROOM TO AN INQUIRY AND ACTIVE LEARNING ENVIRONMENT

Classroom Renovations and Redesign – One of the major project undertakings by the STEM Center for Active Learning was to renovate two class/lab spaces into Math & Science Emporium and SCALE-UP labs on the second floor of the Mathematics Building, and to redesign, or reform 12 math & science gatekeeper courses. The Emporium lab or SCALE-UP lab with workstations would enable redesigned courses or associated lab sessions to be instructed in such a way that active and collaborative learning among students was facilitated. Improving student learning and success is the major impetus for implementing the Emporium or SCALE-UP model in these mathematics and science courses. Research shows that both the Emporium and SCALE-UP models have allowed institutions to improve student learning and performance significantly.

The primary reason for the success of the Emporium Model is that it allows students to work through the course material individually at their pace, obtaining individual feedback and assistance as learning challenges arise (Twigg, 2011). The flexibility in course/topic pace, relative to traditional lecture formats, keeps students focused and less discouraged when they do not grasp the concept as fast as others. In the Emporium Model, the student spends much more time actually working through mathematics problems, in contrast to the lecture model in which the instructor typically works through the problems and the student watches and listens. In addition, in this model, the instructors, graduate assistants, and undergraduate peer tutors are available to answer students' questions. Learning takes place through active engagement in *doing* math and improving engagement and learning.

The main reason for the success of the SCALE-UP model is that it integrates non-standard classroom design and technology-based resources to create a highly interactive learning space, which facilitates student-centered active learning. In the SCALE-UP model, large studio classes are taught with an emphasis on learning by guided inquiry instead of listening, and the students work in carefully structured groups and follow a curriculum intentionally designed to ensure collaborative learning, with instructors serving as guides and facilitators (Beichner, 2006). This method, which has been shown to be highly effective in improving course retention, conceptual learning, and overall performance, has been widely adopted nationally and internationally in various disciplines at two-year and four-year institutions.

The renovation of two large classroom spaces to computer lab-type active learning environments took place in two phases. The first phase was to renovate a 2,750 square-foot area of four rooms on the second floor of the Mathematics Building to create the Mathematics & Science Emporium teaching laboratory. This interactive teaching environment included 10 modular hexagon workstations each equipped with six thin-client terminals with wired network connections, three high-resolution projectors with drop-down screens, two smart boards, a network printer, and two whiteboards. Both projectors and smart boards in this new classroom are controlled by one of the three instructors' podiums through the onboard smart podium. The second phase of renovation included converting another 2,750 square-foot space in the Mathematics Building into a studio-style collaborative SCALE-UP laboratory. Renovations began during the spring semester of 2014 and were completed in January of 2015. This interactive teaching environment included eight 7-foot round tables each equipped with three thin-client terminals with wired network connections, three high-resolution projectors with drop-down screens, two smart boards, two whiteboards, and a network printer. Both projectors and smart boards are controlled by one of the three instructors' podiums through the onboard smart podium. The SCALE-UP teaching lab in this newly remodeled room has a seating capability of 72.

Active Learning Course Redesigns – The SCEAL project engaged STEM faculty in transformative course redesign of 12 selected mathematics and science courses of college algebra and trigonometry (MATH 103 and 104), pre-calculus and calculus I (MATH 110 and 131), Biology I & II (BIOL 101 and 102), General Chemistry I & II (CHEM 106 and 107), and College and General Physics (PHYS 225, 226, 241, and 242). A synopsis of the course redesigns for each department is highlighted in Fig. 2.

In the Department of Mathematics, both MATH 103 – College Algebra and Trigonometry for Engineers and Scientists I and MATH 104 – College Algebra and Trigonometry for Engineers and Scientists II were redesigned using the Math Emporium Model. The Emporium Model eliminates the traditional lecture

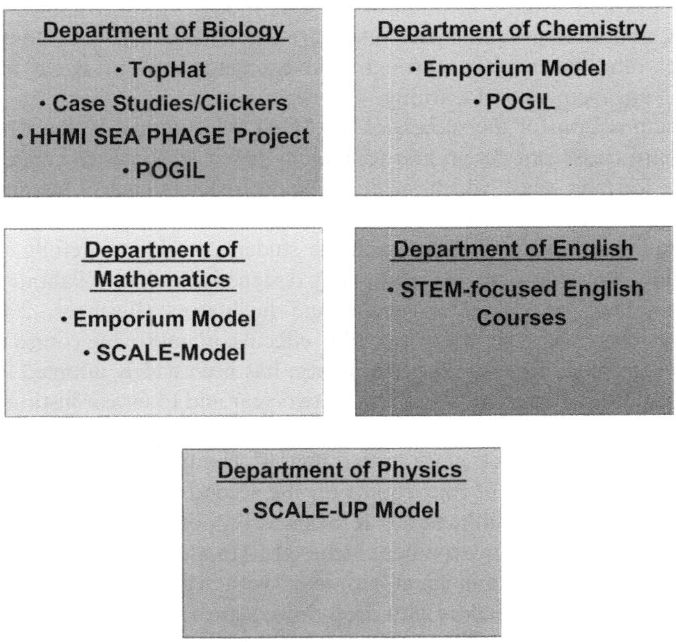

Fig. 2. Active Learning Course Redesign Models and Strategies in the STEM Center of Excellence for Active Learning.

and uses commercially available interactive computer software combined with personalized on-demand assistance and mandatory student participation. In the Math Emporium Model implementation, the Pearson's MyMathLab for course delivery and management was used, and a required attendance course structure was adopted. Students were divided into course sections that met three times per week in the Emporium Lab and one hour per week in a traditional classroom setting. For MATH 110 and 131, the SCALE-UP Model was implemented. The SCALE-UP Model divided students into three-member teams, arranged to have three student teams sit at each round table, and engaged student teams with in-class activities to explore and discover while their instructor roamed – asking questions, sending one team to help another, or asking why someone else got a different answer. Within each team, the following roles were assigned: (1) the researcher who would be in charge of making calculations, (2) the scribe who would be in charge of writing out the group's solutions, and (3) the manager who would be in charge of overseeing the collaboration and making sure everyone remained on task. These roles were fine-tuned during development and rotated every three to four weeks, so that everyone in the active learning classroom participated in each job over the course of the semester. In the implementation of the SCALE-UP model, lecture time was reduced to about 15–20 minutes at the beginning of each 50-minute class period, and the remainder of the class period was used for team-based learning activities and whiteboard presentations.

In the Department of Biology at NC A&T, the use of the active learning environment to provide student-centered instruction and inquiry-based learning, including case studies, problem-based activities, interactive online learning tools, and inquiry-based labs would occur in the BIOL 101 – Concepts of Biology I and BIOL 102 – Concepts of Biology II courses. The students would also meet once a week for two-hour laboratory sessions designed to engage them in collaborative problem-solving using virtual research laboratory activities and case-based studies.

The Department of Chemistry initially redesigned CHEM 106 – General Chemistry VI and CHEM 107 – General Chemistry VII (Parts 1 and 2 of General Chemistry for STEM majors) courses, modeling the Discovery Learning Method (DLM) after the University of Maryland Baltimore County's (UMBC) program. Using the NC A&T-modified DLM model, students met three days a week for a 50-minute session consisting of a five-minute quiz, mini-lecture, and group work. On the fourth day, the students met for a two-hour discovery learning session to complete modules that had been developed for CHEM 106 or CHEM 107. The five-minute quizzes were conceptual and focused on the material to be covered on that day. The mini-lecture highlighted areas that were typically confusing to students or not thoroughly explained in the text. Throughout the lecture, students were questioned in ways which required that they demonstrate not only an understanding of the material but also the ability to analyze and apply the material to new situations. The initial pilot of this method showed great signs of success. However, it was not widely accepted by the Chemistry faculty. This led to the faculty meetings that discussed the best approach to transform the chemistry classroom. As a result, a team of faculty later developed an Emporium Model for Chemistry. This

Emporium Chemistry model takes advantage of the ALEKS software as the main vehicle for active learning and student engagement.

In Physics, the traditional lecture style courses were complemented by adopting the studio model as a method for recitation and laboratory instruction (Wilson, 1994). The studio format is a student-centered learning method that promotes active learning and integrates lecture and laboratory work into one activity. An instructor introduces the physics concepts or some assigned homework problems and engages small groups of students in active and collaborative inquiry-based learning. The presented concepts are explained and reinforced by immediate hands-on laboratory or simulation experiences. By incorporating the experiments/simulations into the flow of the course material, students are more deeply engaged and reflect on the introduced concepts.

TWO TEACHING MODELS FOR STUDENT-CENTERED ACTIVE LEARNING AT NC A&T HIGHLIGHTED

Here, we provide a more in-depth discussion of how two models which emerged from the activities of the CAS which stand out as innovations in student-centered active learning were developed at NC A&T.

The Emporium Model – A group of four faculty members completed course redesign for MATH 103 – College Algebra and Trigonometry for Engineers and Scientists I and MATH 104 – College Algebra and Trigonometry for Engineers and Scientists II sequence in the spring semester of 2013. They developed five interactive modules for MATH 103 and five modules for MATH 104 using Pearson Publishers' *MyLabsPlus* software as a course delivery platform as well as an accompanying workbook, *College Algebra and Trigonometry for Engineers and Scientists Workbook*, for the course. Each course module consisted of a pre-test, instructional videos, workbook examples and exercises, homework problems, a quiz, a final quiz, and a module test. Homework assignments within the module were generated based on the results of the pre-test. During class time, students worked through their assignments and were able to ask the instructors and graduate teaching assistants (GTAs) questions about their work. Sometimes the instructors and GTAs were not able to get to a student immediately and during their waiting time and therefore students were encouraged to ask each other questions. Therefore, in this method, peer tutoring was highly encouraged. For the fourth hour of the week, the students would meet with one of the GTAs in a regular classroom to complete additional practice problems from the course workbook. The GTAs encouraged the students to lead the process in solving the problems while the GTAs facilitated the process. A 460-page workbook was published by the Pearson Learning Solutions over the summer 2013, a second edition was published in the summer 2014, and a third edition was published in the summer 2016.

More specifically, students enrolled in the Emporium instructional sections met for four hours total per week. For three of these hours, students met in an Emporium classroom in which each student sat at a computer and worked through the class assignments. The Emporium classroom, where course delivery occurred, is a computer lab with workstations. A workstation includes a modular desk, a wheeled chair, and a networked computer for a student to work through online instructional modules. Each lab is equipped with thin-client terminals to facilitate students' interactive learning of course modules. Each class is staffed with an instructor and enough GTAs to have a minimum ratio of one instructor or GTA for every 12 students.

The class structure for the Emporium begins with a student entering the module. To start a module in each of the two courses, a student is required to take the module pre-test to assess his/her prior knowledge and mastery of the module content. If the student scores 85%, or higher in the module pre-test, s/he will be allowed to proceed directly to the next module. Otherwise, the student will be required to go through the module by completing the following tasks: watching instructional videos, working through the practice exercises in the workbook, completing online homework assignment, taking online quiz and practice test, and then taking a proctored and timed module test. Satisfactory performance in homework, quiz, and module test will allow the student to move to the next module. When watching online instructional videos, and working through practice exercises, students may ask for assistance from the instructor or teaching assistants. To seek assistance, a student places a plastic cup at his/her workstation on top of the desk to signal to the course instructor or GTAs that s/he needs help. While modular courses for MATH 103 and MATH 104 are self-paced, students are provided with a pacing guide for each course to ensure timely module completion and progression (Table 1).

The SCALE-UP Model – A group of five mathematics faculty members were engaged in MATH 131 – Calculus I course redesign refinement. This group of faculty members started their redesign endeavor in the fall semester of 2012, and they piloted the SCALE-UP model in MATH 131 in the fall semester of 2012. In SCALE-UP model, students worked in carefully structured groups and follow a curriculum intentionally designed to ensure problem-based interactive group learning with instructors serving as guides and facilitators. While the laboratory sizes vary, the method has been standardized to employ 7-feet round tables, each seating nine students to facilitate teamwork, computer projection systems, instructor station, whiteboards, three client computer terminals in each table, Internet connectivity to catalyze hands-on collaborative learning, and a workshop-style curriculum design to guide course activities. The SCALE-UP model enables a collaborative study environment for instructor, GTAs, and students. In this model, students meet five days a week, a total of five hours per

Table 1. Summary of the Enrollment, Satisfactory Academic Performance (ABC) Rates and Passing (ABCD) Rates for College Algebra I & II Courses from Spring 2014 – Fall 2017 using the Emporium Model.

Course	No. Section	Enrollment	No. ABCs	% ABCs	No. Passing	% Passing
MATH 103 – Algebra I	81	2,470	1,476	57	1,725	68
MATH 104 – Algebra II	48	1,351	858	62	996	74

week. In addition to the four weekly meetings in the SCALE-UP lab per week, students are also required to meet for supplemental instruction (SI) laboratory hours per week in a traditional classroom setting with the class GTA. The purpose of the weekly SI meeting is to focus students' attention on the week's upcoming tasks and recap the previous weeks' materials and their relationship to new materials to be learned. Students are expected to read the materials before coming to class. Students are divided into three-member groups. During class time, students listen to a short lecture by the instructor, work through their worksheet assignments, and participate in in-class activities as a group with the assistance of the instructor and GTAs.

The SCALE-UP curriculum was developed such that most of the class time is spent on the tangible and ponderable, which engage students as active learners and encourage them to invent their own knowledge. NC A&T faculty developed a Calculus I workbook by the SCALE-UP faculty team, compiling the concise lecture notes, fully worked out examples, practice exercises, and formula sheets, all in a modular format. This workbook is currently used by all SCALE-UP MATH 131 sections. Central to the SCALE-UP adaptation is commercially available software used in the labs. Pearson's MyLabsPlus software platform was used for online homework delivery and management of the course. MyLabsPlus provided immediate feedback and assistance to students in solving problems through its "Show Me How" and "Work a Similar Exercise" functions. Also, online homework and quizzes are graded immediately by MyLabsPlus upon submission. Much effort was made to ensure that all SCALE-UP Calculus I sections followed the same format. All sections of the SCALE-UP Calculus I course use the same summary notes, same or very similar worksheets, and have a common final with the same grading standards. In the MATH 131 implementation of the SCALE-UP model, lecture time is reduced to about 15–20 minutes at the beginning of the 50-minute class period, and the remainder of the class period is used for team-based learning activities. The course instructor and one or two GTAs were readily available to support student team activities. Moreover, in the SCALE-UP setting, course instructors have implemented a series of standards-based teaching and learning strategies, such as course syllabi with *shared vision and common student learning outcomes, common test items, common pedagogies (active and interactive learning environment as well as integration of problem sessions into the course), and common grading rubrics.* This group of math faculty members also published a Calculus I workbook through Pearson Learning Solutions in the summer of 2014. The Calculus I workbook has been used in all the SCALE-UP sections of Calculus I since the fall 2014 (Table 2).

A group of three mathematics faculty members were engaged in redesigning MATH 110 – Pre-calculus for Engineers and Scientists in the 2013–2014 academic year. They developed six course modules and completed a preliminary version of MATH 110 workbook in the spring and summer of 2014. They started the SCALE-UP pilot in MATH 110 in the fall 2014.

Table 2. Summary of the Enrollment, Satisfactory Academic Performance (ABC) Rates and Passing (ABCD) Rates for Calculus I Courses from Fall 2012 – Fall 2017 using the SCALE-UP Model.

Course	No. Section	Enrollment	No. ABCs	% ABCs	No. Passing	% Passing
MATH 131 – Calculus I	107	3,852	2,151	56	2,642	69

Table 3. Summary of the Enrollment, Satisfactory Academic Performance (ABC) Rates and Passing (ABCD) Rates for MATH 110 from Fall 2014 – Fall 2017 using the SCALE-UP Model.

Course	No. Section	Enrollment	No. ABCs	% ABCs	No. Passing	% Passing
MATH 110 Pre-Calculus	16	568	344	61	418	74

The following table summarizes the enrollments, satisfactory academic performance (ABC) rates, and passing (ABCD) rates for the SCALE-UP sections of MATH 110 from fall 2014 through fall 2017 (Table 3).

STEM CENTER OF EXCELLENCE FOR ACTIVE LEARNING LESSONS LEARNED

Faculty Development – The former College conducted an impact analysis of these efforts that included a survey of faculty throughout the college to gauge perceptions of success of the program and an analysis of new collaborations formed because of participation in the program. Forty-nine faculty members (representing 24.3% of this group) responded to the survey. Respondents indicated the following:

- 70% of recipients reported that the Fund had a positive impact on their ability to collaborate within the CAS.
- 74% reported an increase in inter-departmental collaborations.
- 60% stated that the Fund allowed them to increase their collaboration across the University.
- 75% of grant recipients stated that the Innovation Ventures Fund had a positive impact on their scholarly activity.

All of the departments in the former CAS participated in these efforts to varying degrees, and it was found that the incentives and supports available through College and SCEAL worked together to foster deep collaborations that are transforming what faculties are doing in the classroom in ways that will have long-term impacts on the scholarly productivity of faculty. Additionally, increased dialogue among departments that have traditionally operated as siloes and most importantly offer a strong foundation for developing and integrating student success initiatives was brought into the culture of the College.

Course Redesign Lessons Learned (Emporium Model) – There were several observations that may shed light on student performance in the Emporium sections of MATH 103 and MATH 104. Math faculty who taught the Emporium sections of MATH 103 and MATH 104 observed that the students who performed well bought into the system. They worked outside of class and were eager to move on. The students who have been really successful in the Emporium have not necessarily been the most academically gifted students, but rather the students who have a strong work ethic. The majority of the students who were able to finish the course early spent a considerable amount of time working outside of class, and they were not afraid or embarrassed to ask for help. They never complained about the style of the class. In general, they accepted the Emporium style and made the most of the opportunities that were presented to them. These students embraced the Emporium's goals and objectives to achieve as a challenge, had excellent class attendance, and were rarely distracted from their work. They also viewed "learning math by doing math" as an opportunity to demonstrate and enhance their academic skill set, listened well, and followed directions without procrastination, seemed to enjoy the push of high expectations and the challenge of constructive criticisms. It was noted that they also asked specific questions and took detailed notes in a separate notebook. The Emporium was there for them, offering the academic flexibility for them to excel. However, the majority of the students who performed unsatisfactorily in the Emporium class resisted putting the required time into the class. They regularly missed class, and/or they spent little to no time outside of class working on class material. Faculty members gave students three chances on all of their tests, provided that the attempts were made prior to the test deadline. The majority of the students who failed the course only took each test once (if at all) because (again) they wouldn't work outside of class to get to the point where they could take the test early.

Faculty members involved with the implementation of the Emporium Model in MATH 103 and 104 agreed that the Emporium Model is a very good and effective teaching and learning model for improving student success. It provides ample opportunities for our students to succeed. The on-demand nature of the instruction in the Emporium classroom allows each student to ask questions and receive assistance on specific topics with which they may be struggling. This style of instruction is especially efficient for students who have some prior knowledge of the content, or those who have an increased aptitude for quantitative reasoning. It is also very good for those students who may struggle with mathematics in general but are willing to extend an effort (both in class and out of class) to learn and practice the material. The Math Emporium also allows instructors to witness students working problems. One of the major strengths of the Emporium Model is that it allows instructors to meet and challenge students, individually, at where they are. Instructors can challenge accelerated students and allow them to move forward and stay engaged within the course, while providing support to students with deficiencies both in content and in preparation. The Emporium Model also affords instructors better and more frequent one-on-one time with students and the ability to help address their root problems in learning math and not just address the concrete issues that they have with a specific problem. It has improved the students' understanding of algebraic concepts because they spend class time working through the math problems. It will result in stronger students overall, but only if we can interact with the students early in their academic careers. First- and second-year students in the Emporium classes typically accepted the course more readily than third- and fourth-year students. The Emporium Model of instruction is also beneficial to the instructor in terms of monitoring student success. While in a traditional classroom, a majority of the class time is devoted to a "blanket" lecture for everyone. This may cause problems for the students who are already familiar with the material as well as those who are behind and struggling to grasp the material. In the traditional lecture, it is difficult for the instructor to ascertain how the students are proceeding in the class until a test is taken, which is often too late. Additionally, the interaction between instructor and students is severely limited in a traditional lecture class. However, with the Emporium setting, the instructor has the opportunity to work with each student individually which allows for a better understanding of where each student is with regard to the material. The software used in the Emporium also allows the instructor to closely monitor student progress.

Lessons Learned (SCALE-UP Model) – There were a couple of observations, which may shed light on student performance in the MATH SCALE-UP sections. Math faculty who taught the SCALE-UP classes observed that the students who performed well were generally motivated students known as willing students. This group of students included both prepared students (able and willing students) and underprepared but willing to try students (less able but willing students). They appeared to enjoy the challenge of pre-reading and trying concept and skill mastery on their own. They came to class with completed pre-class reading assignments, and their questions reflected that they had contemplated the concepts and problems. These students were not necessarily all high-achieving students, but they seemed to understand the purpose of being in class each day and trying to learn a new concept or master a new skill. We did see some borderline students (D average students) moved up to get a C or better for the course as well as the average C or B students moved up to B and A columns. It appears that not solely student ability but rather the motivation to learn is what determines success. However, the majority of the students who performed unsatisfactorily in the SCALE-UP classes were in the categories of less able and unwilling students as well as able but unwilling students. They somehow viewed the SCALE-UP teaching and learning method as a burden, resisted putting required time into the class, and were not motivated to even try. They regularly missed class, and/or they spent little to no time outside of class working on class material. When they sometimes came to class, they appeared unprepared for the class, not knowing what topic or section was to be covered as this was readily available in their course syllabus. They simply expected to sit there and have someone work out problems on the board, the exact ones to be tested on. The biggest challenge for this group of students was to comprehend what is required in order to learn and master Calculus and develop a strong work ethic. The biggest challenge for faculty, as well as the institution as a whole, was to get them motivated to learn and cultivate a culture of responsibility and commitment to inquiry, ownership, and engaged learning.

Faculty members involved with the implementation of SCALE-UP model in Calculus I mostly agreed that the SCALE-UP model is a good and effective teaching and learning model for improving student success when students are motivated and have a desire to learn. It seems that the average students gained the most with the additional work, while strong students mostly did well. However, it did not impact less prepared and unwilling students as much because they often did not come to class or did not really work in class. It was difficult for instructors and TA's to keep classes at the same place and coordinate exams when having common exams in the evening. While interactive group learning affords an opportunity for more students working on problems and spending more time on task, it also poses a challenge for faculty to maintain the classroom order and engage everyone into active learning.

The SCALE-UP teaching and learning methodology does impact the intellectual climate in a positive way. Faculty members now see more students are engaged in determined and purposeful collaboration and problem discussions in class, and students are aware that the instructor is there to assist what they themselves are trying to learn, as opposed to just sitting there waiting to copy something off the board. Besides the in-class group work, some students began to form study groups that met evenings and weekends. Some students seemed to be aware of their responsibilities beyond the classroom, while other students learned the power of peer learning as evident by the student comments listed above. It seems some of them have learned the importance of discussing mathematics and learning mathematics. Many learned that more was required of them for learning to occur.

STEM CENTER IMPACT

Through the use of funds from the University and NC GlaxoSmithKline Foundation, two existing classroom/laboratory spaces were renovated into Mathematics & Science Emporium and SCALE-UP laboratories, for active learning instruction. Course redesigns were completed for 12 Biology, Chemistry, and Mathematics gateway courses (BIOL 101/102, CHEM 103, CHEM 106, MATH 103/104, MATH 110, and MATH 131). Efforts are ongoing in all of the departments highlighted in this book chapter. Thus far, the STEM Center since inception has impacted over 7,500 students through course redesigns in Mathematics, Biology, Chemistry, and Physics. Building a "community of practice" which consists of multiple inter- and intrafaculty collaborations now exists on the NC A&T campus. These collaborations along with other campus partners and administrators have been the key to the STEM Center's success. The STEM Center has supported faculty course implementation and student success by providing several professional development workshops and the Innovation Ventures Fund project awards, which provided seed funds to support faculty student success projects. Faculty have published several mathematics workbooks, refereed journal articles, and a book chapter. The STEM

Center and its affiliated faculty realize that in order to continue sustaining its efforts to drive STEM student success it must obtain external funding. Faculty in Physics, Mathematics, and Chemistry have garnered over US$3.21 million dollars in funding from the National Science Foundation (NSF) to continue implementation, assessment, and refinement of active learning pedagogies in Mathematics, Chemistry, and Physics and implement opportunities for faculty development.

STEM CENTER LONG-TERM IMPACT

Given NC A&T's very transparent focus on STEM, it has been an imperative to provide the necessary avenues for strengthening teaching and learning as well as faculty development and scholarship in STEM that will increase the numbers of STEM graduates to enter the global workforce for the future. Establishing a SCEAL is certainly consistent with the national agenda to diversify and broaden the STEM workforce. It is also consistent with NC A&T's strategic plan, *A&T Preeminence 2020*, and especially goal 3 of the plan, to position the university to become a national, premier research-intensive doctoral, and Science and Technology-focused learning institution. The SCEAL has provided a base for introducing, engaging, and advancing students in STEM at the foundational levels. Likewise, it engages and advances faculty members who provide instruction for STEM courses in ways that improve pedagogy, scholarship, and confidence in twenty-first-century STEM instruction. Creating new and innovative collaborative student instructional spaces and matching them with a quality STEM-focused and engaged faculty will certainly broaden participation in STEM. Furthermore, developing courses that move beyond the traditional way of teaching is crucial for engaging the new millennial STEM student. The efforts of the STEM Center's initiatives will lead to STEM student success for years to come.

REFERENCES

Adelman, C. (1999). *Answers in the tool box: Academic intensity, attendance patterns, and bachelor's degree attainment.* Washington, DC: U.S. Department of Education.

Adelman, C. (2006). *The toolbox revisited: Paths to degree completion from high school through college.* Washington, DC: U.S. Department of Education.

Anderson, W. A., Banerjee, U., Drennan, C., Elgin, S., Epstein, I., Handelsman, J. ... Olivera, B. (2011). Science education. Changing the culture of science education at research universities. *Science (New York, NY)*, *331*(6014), 152–153.

Acrum, R., & Roksa, J. (2011). *Academically adrift: Limited learning on college campuses.* Chicago, IL: University of Chicago Press.

Austin, A. E. (2011). *Promoting evidence-based change in undergraduate science education.* A Paper Commissioned by the National Academies National Research Council Board on Science Education 1–25. Retrieved from http://citeseerx.ist.psu.edu/viewdoc/download?doi=10.1.1.675.9703&rep=rep1&type=pdf

Beichner, B. (2006). Chapter 29. North Carolina State University; SCALE-UP. In D. Oblinger (Ed.), *Learning spaces: An EDUCAUSE E-book.* EDUCAUSE.

Eble, K. E., & McKeachie, W. J. (1985). *Improving undergraduate education through faculty development. An analysis of effective programs and practices*. ERIC. San Francisco, CA: Jossey-Bass Inc., Publishers.

Handelsman, J., Ebert-May, D., Beichner, R., Bruns, P., Chang, A., DeHaan, R. ... Tilghman, S. M. (2004). Scientific teaching. *Science*, *304*(5670), 521–522.

Higher Education Research Institute. (2010). *Degrees of Success: Bachelor's degree completion rates among initial STEM majors*. Higher Education Research Institute at UCLA, HERI Report Brief, January 2010. Retrieved from http://www.heri.ucla.edu/nih/HERI_ResearchBrief_OL_2010_STEM.pdf

National Academy of Sciences, National Academy of Engineering, and Institute of Medicine. (2007). *Rising above the gathering storm: Energizing and employing America for a brighter economic future*. Washington, DC: National Academies Press.

National Research Council. (2012). *Discipline-based education research: Understanding and improving learning in undergraduate science and engineering*. Washington, DC: National Academies Press.

National Science Foundation. (2003). *National science board. The science and engineering workforce: Realizing America's potential. NSF 03–69*. Arlington, VA. Retrieved from http://www.nsf.gov/nsb/documents/2003/nsb0369/nsb0369.pdf

National Science Foundation. (2010). *Division of science resources statistics. Science and engineering degrees, by race/ethnicity of recipients: 1997–2006. Detailed Statistical Tables NSF 10–300*. Arlington, VA. Retrieved from http://www.nsf.gov/statistics/nsf10300/

National Science Foundation, National Center for Science and Engineering Statistics. (2017). *Women, minorities, and persons with disabilities in science and engineering: 2017. Special report NSF 17–310*. Arlington, VA. Retrieved from www.nsf.gov/statistics/wmpd/

NSSE Annual Results. (2011). *Fostering student engagement campus wide. National survey of student engagement*. Retrieved from http://nsse.iub.edu/html/annual_results.cfm

President's Council of Advisors on Science and Technology. (2012). *Engage to excel: Producing one million additional college graduates with degrees in science, technology, engineering, and mathematics*. Retrieved from http://www.whitehouse.gov/sites/default/files/microsites/ostp/pcast-executive-report-final_2-13-12.pdf

Roksa, J., & Arum, R. (2011). The state of undergraduate learning. *Change: The Magazine of Higher Learning 43*(2), 35–38.

Rouseff-Baker, F. (2002). Leading change through faculty development. *New Directions for Community Colleges*, *2002*(120), 35–42.

The College Board. (2005). *2005 College-bound seniors: Total group profile report*. Retrieved from http://www.collegeboard.com/prod_downloads/about/news_info/cbsenior/yr2005/2005-college-bound-seniors.pdf

Twigg, C. (2011). *The Math emporium: Higher education's silver bullet. Change: The Magazine of Higher Learning*, May-June 2011 Issue. Retrieved from http://www.changemag.org/Achives/Back Issues/2011/May-June-2011/math-Emporium-full.html

Wieman, C. (2007). Why not try a scientific approach to science education? *Change: The Magazine of Higher Learning*, *39*(5), 9–15.

Wieman, C. (2012). Applying new research to improve science education. *Issues in Science and Technology*, *29*(1), 1–8. Retrieved from https://issues.org/issue/29-1/

Wilson, J. (1994). The CUPLE physics studio. *The Physics Teacher*, *32*(9), 518–523.

TRANSFORMING STEM DEPARTMENTS FOR INCLUSION: CREATIVE INNOVATION, CHALLENGES, ADAPTATION, AND SUSTAINABILITY AT THE UNIVERSITY OF ARKANSAS-FORT SMITH

Sayo O. Fakayode, Jennifer Jennings Davis, Linus Yu, Paulette Ann Meikle, Ron Darbeau and Georgia Hale

ABSTRACT

Strengthening the nation's technological workforce, competing and expanding its relevance in the global economy, and maintaining personal as well as homeland security will be highly dependent on the quantity, quality, and diversity of the next generations of scientists, engineers, technologists, and mathematicians. Production of a diverse generation of human resources with relevant, competitive skills is critical. However, so too is the need to raise an enlightened citizenry with cross-cultural experience and cultural awareness competency, with a broad worldview and global perspectives. These requirements are critical to understanding the challenges and opportunities of scholarly activity in a pluralistic global environment and positioning ourselves to capitalize upon them. Scholars with cross-cultural experience and competency are empowered to adapt and work collaboratively, nationally and globally, with scholars of different races, geopolitical, socioeconomic, and cultural backgrounds. Development of effective strategies to transform science,

Broadening Participation in STEM: Effective Methods, Practices, and Programs
Diversity in Higher Education, Volume 22, 73–105

ISSN: 1479-3644/doi:10.1108/S1479-364420190000022004

technology, engineering, and mathematics (STEM) departments for inclusion and to broaden the participation in STEM across cultures, socioeconomic standing, race, and gender in higher education has been a dominant topic of pedagogical interest of national priority in the last several decades. However, success in these endeavors is achievable only through systemic change and a cultural shift to address the underlying root causes of socioeconomic disparity, gender, and racial disparities and a paucity of cultural awareness among all educational stakeholders. STEM departments can only be truly transformed for inclusion through the development of sensitive, creative, and student-engaging curricula and targeted recruitment and retention of underrepresented minorities in STEM. Formation of well-coordinated alliances spanning educational sectors, governmental and non-governmental organizations, and community engagement and outreach are also critical to promoting inclusive and broad participation in STEM education.

The first section of the chapter gives an introduction to various challenges, obstacles, and hindrances that prevent a successful transformation of K–12 science education as well as STEM departments in higher education for inclusion. The second section discusses historical perspectives of the University of Arkansas-Fort Smith (UAFS) – the institutional profile, missions, and visions of UAFS as a regional university. Policies and strategies for addressing the socioeconomic disparity, faculty gender, and racial disparities and cultural competency awareness at UAFS are also highlighted in this section. Other approaches including targeted efforts to recruit and retain underrepresented minority students, provision of financial assistance for students from low-income families, and a creative "Math-up" curriculum innovation to promote inclusive and broad participation in STEM at UAFS are highlighted in the latter section of the chapter. Formation of alliances between UAFS, local K–12 school districts, and governmental and non-governmental agencies to promote broad participation in STEM at UAFS are discussed. The last section of the chapter provides recommendations for adaptation and sustainability of strategies and efforts aimed at transforming national STEM departments for inclusion.

Keywords: University of Arkansas-Fort Smith; inclusive transforming STEM departments for underrepresented minorities in STEM; inquiry learning creative curriculum innovation; cultural awareness competency; STEM faculty professional development; partnerships with local K-12 school districts

INTRODUCTION

Strengthening the nation's technological workforce, competing and expanding its relevance in the global economy, and maintaining personal as well as homeland security will be highly dependent on the quantity, quality, and diversity of the next generations of scientists, engineers, technologists, and mathematicians. Production of diverse generations of human resources with relevant, competitive skills is critical. However, so too is the need to raise an enlightened citizenry with cross-cultural experience and cultural awareness competency, with a broad worldview and global perspectives. These requirements are critical to understanding the challenges and opportunities of scholarly activity in a pluralistic global environment and positioning ourselves to capitalize upon them. Scholars with cross-cultural experience and competency are empowered to adapt and work collaboratively, nationally, and globally with scholars of different races, geopolitical, socioeconomic, and cultural backgrounds.

It is therefore imperative to develop robust, effective strategies to transform higher education in general, and STEM departments in particular, for inclusion. It is also crucial to develop sustainable, long-term strategies that will broaden the participation in STEM education and to reduce achievement gaps across all races, genders, cultures, socioeconomic backgrounds, and disability status. Transforming STEM departments for inclusion and broadening the participation across the entire human spectrum is also essential not only to maintain but to strengthen the United States' future STEM and STEM-derivative (for example, health care) workforce, especially with the current retirement of a massive number of "*baby boomers*" (AARP, 1996; Bureau, 2014; Costa & Di Milia, 2008; Kemmlert & Lundholm, 2001; Klug, 2009; Kneeland, Kneeland, & Wachter, 2010; Hopkins, Nestleroth, & Bolick, 1991).

The rhetoric of inclusive strategies to broaden the participation in STEM education across cultures, race, and gender has been a dominant topic of pedagogical interest over the past few decades. Strategies including the implementation of new curriculum innovation, the formation of alliances between higher education, local school districts, community outreach, governmental and non-governmental agencies, and other educational stakeholders have been strategically implemented, employed, and adopted to address these challenges and to promote inclusive and broader participation in STEM (Adams & Lisy, 2007; Amaral, Shank, Shibley, & Shibley, 2013; Craig, 1999; Fakayode, 2014; Fakayode, 2015; Fakayode, Abel, & Pollard, 2014; Fakayode, Mayes, Kanipes, Johnson, & Cuthbertson, 2016; Fakayode, Pollard, Mohammed, Adeyeye, & Yakubu, 2014; Fakayode, Snipes, & Kanipes, 2016; Fakayode, Snipes, Kanipes, Mohammed, & Wilson, 2016; Hutchison & Atwood, 2002; Lindsay & McIntosh, 2000; Mabrouk, 2009; Slezak, 1999; Wilson, et al., 2012; Wilson, Iyengar, Pang, Warner, & Luces, 2012).

Although these initiatives have made some strides, there is an organic consensus that our national educational systems in the twenty-first century are far less diverse and inclusive. Multiple, diverging, and often competing factors and barriers, including social stratification, economic disparity, gender and racial inequality, and cultural awareness incompetency that historically undermined and threatened inclusive and broad participation in K–12 science education and STEM majors in higher education, continue to be major endemic challenges of national concern. Racial disparities and economic inequalities permeate every fabric and aspect of human lives, with negative consequential effects on K–12 science education as well as higher education. Many of the current initiatives, alliances, and educational outreach often lack coherent coordination among key educational stakeholders to make significant, lasting impact or to achieve desirable outcomes with respect to inclusive and broad participation in STEM. Efforts are also often duplicated, redundant, and sometimes counterproductive, resulting not only in poor and undesirable outcomes but also leading to a waste of valuable resources and systemic frustration among the spectrum of stakeholders.

Socioeconomic disparity, cultural awareness incompetency, gender and racial inequality, and inadequate K–12 preparation are among the major barriers, contributing to the underperformance of underrepresented minority students not only in K–12 science education but also in STEM disciplines in college. Our national K–12 school districts are still predominantly structured, zoned, and stratified along social–economic lines and racial boundaries. For example, children of affluent parents are likely to attend high-achieving and well-funded schools fully staffed with qualified science and mathematics teachers. Such high-achieving schools are also equipped with state-of-the-art technology, resources, and infrastructure necessary for effective classroom instructions to stimulate students' learning in a conducive environment. Also, students that attend high-achieving schools often take AP courses in Chemistry, Physics, Biology, and Mathematics that fully prepared them for the rigor of STEM majors in college. Furthermore, these students likely had early and frequent (or maybe multiple) exposure(s) to STEM-related (extra-curricular) enrichment activities. Expectedly, students who attend well-funded K–12 schools with adequate resources and infrastructure are at an advantage and usually perform better in terms of GPA, SAT, or ACT scores. Most importantly, such students are more likely to enroll in College Mathematics or Calculus I courses in their first semester in college, enhancing the likelihood of their graduation in four years or sooner. Such students are also more likely to be offered admission and competitive scholarships to highly selective and prestigious colleges. It is also likely that their parents are able to afford the high tuition and fees at prestigious colleges and universities even without scholarships or awards.

On the contrary, children from low-income families often attend low-performing, underfunded K–12 schools. Such schools are typically poorly equipped with aging, outdated classroom technology and dilapidated infrastructure. It is also very challenging to recruit and retain qualified teachers in such school districts (National Education Association, 2011). Low-performing K–12 schools often hire inexperienced and/or underqualified science and mathematics teachers due to a limited budget, resources, and prestige. Additionally, professional development opportunities are limited for many of these teachers affecting their teaching effectiveness. The study has also shown that compared with teachers in high-performing and well-funded schools, teachers in low-performing schools have low job satisfaction and are more likely to leave their school district for greener pastures (Dilworth & Coleman, 2014).

Sadly, a larger percentage of minority students are structurally destined to attend these ill-equipped schools. As a result, students who attend low-achieving and poor-performance schools tend to be inadequately prepared (with relatively low GPA, SAT, or ACT scores) for the rigor of STEM majors and are likely to avoid or dissociate from STEM majors in college. In general, minority students with relatively low GPA, SAT, or ACT scores are more likely to attend community colleges, or minority-serving institutions, where the tuition and fees are more affordable. Most importantly, students with inadequate mathematics preparation are likely to take remedial mathematics courses in their first year of college, prolonging their time to graduation to five, six, or more years, thereby suffering lost potential wages while incurring additional financial burdens.

It is also true that minority students in K–12 are disproportionately suspended and are more likely to be expelled for minor offenses, with direct and/or indirect negative impacts on their educational success in K–12 (American Civil Liberties Union, 2012; Arcia, 2006; Balfanz, Byrnes, & Fox, 2014; Balfanz, Herzog, & Mac Iver, 2007; Krezmien, Leone, & Achilles, 2006; Losen & Martinez, 2013; Mizel, Miles, Pedersen, & Tucker, 2016). Consistently, there is racial disparity in high school dropout rates, which is more prominent among the African American and Hispanic communities as compared with the White community for various and diverging reasons (Campbell, 2015; Carbonaro & Workman, 2013; English, Lambert, & Ialongo, 2016; Ewing, D'Amico, & Humor, 2014; Fortin, Lessard, & Marcotte, 2010; Leventhal-Weiner &Wallace, 2011; Oyserman, 2013; Ricard & Pelletier, 2016; Robison, Jaggers, Rhodes, Blackmon, & Church, 2017). For instance, an estimated 7% dropout rate for those aged 16–24 years was reported the US in 2012 as a result of socioeconomic factors, family instability and household dysfunction, substance abuse, neglect, child abuse, attitudinal problems, adverse peer influences, and lack of self-efficacy among underrepresented minority students. The "*pipeline*" delivering underrepresented minority students to college and beyond is, therefore, "*bleeding*" at an alarming and unprecedented rate.

Given that students start deciding on a college path somewhere around sixth grade, the need for racial diversity and gender equality in STEM education in K–12 is also high. However, there are insufficient minority teachers and high racial disparity in K–12 teachers. Many K–12 teachers also lack cultural sensitivity (Bitterman, Goldring, & Gray, 2013). Whites constitute an overwhelming percentage of K–12 teachers nationally. For instance, in the state of Arkansas, 90% of K–12 teachers are White compared to only 10% of minority teachers (Dilworth & Coleman, 2014). The trend of racial disparity in K–12 teachers nationally is similar to the observed disparity in Arkansas. The high prevalence of race disparity and gender inequality in national teacher preparation programs also compound the racial disparity challenge. Out of the 728,310 in teacher preparation programs nationally in 2013, 74.3% are female and 24.7% are male. Also, the majority (68.1%) of potential teachers in preparation programs are White. However, only 10.6% are Hispanic or Latino, 8.6% are Black or African

American, 2.6% are Asian or Pacific Islander, and 0.8% are Native American or Alaska Native (Dilworth & Coleman, 2014).

The highlighted challenges, dichotomy, disparity, and inequality in K–12 school districts systems also have serious ramifications and cumulative effects in higher education. Many colleges and universities across the spectrum struggle with meaningful inclusivity and diversity. Racial disparity, gender inequality, and the shortage of minority faculty as mentors continue to be challenges and problematic attributes of higher education. A recent survey of 412,115 professors in over 1,500 universities revealed that 75% are White faculty. However, Asian, Black, and Latinx faculty only constitute 10%, 5%, and 4%, respectively, of all Carnegie classifications professors (Myers, 2016; Taylor, Apprey, Hill, McGrann, & Wang, 2010). The trend is not significantly different at highly selective research-intensive universities, large master's universities, medium master's universities, or at Liberal Arts & Sciences baccalaureate colleges, where the percentage of Black and Hispanic faculty is consistently less than 10%. The Black female professors also constitute only approximately 6%, and Latina professors are roughly 4% of the faculty across the spectrum from research-intensive universities to arts and sciences baccalaureate colleges. On the contrary, White women constitute 75% of female faculty members across the board.

Inequality and disparity in faculty ranks are also notable. Less than 5% of Black women and Latina faculty attained the rank of full professor. Indeed, the student body is usually more diverse at research-intensive universities than the faculty body. The shortage of minority faculty in STEM follows the observed overall pattern of minority faculty at colleges and universities. For example, less than 4% of chemistry professors in the US are minorities (Widener, 2017), with similar percentages in other STEM fields. The monolithic ethnicity, racial disparity, and gender inequality among faculty are problematic in higher education and constitute hindrances to attracting and recruiting minority students into STEM majors. Research and experience have shown that minority students are more receptive, inspired, motivated, and willing to attend institutions and STEM departments where they see minority faculty who share their cultural values. Minority faculty members are excellent resources to attract and retain minority students in STEM departments – serving as mentors, role models, and advisors for underrepresented minority students to succeed in the STEM disciplines.

The challenges of educational inequality and racial disparity in K–12 education combined with the lack of faculty diversity in colleges and universities have notable reverberations and repercussions in STEM higher education and beyond. Inadequate preparation of minority students and students from low-income families has resulted in a prevalent disparity between the motivation, retention/persistence, graduation rates, achievement, and success of underrepresented minority (URM) students in STEM fields (Adams & Lisy, 2007; Claudio, 1997; Foundation, 2004; Galama & Hosek, 2007; Gladieux & Kahlenberg, 2004; Hemphill & Vanneman, 2011; Hrabowski III, 2003; Huang, Taddese, & Walter, 2008; National Science Foundation, 2013; Poodry, 2003; Rohrbaugh &

Corces, 2011; Science and Engineering Indicators, 2010). According to the available data from the National Center for Educational Statistics, the total number of bachelor's degrees awarded from 1993 to 2007 has increased from 180,000 to 245,000 – an increase of 36%. However, the proportion of those degrees that are awarded to URMs in the STEM disciplines has only risen from approximately 13–17% (Chen, 2009). The demography of the United States is also rapidly changing, and it is projected that the proportion of underrepresented minorities in the US population will significantly increase by the year 2050. For instance, the Hispanic underrepresented minorities only account for 7% of awarded STEM PhD degrees. The Hispanic population is projected to almost double from the current 16% to approximately 30% by the year 2050 (Bureau, 2014). Meanwhile, the African American population is projected to stay about the same, at approximately 13% during the same period.

Racial disparity and gender inequality in our national educational system are problems that need immediate attention. For instance, racial disparity and educational inequality are directly related to income disparity and economic inequality. There is also a strong correlation between the levels of educational attainment and income earned over a lifetime and quality of life. Poorly educated individuals (often the minorities) are more likely to have serious social, economic, and behavioral issues and health challenges. Resolving national socioeconomic issues and reducing the opportunity and achievement gap and community safety can only be achieved by providing equitable access to quality education across cultures for all citizenry. Promoting the nation's technological workforce, competing, and expanding its relevance in the global economy and homeland security are only achievable by eliminating racial disparity and gender inequality in the next decades.

The US prospective racial diversity, ethnicity pluralism and "*brown America*," is a wakeup call to all educational stakeholders to promote a more inclusive and diverse future STEM workforce. The overarching goal of this chapter is to discuss and to highlight various strategies and create innovative approaches that are being utilized to transform STEM departments for inclusion at the University of Arkansas-Fort Smith (UAFS). Our conceptual framework for transforming STEM for inclusion at UAFS is shown in Scheme 1.

The first section of the chapter examines general challenges hindering this transformation. The second section discusses the historical perspective of UAFS – its institution profile, mission, and vision as a regional university. Strategies for addressing socioeconomic disparity, gender and racial inequality, and cultural competency awareness including an institutional/administrative initiative toward recruitment of more minority faculty, flexible teaching schedules for female faculty, offering of professional development and mentoring plans for junior faculty to promote inclusive and racial and gender equality at UAFS are discussed in the second section of the chapter. Other strategies such as the use of the creative Math-up curriculum innovation, the introduction of inquiry-based learning pedagogies and the formation of alliances between UAFS, local K–12 school districts, and governmental and non-governmental agencies to transform STEM departments for inclusion at UAFS are discussed next. The last section

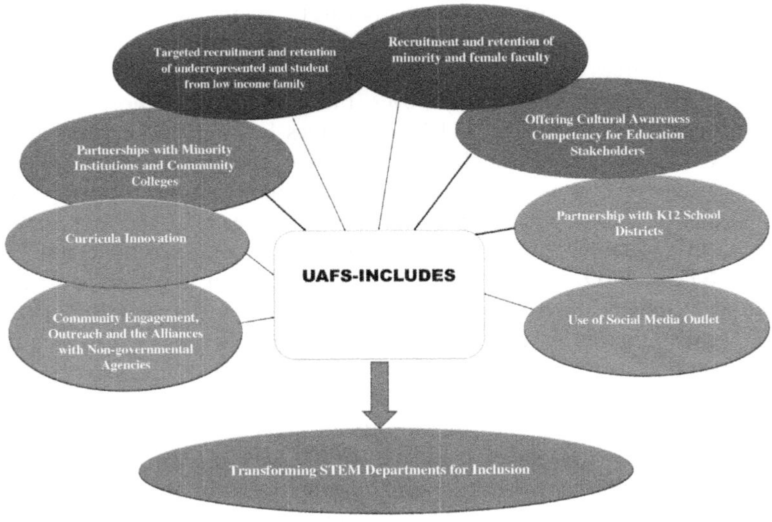

Scheme 1. Conceptual Framework of Transforming STEM Departments for Inclusion at UAFS.

of the chapter provides recommendations for adaptation and long-term sustainability of efforts toward the stated objectives.

TRANSFORMING STEM DEPARTMENTS FOR INCLUSION AT THE UNIVERSITY OF ARKANSAS-FORT SMITH

University of Arkansas-Fort Smith – Institutional Profile

UAFS began as Fort Smith Junior College with 34 students in 1928. It successfully transitioned to a four-year regional university in 2002 after going through two other iterations as WestArk Junior College (1966) and WestArk Community College (1972). The institution has expanded in size and vision over the years – finally taking on its current name and identity as a premier regional university when it joined the University of Arkansas System in 2002 to meet the educational need and demands of the workforce of the region. Table 1 shows the UAFS student enrollment and race/ethnicity composition over the past five years. The total student enrollment at UAFS is approximately 6,700 students, with 76 academic programs comprising one master's degree, 33 bachelor's degrees, 14 associate degrees, and 28 certificates. UAFS contributed over 300 million dollars to the regional economy in 2016.

The majority of UAFS students are first generation, and approximately 80% of them qualify for Pell Grants. The institution's strategic plans are to develop as a center for intellectual, artistic, social, and cultural advancement, to maximize real-world experiences for its students, to solidify and expand its position

Table 1. UAFS Student Enrollment Race/Ethnicity Composition, 2013–2017.

Race/Ethnicity	2013		2014		2015		2016		2017	
All Students	#	%	#	%	#	%	#	%	#	%
American Indian or Alaska Native	211	2.9	186	2.7	155	2.3	168	2.6	148	2.2
Asian	309	4.3	299	4.4	333	5.0	325	4.9	349	5.3
Black or African American	314	4.4	300	4.4	260	3.9	246	3.7	275	4.2
Hispanics of any race	575	8.0	663	9.7	758	11.3	748	11.4	819	12.4
Native Hawaiian or Other Pacific Islander	4	0.1	4	0.1	7	0.1	7	0.1	8	0.1
Non-resident Aliens	82	1.1	104	1.5	90	1.3	92	1.4	81	1.2
Race and ethnicity unknown	0	0.0	10	0.1	4	0.1	4	0.1	0	
Two or more races	365	5.1	359	5.3	414	6.2	476	7.2	532	8.1
White	5,294	74.0	4,900	71.8	4,683	69.9	4,503	68.5	4,396	66.5
Total	7,154		6,825		6,704		6,569		6,608	

as a primary contributor to economic development, to lead the innovative use of technology, and to promote global learning initiative. Through partnerships with the state and with local industry, UAFS is consistently anticipating the future demands of the workforce to provide premier education and training to students and the community. Although UAFS remains a predominantly teaching institution, it has begun to place greater emphasis on a commitment to excellence in research-training and scholarly activity in order to better prepare students for the global challenges and opportunities of the twenty-first century in academia, professional schools, industrial and governmental jobs, private business, entrepreneurship, and the STEM workforce. UAFS has several institutional initiatives, resources, and platforms highlighted and discussed below to transform STEM departments for inclusion and to promote broad participation in STEM across cultures, race, and social–economic backgrounds.

College of STEM at UAFS

The College of STEM, established in July 2008, initially encompassed four departments: Sciences, Information Technology, Engineering, and Mathematics. After some restructuring, the College of STEM now houses departments of Biological Sciences, Physical Sciences, Computer & Information Systems & Engineering, and Mathematics (combined with DevEd Math). It has also added the School of Education in 2014. Table 2 shows the College of STEM student enrollment race/ethnicity composition.

The departments of Physical Sciences, Biological Sciences, and Mathematics are located in the Math–Science building and share infrastructure and resources. The Department of Physical Sciences offers bachelor's degrees in Chemistry (with a concentration in Biochemistry) and Geosciences as well as minors in Physics, Chemistry, and Geology. The department has approximately

Table 2. College of STEM Student Enrollment Race/Ethnicity Composition, 2013–2017.

Races	Gender	2013	2014	2015	2016	2017
American Indian	Female	16	22	16	13	8
	Male	36	29	21	21	14
	Total	52	51	37	34	22
Asian	Female	35	38	41	39	37
	Male	55	55	54	43	58
	Total	90	93	95	82	95
Black	Female	35	31	19	11	24
	Male	18	25	11	19	18
	Total	53	56	30	30	42
Hispanic	Female	46	50	60	75	85
	Male	54	63	58	72	67
	Total	100	113	118	147	152
Native Hawaiian/PI	Female	1	1	0	0	0
	Male	1	0	0	1	0
	Total	2	1	0	1	0
Non-residence Alien	Female	13	13	7	5	6
	Male	19	20	21	19	19
	Total	32	33	28	24	25
Two or more Races	Female	43	39	51	55	60
	Male	44	32	42	54	67
	Total	87	71	93	109	127
White	Female	692	608	529	463	440
	Male	604	559	496	439	394
	Total	1,296	1,167	1,025	902	834
	Female	881	802	723	661	660
	Male	831	783	703	668	637
STEM Total		1,712	1,585	1,426	1,329	1,297

225 majors. The Department of Biological Sciences has ~ 350 majors and a single concentration. The Department of Mathematics has approximately 100 majors. Our Associate's Degrees in electrical engineering and mechanical engineering are offered in coordination with the University of Arkansas at Fayetteville to which AS graduates matriculate upon graduation. During the final two years of study, program students are taught by UAF faculty on the UAFS campus and receive their BS degrees from the ABET-accredited UAF. The College of STEM also houses the School of Education, which allows synergy, efficient interaction, and collaboration between STEM and Education faculty. Teacher licensure degrees are also available for biology and mathematics as well as other non-STEM disciplines in secondary education (7–12), as are

bachelor's degrees in elementary education (K–6) and middle childhood education (4–8). The College of STEM has 63 faculty and 17 staff. Faculty and staff receive grants from the National Science Foundation, Arkansas Department of Higher Education, Arkansas Department of Education, and some private funding. The overall goals of the College of STEM are to provide a quality education, inquiry-based learning, and hands-on experiences for students on cutting-edge ideas and methodologies. Our programs and courses are specifically designed to promote independence, analytical thinking, problem-solving, communication teamwork, and leadership skills for our graduates. The College of STEM's top priority is to adequately prepare our graduates for the STEM workforce, graduate studies, teaching, other professional work, or medical/healthcare fields.

Institutional/Administrative Initiative

UAFS recognizes the importance of cultural diversity and gender equality among the faculty, staff, and student body. It values and understands the importance of an inclusive academic environment that nurtures and cherishes students – providing them with a culturally relevant education. UAFS has therefore intensified efforts to promote inclusivity, faculty diversity, as well as gender and racial equity. For instance, a considerable number of minority faculty and minority female faculty have been hired in the past few years to reduce the faculty racial disparity and gender inequality UAFS campus-wide and in STEM departments. A significant number of minority faculty have also been hired in key strategic administrative positions in STEM departments and campus-wide. To increase the minority applicants' pool, the College of STEM now post faculty job advertisements in National Organization for Black Engineers and Chemists, Journal of Blacks in Higher Education, Women in Higher Education, and Society for Advancement of Chicanos/Hispanics and Native Americans in Science. It has also mandated that all search committees possess both gender and ethnic diversity. It is anticipated and envisioned that these changes would see a rise in faculty diversity and that the enhanced presence of minority faculty in STEM and other departments will recruit support and retain more underrepresented minority students in STEM majors at UAFS. These minority faculty will also serve as crucial mentors, role models, and advisors for underrepresented minority students, promoting increased retention, graduation, and success.

Balancing teaching, scholarly activity, and service (departmental, college, campus, and community) with family responsibilities is a challenge for faculty. This is especially true during pregnancy and in families with young children. The professional work expectations and challenges of pregnancy, childcare, and other family responsibilities may significantly affect the output and research productivity of faculty, impeding their advancement in ranks. UAFS recognizes these challenges and the importance of achieving a healthy balance between work and family responsibilities. Department Heads in the College of STEM at UAFS have developed more flexible teaching schedules and committee assignment responsibilities for faculty with children during the academic year in order to accommodate their domestic needs. These flexibilities have allowed faculty to drop off and pick up their children and attend cultural and educational events at school. Efforts are also currently being made to establish a Child Development Center on the UAFS campus. This will further allow the faculties to bring their children on campus in the morning and take them home after work.

Gender inequality in promotion and ranks is another issue that is being addressed on the UAFS campus. Several strategies including effective mentoring of female faculty and development of competitive promotion portfolios to make it easier for female faculty to advance through ranks from Assistant to full professor are being implemented. UAFS is also exploring external grants such as NSF-ADVANCE and NSF-INCLUDES grant proposals to obtain additional resources and infrastructure for female and junior faculty members to advance their research and scholarly activity to promote their professional development and academic growth.

Offering Professional Development Programs for STEM Faculty

UAFS recognizes the importance of providing professional development opportunities for faculty members as a catalyst to advancing their career goals. UAFS also recognizes the effects of provision of mentoring strategies, particularly for junior faculty to strike a balance between the responsibilities of teaching, scholarly activity, and service to the department, college school, and community. Accordingly, various professional development programs and mentoring initiatives have been instituted, implemented, and strategically developed for junior faculty members to enhance their teaching effectiveness and to assist them in advancing their professional careers at UAFS. There is also a plan to assign reduced teaching loads as much as possible to future research-oriented junior faculty hires during thier first few years in order to allow them time to establish themselves.

The university also recognizes the importance of the provision of professional development programs for faculty to promote the use of student-centered and student-engaging innovative classroom pedagogy. Professional development program workshops are therefore organized and offered on campus for junior faculty on creative curriculum innovation pedagogy development and classroom instructions to enhance their teaching effectiveness. For instance, the College of STEM recently organized a Process Oriented Guided Inquiry Learning (POGIL) professional development workshop for targeted STEM faculty in fall 2017 to promote the use of active learning pedagogy in STEM curricular. We have also hosted workshops on Flipped Classes and Generational Differences in Learning in spring 2017. Also, a seminar on Fixed vs. Growth Mindsets has been scheduled for spring 2018. Junior faculty members are also encouraged and provided fiscal support to attend professional development program workshops at regional and national conferences through Faculty Development Funds to stimulate their research, promote their teaching effectiveness, and to facilitate networking with their peers at other institutions.

Targeted Recruitment, Retention, and Graduation of STEM Majors across Cultures

Continued budget cuts to higher education, demand a proactive and sustainable strategy to recruit, retain, and graduate students in a timely manner. The current funding formula in the state of Arkansas has shown the importance of STEM majors as the largest contributor in the productivity-based funding formula. Most importantly, the current funding formula places a significant emphasis on the progression through graduation of STEM majors and minority students. This model is based on institutional effectiveness (80%) and affordability (20%). The effectiveness component of the funding formula is based on students' earned credentials (certificates of proficiency, technical certificates, associate's degrees, bachelor's degrees, or master's degrees), progression, transfer success, success of student in gateway courses success (Mathematics, English, and Reading), post-completion success, and non-credit workforce training. The affordability component of the Productivity Funding Formula is made up of student time to degree and credits to completion. More weight is assigned to retention, progression, and graduation of STEM majors, Hispanics and African Americans, and students from low-income families. Detailed information about the Productivity Funding Formula is available at Arkansas Department of Higher Education website and can be accessed at: https://static.ark.org/eeuploads/adhe/Productivity_Funding_Policy_-_Universities1.pdf. Accordingly, recruitment, retention, and graduation of more and more diverse STEM majors are imperative and a top priority for the future financial stability of any university in the state of Arkansas system. UAFS has therefore taken strategic steps and approaches to facilitate targeted recruitment, retention, and graduation of STEM major across social–economic, gender and racial spectra. We have strong partnerships with K–12 school districts and K–12 councilors in Sebastian and, adjoining counties as well as in neighboring Oklahoma counties. We are also working to increase our recruiting footprint across the state and in the other states that border our own. STEM departments distribute booklets, pamphlets, and flyers during our regular open house events to create awareness and to disseminate the resources, scholarships, and financial aid available to attract students. STEM faculty also regularly attend local K–12 school districts events to recruit students and to further create awareness about the application process. The College of STEM also has strong student clubs including The Chemistry Club, Geology Club, Premed student club, and Tri-Beta club. These club members and their faculty advisors are also excellent resources for student recruitment and retention. The club members have served and participated in UAFS open house events, visited several K–12 classrooms, and have also served as peer-mentors/councilors for several high school students as part of their annual community service and educational outreach.

Promoting STEM among Low-income and First-generation Students: UAFS Upward Bound Programs Initiatives

Upward Bound Math–Science and the Upward Bound "Classic" program on the UAFS campus serve 124 first-generation, low-income students in the Fort

Smith region. The programs recruit at seven area school districts. The goal of the programs is to increase secondary school graduation and post-secondary enrollment and graduation of URM students. The programs at UAFS achieve these goals by providing academic tutoring, advising, cultural enrichment opportunities, and college visits. Participants come to the UAFS campus for Saturday Scholars Academy, Summer Academy, and a variety of other events and activities. Upward Bound Math–Science at UAFS serves participants that have expressed a strong interest in a career in the fields of mathematics and/or science. Upward Bound Classic serves participants that show an interest in any field, including mathematics and science. Student Support Services on the UAFS campus serves 140 students that are identified as first-generation, low-income, and/or registered with the disabilities office. The program works with participants to ensure they graduate college within six years. It provides academic tutoring, advising, career exploration, cultural enrichment, and professional/graduate school admissions preparation. All students in the program must be seeking a bachelor's degree from UAFS to be eligible.

Student Emergency Fund through Great Lakes Dash Grant

Approximately 80% of UAFS students are Pell-eligible. A significant fraction of them work – some only part-time, but large numbers work at least 40 hours a week while still enrolling full-time. Many non-traditional students have families and associated responsibilities in addition to working full-time. Balancing school work, family, and off-campus jobs is daunting and often overwhelming. Some students may also face financial emergency situations during the semester which may cause academic stress, poor performance in course work, or result in their outright withdrawal from college if not addressed appropriately. UAFS was awarded a Great Lakes Community Investments Dash grant allowing us to intervene immediately and to address students' emergency financial situations. The Dash fund is available for students with an expected family contribution of US$7,000 or less. Students can apply for and receive a one-time, US$1,000 award. The fund is intended to meet unexpected or emergency costs (not tuition and books) that arise and may keep a student from staying enrolled in college. Medical bills, car repairs, and/or changes in childcare costs are covered by the Great Lakes Community Investments Dash Grant. Also, each emergency grant is used as an avenue to motivate and to encourage students to stay the course with their academic programs. Besides, students are provided with additional campus and community resources that can address the root causes and contributing factors leading to their unforeseen expenses. Furthermore, UAFS provides economic and financial literacy resources as a follow-up to a grant award for especially strong student outcomes.

In 2016, the College of STEM secured a separate Great Lakes Community Investments Grant specifically used to address financial barriers to retention. The objective of this grant is to identify struggling STEM seniors (with 90+ hours) and provide financial and mentoring assistance to enable them to successfully complete their degree. It is designed to provide up to two semesters of assistance – paying for those personal and academic expenses that would otherwise see them derailed. It also makes available one-on-one mentoring and advising to help seniors persist to graduation. Students complete an application that confirms their academic standing and demonstrates their specific financial needs. The Director of Financial Aid verifies their eligibility for receipt of aid. Applicants must also be recommended by their Department Head. A graduation application is included to verify the student is on track for completion and for each applicant to be keenly aware of their remaining academic requirements. Once the application and forms are complete, they are scanned and emailed to the Program Advisory Team for voting. If approved, the student is required to come in and sign a contract confirming their understanding of their requirements for continuing on the Grant and the amount granted for the upcoming semester. Students and their faculty mentors are introduced to each other at a celebration at the start of each semester.

In general, students usually need financial assistance only to cover the cost of their tuition and fees, which ranges from US$2,400 up to US$4,000 per term. However, additional assistance has been provided to cover utilities, food, groceries, rent or mortgages, childcare, parking passes, fuel cards, and so on. Financial assistance provided for the students has ranged between US$600 and US$5,500 per semester. Students have completed the majority of their courses with great success. Thus far, the College of STEM Great Lakes Community Investments Grant has assisted 13 STEM majors. Of these recipients, four of them graduated in December 2017. We are excited, and we look forward to hearing about their successes in their future endeavors. Eight of the remaining recipients are on track to graduate in spring 2018. Unfortunately, one recipient failed to meet the expectation of the contract and funding was withdrawn. The College of STEM Great Lakes Community Investments Grant offered assistance to three additional STEM students for the spring 2018 term.

Curriculum Innovation: Math-up Initiatives

High attrition rates in gateway courses (especially mathematics, chemistry, and physics) is of grave concern and is a major reason leading to low persistence, retention, and graduation of students in STEM programs. For example, the typical success rate, nationally, in general chemistry is approximately 60%. Studies have also shown a continued large disparity between the persistence, retention, and graduation rates of URM students majoring in STEM compared with non-minority STEM majors as a result of cumulative challenges (*vide supra*) (Adams & Lisy, 2007; Claudio, 1997; Foundation, 2004; Galama & Hosek, 2007; Gladieux & Kahlenberg, 2004; Hrabowski III, 2003; Huang et al., 2008; National Science Foundation, 2013; Poodry, 2003; Rohrbaugh & Corces, 2011; Science and Engineering Indicators, 2010). Many K–12 students, especially minority students, often have low SAT/ACT scores in mathematics, preventing them from enrolling in the college-level Math courses required for their course of study. Such students must enroll in remedial mathematics courses in their first year in college, spending an extra semester(s) to take an entire course (or series of courses) for which they may only need a small portion to build their skill set. Prolonged time to graduation also means that these students are likely to exhaust their financial aid and incur additional financial burdens from loans or even spend extra money out of their pockets. The latter often means working or working extra hours, which negatively impacts their already tenuous ability to do well in these courses increasing the likelihood of their dropping out of college.

UAFS has developed a *Math-up* initiative that attempts to meet the needs of students with inadequate mathematics preparation and/or low SAT/ACT mathematics scores. Math-up gives these students an opportunity to improve their mathematics skills during the summer (or at other times before the start of a semester), ultimately allowing them to enroll in entry level credit-bearing or even higher-level mathematics courses that fulfill requirements in their degree plan. In addition to higher student retention and on-time student graduation, the Math-up program also alleviates extra student costs that would have been spent on remedial mathematics coursework. The Math-up program was initiated and piloted in summer 2016 for 124 pre-college-level students STEM majors. The students were allowed to take the first Math-up diagnostics test during the New Student Orientation (NSO) to evaluate their mathematics skills and their readiness to take college mathematics or calculus courses. The Math-up diagnostics test is provided by Assessment and Learning in Knowledge Spaces (ALEKS) software program. ALEKS is a Web-based, artificially intelligent assessment and learning system. Detailed information about ALEKS may be accessed at www.ALEKS.com. Students must score at least 35 points on the Math-up diagnostics test to register for college mathematics courses and 70 points to be eligible to enroll in Calculus I. Students with lower Math-up scores are invited over the summer for continued assessment/improvement of their math skills set. The Department of Mathematics, often in collaboration with the UAFS Education Renewal Zone, paid for student's ALEKS registration as an incentive and to reduce the financial burden on students. Students also have the flexibility to work off-campus at their own paces on their ALEKS Math-up exercises. Math-up was designed and customized to meet the specific content needs of each student and to address their individual mathematics shortcomings and college-level readiness.

The first assessment is the most crucial part of the Math-up program. The students who signed up for the Math-up program were contacted via phone during the summer and encouraged to practice more Math-up excises and to come back to campus for a second assessment. After students have practiced for at least 10 hours, they are invited back to campus for a second assessment to evaluate how much progress has been made with their mathematics skills. If a student has improved sufficiently, they are encouraged to practice more Math-up exercises and come back for a third, fourth, or fifth assessment until they attain the necessary skills to register for the appropriate credit-bearing course. The completion of assessments for students who completed Math-up overall is as follows: second assessment – 44%, third assessment – 20%, fourth assessment – 30%, and fifth assessment – 6%.

Most of the Math-up participants attained an increase of over 50% in their Math-up scores (from 22 to 33 points) and math skill set over a period of one month and by the second assessment. Incredibly high increases in Math-up scores (from 22 to 70 points), an average of 218% increase, were observed for some students after two months of the summer Math-up exercise and the third assessment. As a result, these students, many of whom were slated for remedial math, were able to bypass College Algebra and were enrolled in the Survey of Calculus course. Also, two months of Math-up saved such students two semesters and the expense of seven hours of non-credit coursework. Some biology majors who participated in the Math-up program have also been able to increase their math skills and therefore enrolled in higher mathematics courses such as Probability & Statistics without the need to take a College Algebra prerequisite. This also allowed these students to enroll in College Chemistry I (which has a College Algebra prerequisite). Most importantly, such students are offered a guaranteed 8-semester degree plan after Math-up, meaning they can graduate in four years, or sooner. This program may well represent one of the best ways to retain our students in the STEM fields. Non-traditional students and veteran students have also benefited from the Math-up program. They used it to refresh their math skills and to get back on track with no extra cost and/or time spent.

The summary of the final grades of the Math-up participants in their enrolled mathematics courses in fall 2016 is shown in Fig. 1.

Approximately 8% of students who participated in the Math-up program could not enroll in a math course due to restrictions in their financial aid. An

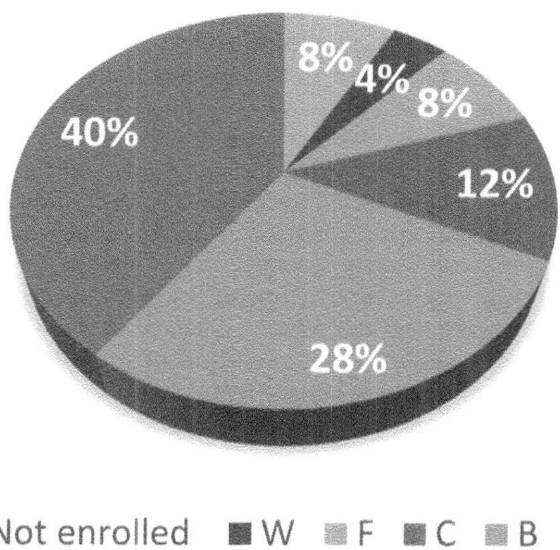

Fig. 1. Final Grades of Math-up Participants in their Enrolled Mathematics Courses in Fall 2016.

overwhelming 87% of the participants earned passing grades of "C" or better in their mathematics courses. On the contrary, only 70% of non-Math-up participants earned passing grades of "C" or better in their mathematics courses. Better performance of the Math-up participants may be attributed to their motivation to work more independently and to spend considerable time solving mathematics problems outside of the classroom, a skill set they learned over the summer during their Math-up practice excise. Most importantly, students who completed the Math-up program during summer 2016 saved a total of 202 credit hours of coursework, $40,400 in tuition costs, and 56 semesters toward graduation. The total direct cost to the institution, at $25 per student, was $3,100.

Offering Math Reviews to Strengthen Student Success in Introductory College Chemistry Courses

The Department of Physical Sciences at UAFS has also developed a set of "*mathematics review*" questions, covering basic topics in college algebra and pre-calculus, including ratios and proportions unit conversions and dimensional analysis. The overall objective of the "*mathematics-review*" problems initiatives is to allow students to practice individually and later to solve problems in groups in class and thus be better prepared for future calculations and data analysis in College Chemistry courses and laboratory experiments. Some instructors are further using a Calculation in Chemistry book (Dahm & Nelson, 2013) and Mastering in Chemistry problem tools as additional resources to strengthen students' understanding of math in chemistry.

One of the challenges faced by students is the lack of adequate preparation before the class or laboratory activity in STEM. To address this issue, course materials, PowerPoint® lecture notes, pre-lab assignments, and assessments are often posted on Blackboard® for completion prior to coming to the class or laboratory. The students are quizzed on these materials to evaluate their readiness and to assess where they have difficulties. Scheduling exams and their reviews outside of the class for success marker courses allows faculty and students to have more class time to go over problems. In organic chemistry, instructors schedule six cumulative exams during the semester, and the best four scores are taken. In this way, students are encouraged to persist in the course even if they may have scored poorly on one or two early exams and even if their mid-term grade is low. In this model, even at mid-term, as much as 75% of a student's grade may still be undetermined and, hence, influenced by a change in their strategy and increased focused effort. Other initiatives such as holding recitation sections, extra tutorial sections, an in-class activity, and teamwork have also been used to meet the needs of students who need additional help in STEM courses.

FORMATION OF ALLIANCES AND PARTNERSHIPS WITH KEY EDUCATIONAL SHAREHOLDERS

UAFS recognizes that broadening participation, resolving the many challenges associated with majoring in STEM, and reducing achievement gaps across all cultures, requires a coherent, well-coordinated effort and the formation of productive alliances spanning educational sectors among key educational stakeholders. Formation of strong alliance among UAFS-STEM educators, staff, and students to promote STEM education is our top priority. Scheme 2 is the overview of our alliance-oriented model for building and sustaining strategic associations and participation at UAFS.

UAFS recognizes that the broader scientific community is rapidly becoming interdisciplinary and is increasingly functioning as a global village. It also recognizes the importance of offering more robust and broad-based interdisciplinary programs that will provide enriched and diverse experiences for students. Development of interdisciplinary programs will not only promote cross-pollenation of ideas among the faculty from diverse backgrounds, expertise, and professional experiences, but it will simultaneously provide teamwork, interdisciplinary, and experiential learning opportunities for students. One of our immediate goals in the Department of Physical Sciences is also to establish an interdisciplinary curriculum to promote Forensic Science and Crime Scene Investigation in the College of STEM. Forensic chemistry is in demand and represents a very broad, interdisciplinary exposure, involving teams of experts from various disciplines, STEM, non-STEM, and non-academic (Amendt et al., 2007; National Institute of Justice, 2000; Rausch, 2015; Snow, 1982). Our previous study demonstrated the positive impacts of interdisciplinary forensic science in promoting the success, teamwork, and collaboration between STEM and Criminal Justice (Fakayode et al., 2016). Our forensic chemistry program will, therefore, be interdisciplinary involving the Department of Criminal Justice, Department of Biology, Department of Physical Sciences, and Department of Computer & Information Science and Engineering in the College of STEM. It will also enjoy partnerships with the local judiciary system including the Police Department and attorneys.

Because forensic chemistry will naturally involve and require collaborative approaches, it will support collective learning, communication/deliberation of concepts, personal confidence, and interdisciplinary problem-solving. The forensic chemistry curriculum will also provide an experiential learning environment, which facilitates student mastery of concepts through visualization, application, and measurement. Plans are underway to initiate Internship opportunities with the local Police Department, Arkansas Crime Laboratory (Little Rock), and Federal Bureau of Investigation (Little Rock). Student interaction with subject matter experts will provide them with excellent mentoring opportunities and expose them to professional standards and the rigors of scientific and legal reviews. The program will also provide students with experience in producing professional grade work/research. Forensic chemistry, forensic biology, and forensic geology exercises will broaden the research theory and methods options available to students by inserting them in a problem-solving environment with a team of problem-solvers from multiple disciplines.

Our forensic chemistry program will focus on the principles of forensics that relate to the chain of custody, physical evidence collection and storage, laboratory analysis of evidence, and ethical considerations in data interpretation and

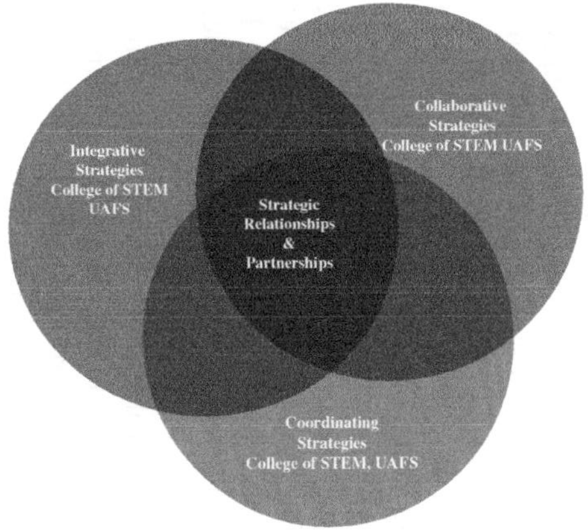

Integrative Strategies College of STEM, UAFS	Collaborative Strategies College of STEM, UAFS	Coordinating Strategies College of STEM, UAFS
People, their Dispositions, Organizational Interactions, and Inter-organizational relationships	Internal organizational structures/departments and external affiliates, partners, communities, and organizations	Implementation, Project Outcomes, Viability, and Sustainability of Partnerships
1. Teamwork, supportive attitudes, and organizational dispositions 2. Trust building 3. Role/responsibility definition 4. Respect for persons 5. Role support/empowerment of team members 6. Build credibility and inspire support 7. Embrace unique talents, gift, and abilities of educational stakeholders 8. Unity and inspiration 9. Purposive and open communication among College of STEM constituents 10. Shared objectives among College of STEM constituents 11.Cross-cultural teams and networks 12.Rewards among College of STEM constituents 13.Action, Accountability, and Consequences among College of STEM educational stakeholders 14.Clear communication of College of STEM core values, ethics, and principles 15. Support for a culture of learning and discovery and the power of ideas	1. Definition of goals 2. Expectations defined 3. Understanding for local contexts 4. A culture of dynamic discourse and dialogues 5. Support for scientific innovation and initiatives 6. Understanding of common goals 7. Mutuality and consideration for local context 8. Service to others and the community 9. Inclusive/participatory decisionmaking 10. Identification of and use of common threads 11. Teamwork (horizontal groups, vertical groups, and combined horizontal and vertical groups) 12. Common understanding of College of STEM core values, ethics, and principles 13. Trust among diverse College of STEM constituents 14. Confidence in products and outcomes 15. Common adherence to regulations of the UAFS	1.Protecting the core values, ethics, and principles of the College of STEM. 2. Accountability to College of STEM stakeholders 3. Stakeholders' action pegged to College of STEM operational standards 4. Internal task assignments are harmonized with external partners 5. Quality assurance 6. Weaving of common threads among stakeholders 7. Effective implementation of goals and objectives 8. Consideration for local contexts 9. Sharing of fruits of labor 10. Evaluation and assessment of outcomes 11. Precise feedback 12.Strengthened confidence in College of STEM organizational efficacy and efficiency

Scheme 2. An Alliance-oriented Model for Building and Sustaining Strategic Associations and Participation: University of Arkansas at Fort Smith.

reporting. The initiative is expected to improve the recruitment, retention, and graduation rates of Criminal Justice, chemistry majors, and other STEM majors at UAFS. The initiative will likely have a positive long-term impact in motivating future generations into criminal justice, forensic science, law enforcement, and homeland security.

UAFS has built on the unique strength of its identity as a regional university to form a system of robust alliances with key educational stakeholders. Among these are minority-serving institutions, community colleges, local school districts, policymakers, governmental and non-governmental agencies, and local businesses with positive complementary and synergistic effects to promote inclusion, diversity, and culturally relevance in education and to broaden participation in STEM education. Some of these partnerships are highlighted below:

Partnership with K–12 School Districts: Office of P-20 Collaboration

Our research (Fakayode et al., 2016; Fakayode, Pollard, & Snipes, 2014; Fakayode, Snipes, & Kanipes, 2015) and that from other educators (Blasie & Palladino, 2005; Bundschuh, El-Awady, Hardin, Hughes, & Shelton, 1982; Fuller, Lefler, Panush, & Will, 1971; Herrington, Luxford, & Yezierski, 2012; Howell, 2001; Sarquis, 2001; Sweeney, 2003) has shown the importance of partnerships between STEM faculty in higher education and school districts in promoting grass-roots science education through provision of resources or offering of a professional development program for teachers. The UAFS-STEM Education Center is an integral part of the Arkansas Network of STEM Education Centers. It provides educational outreach to 22 school districts (over 110 schools) in Sebastian, Crawford, Franklin, Scott, Logan, and Johnson Counties in the state of Arkansas. At UAFS, both the Education Renewal Zone (ERZ) and the STEM Education Center fall under a single, coordinating Executive Directorship. The prime objective of the STEM Education Center is to provide educational outreach focusing on teacher and administrator professional development within the STEM fields. The overarching objective of the Education Renewal Zone is to connect higher education and K–12 through initiatives in any field of study that advance student achievement. Both programs have the same ultimate goal of enhancing and increasing learning outcomes for Arkansas Pre-K–12 students. Arkansas hosts six ERZs and 12 STEM Education Centers, all housed at four-year institutions of higher education. UAFS is fortunate to house both programs. Each program is funded with separate, annual grants from the Arkansas Department of Education. Additional annual grants supporting the work of our programs are secured from additional state-level sources. The resources and facilitates in the UAFS-STEM Center include a well-equipped library, and books, for K–16 science education. The STEM Education Center also has over US$500,000 in math and science manipulatives, 30 laptop computers, materials, kits, field equipment, virtual reality instruments, and print resources available for checkout to regional K–12 educators. The STEM Education Center has conducted professional development programs for thousands of K–12 teachers, UAFS student/preservice teachers, UAFS faculty and staff, and parents in the past three years.

UAFS is uniquely positioned as it is the only campus in the state of Arkansas that has both programs conveniently located in one building and organizationally structured as its own unit. This unit is managed by the Executive Director of the ERZ and STEM Education Center and is under the purview of the College of STEM, yet collaborates with all colleges and departments campus-wide. Our unique model not only increased the collaborative, synergistic effect of the two programs but also enhanced the productivity of each. Both the physical and organizational structure of these programs have leveraged state resources by maximizing financial benefits and strengthening the educational experience for our internal (university faculty, administrators, and students) and external (K–12 administrators, teachers, students, and community members) constituents. Some of the initiatives including Adopt-a-Professor, Festival of Science, and Girls in STEM Leadership Conference that are currently being implemented by the STEM Education Center are highlighted below:

Adopt-a-Professor Initiative

The overarching goal of the Adopt-a-Professor program is for the ERZ to pair volunteer Pre-K–12 teachers with volunteer UAFS faculty and staff to collaborate and develop three "learning engagements" over the course of a semester. These learning engagements can take place in the teacher's classroom, on the UAFS campus, or in the greater community, as appropriate. The purpose of the program is twofold: (1) to build collaborative relationships between Pre-K–12 and higher education and (2) to strengthen learning outcomes for Pre-K–12 students. The Adopt-a-professor program started on the UAFS campus in the spring of 2014 and, with one exception, has consistently operated every semester since. Over 64 UAFS faculty and 67 Pre-K–12 teachers in seven school districts have participated in the Adopt-a-professor program, to date. Some of the projects and science demonstration that have been conducted in the Adopt-a-professor program for Pre-K–12 students and teachers includes the following: (1) Zombie autopsy, (2) Structures and earthquake project, (3) Coding and video game project, (4) Introduction to industrial robots, (5) Programmable logic controllers, (6) Electronic Circuit Building, (7) Digital Multimeters, and (8) Resistor Color Codes. These demonstrations have generated a lot of excitement and motivation, stimulating the interest of Pre-K–12 students in science education.

Festival of Science

The Festival of Science was initiated in fall 2012 as a collaboration between UAFS-STEM faculty, the ERZ , and the STEM Education Center and has been held every fall since. This annual event is targeted for area third graders and is held in the on UAFS campus with the purpose of cultivating young students' interest in STEM fields by engaging them in exciting mathematics and science experiences. STEM faculty from all STEM departments open their labs for two hours on a Saturday afternoon and rotate students through a series of 20-minute, hands-on, lab experiences. UAFS students and pre-service teachers have also served as laboratory assistants or as team leaders for the third-grade teams. In addition, the College of STEM also has student clubs: Chemistry Club, Geology Club, Premed student club, and Tri-Beta club. These club members and their faculty advisors also participate in Festival of Science. Each student, faculty, staff, and volunteer was given a Festival of Science T-shirt that was worn during the Festival of Science event.

Some of the selected fun experiments and demonstrations that were conducted for the students during the Festival of Science include the following: (1) Strawberry DNA extraction, (2) Microscopic examination of cells, (3) Pond Life, (4) Mirras, (5) Helicopters, (6) Robotic car, (7) Tanagrams, (8) Slime & Oobleck, (9) Dry ice show, and (10) Gold Panning. A Festival of Science T-shirt was also made and given to each student, faculty, and volunteer. Over the six years, 678 third-graders have participated, joined by 139 parents/family members. The highest number of STEM faculty volunteers, 27, was obtained in 2017. At each laboratory exercise section, the students were fully engaged and participated in the laboratory activity. Also, the students were given some fun door prizes as motivation and as mementos.

Girls in STEM Leadership Conference

A *Girls in STEM Leadership Conference* event has been hosted on the UAFS campus for the past two years and is a collaboration between the ERZ, STEM Education Center, College of STEM, and community and regional partners. We have secured funding and are currently planning our third annual event for spring 2018. This female-focused program aims to increase the girls' awareness in three areas: (1) identify STEM careers and their relationship to female interests, (2) recognize their potential as females to be successful in STEM careers, and (3) view university faculty and STEM professionals as academic resources. Female UAFS-STEM faculty members volunteer to host 90-minute, interactive labs during this one-day conference. The highlight of one conference was hosting two NASA scientists from the US Space & Rocket Center, Huntsville, Alabama. These representatives conducted laboratory demonstrations on engineering and designing a Mars Rover machine. Girls in STEM Leadership conference hosted 135 girls (9th grader, 2; 10th grader, 29; 11th grader, 63; 12th grader, 39) from six school districts in spring 2017. The ethnicity and demography of the Girls in STEM Leadership participants are made of American Indian/Alaskan Native (2%), Asian (4%), Black or African American (5%), Hispanic/Latina (13%), Native Hawaiian/Pacific Islander (1%), White (72%), and two or more (2%). In the post-conference, 126 out of the 135 participants indicated that they were more confident in pursuing a STEM career after attending the conference.

Overall, the Adopt-a-Professor, Festival of Science, and Girls in STEM Leadership Conference initiatives conducted by UAFS-STEM Education Center have been a tremendous success. It is envisioned that Adopt-a-Professor, Festival of Science, and Girls in STEM Leadership Conference initiative experiences will have a long-term and positive impact on pre-K–12 students. These experiences will stimulate pre-K–12 students' interest in science and promote a quality "*pathway*" for STEM majors in college. This will have a broader impact on the production of next the generation of STEM professionals. Adequate preparation of an inclusive and diversified next generation of STEM graduates will also have ripple effects that positively strengthen the future STEM and technological workforce. This will promote self-reliance, sustainability, and US competitiveness in global economy. These well-prepared students will have the potential to become leaders in science, medicine, and engineering, in academia and industry. They will also enhance national research laboratories and take their places as STEM educators in K–12 and community colleges.

Partnerships with Minority and HBCUs Institutions

Faculty members from UAFS have developed strong collaborations in STEM education research as well as in fundamental research with peers at several minority-serving institutions including Historically Black College and University (HBCUs). Specifically, some faculty from UAFS have developed collaborations and alliances with faculty at University of Arkansas at Pine Bluff, North Carolina A&T State University (NC A&T), North Carolina Central University, and Winston-Salem State University, NC.

North Carolina A&T State University is the largest HBCU in the nation, with over 11,000 students enrolled. NC A&T is also recognized as a leader in the production of science, and engineering bachelor's degrees conferred to minorities, proudly ranking seventh nationally as the baccalaureate origin of Black science and engineering doctoral recipients from 2002 to 2011. Approximately 59% of all students receive Federal Pell Grants to support their educational expenses, and 78% receive Federal student loans.

North Carolina Central University (NCCU) located in Durham, NC near the world-renowned Research Triangle Park, is one of the 17 constituent institutions in the University of North Carolina (UNC) System and is one of the five public HBCUs in the state. Established in 1910, it has grown into a major regional university with doctoral and master's level programs and a diverse enrollment of almost 8,200 students majoring in 43 baccalaureate degree programs and 12 graduate programs. North Carolina Central University's mission is to provide its students with necessary skills and values needed to contribute to the changing economy of the twenty-first century. NCCU has biomanufacturing and biomedical research institutes and has established a research core in the study of health disparities through collaborations with biotechnology and pharmaceutical companies based in Research Triangle Park and across the state.

Winston-Salem State University is an HBCU, Institution of higher learning, located in Winston-Salem, North Carolina, with nearly 6,000 student enrollment. The primary mission is to prepare diverse students for success in the twenty-first century. A large number of WSSU students are first generation.

WSSU students are particularly engaged and challenged through active and experiential learning to become independent thinkers, problem-solvers, and scholars of high reputation and integrity. They are also dedicated to the development of students through excellence in teaching, scholarship, and service. The vision of WSSU is "to develop graduates of distinction known for leadership and service in their professions and communities."

Faculty from UAFS and HBCUs have also submitted joint institutional grant proposals on areas of mutual need involving student training and faculty development programs. It must be highlighted that the leading author of this chapter has over 10 years of teaching and research experience at WSSU and NCA&T. He has also trained and mentored over 85 URM students. Collaboration between faculty members from UAFS and minority institutions will facilitate networking of HBCUs students and faculty with their peers at UAFS. This collaboration is valuable and is envisioned to strengthen joint inter-institutional student summer research training and scholar exchange program that will broaden participation in STEM education across cultures and diversification of the next generation of scientists and engineers.

Development of Associate Degrees and Certificate Programs

As a former community college that has successfully transitioned into a regional university, UAFS recognizes the importance of the development of educational programs that meet the needs of the local community and regional workforce. As a result, the College of Applied Science and Technology (CAST) at UAFS offers career-based degree and certificate programs that combine academic coursework and hands-on learning. In addition to BS degrees, the CAST also offers associate degrees and certificate programs targeted to the local and regional workforce. Specifically, CAST offers BS degrees in Animation Technology, Applied Science, and Organization leadership. CAST also offers associate degrees and certificates in Automotive Technology, Computer Graphic Technology, Electronics Technology, General Technology, Office Management Technology, Robotics Technology, Welding, and Workforce Leadership. Students gain valuable preparation for the technical and occupational workforce while connecting with employers through internships, cooperative education, and other opportunities along the way.

These associate degree programs and curriculum are designed to promote, where possible, easy transfer of students from AS degrees into BS degree in CAST and STEM majors. For instance, our associate programs, curriculum course content, prerequisites, and schedules of courses at associate degrees are designed and aligned with the corresponding courses and curriculum at BS degrees program to promote easy course credit and student transfer from associate degrees to BS degree in STEM. The offering of associate degrees and certificate programs has further allowed us to meet the educational need of students of low-income families and minority students in STEM. In addition, there are several ongoing initiatives including joint proposal writing, joint student-mentoring, and training in research activity between STEM faculty and faculty from CAST to further promote the inclusive and easy transfer of students from associate degrees into STEM majors at UAFS.

The Department of Computer and Information Sciences and Engineering in the College of STEM also has a [2 + 2] arrangement with the University of Arkansas, Fayetteville College of Engineering. In this arrangement, engineering students spend the first two years of their engineering program at UAFS. The students are awarded an associate degree in engineering after two years. Students are subsequently taught by engineering faculty from the University of Arkansas, Fayetteville to complete their Bachelor of Science in while still in residence Engineering on UAFS campus in last two years. This arrangement promotes interaction between engineering faculty from UAFS, a predominantly undergraduate institution, and faculty from University of Arkansas, Fayetteville Engineering, a research-intensive institution.

ADAPTATION, SUSTAINABILITY, AND RECOMMENDATIONS

Several initiatives and efforts have been made to promote diversity and inclusiveness and to broaden participation in higher education, particularly in STEM, in the past decades. However, we still have a long way to go, and these initiatives must be intensified and sustained to achieve the desired outcomes. Research-intensive and primarily White institution should do more to hire more qualified minority and female faculty to promote faculty diversity and racial and gender equality. Efforts should be made to post faculty job positions in sites frequented by underrepresented minority professionals to increase the pool of minority applicants. Effective mentoring plans must also be implemented to mentor, nurture, and retain minority and female faculty in STEM. More minority and female faculty in STEM will motivate and inspire more minority students to attend and be successful in STEM disciplines at White and research-intensive colleges and universities.

Continued partnerships between predominantly White institutions, minority-serving institutions, and community colleges are required and are recommended to promote inclusivity so as to diversify the next generation of STEM professionals. These collaborations should focus more on joint curriculum development, joint institutional grant proposal writing, and submission in areas of common need. Provision of scholarships to qualified minority and disadvantaged students is critical. Development of faculty and scholar exchange programs between research-intensive institutions and minority-serving institutions should be encouraged and supported. This will facilitate cross-pollenation of pedagogy and research ideas that will promote networking between the faculty scholars from these institutions.

More minority students should be recruited into Summer Undergraduate Research Experiences (SURE), Research Undergraduate Experiences (RUE) and equivalent program to further provide hands-on research experience for underrepresented minority students. Development of an enduring bridge-program between research institutions and minority institutions is valuable and is recommended. For instance, the College of STEM at UAFS collaborates with Arkansas State University's Biomedical Institute on summer residential programs for biology and chemistry majors. Each year underrepresented minority students in STEM participate in a six-week residential research program at ASU in Jonesboro. Such bridge-programs facilitate undergraduate students interest in advancing to graduate program in STEM at research-intensive institution, promoting inclusive and broad participation in STEM across cultures, races, and genders. Concerted efforts should also be devoted to targeted recruitment, retention, and graduation of underrepresented minorities in STEM. Predominantly White institutions should develop strong collaboration with minority associations such Annual of Biomedical Research Conferences for Minority Students (ABRCMS), The National Organization for the Professional Advancement of Black Chemists and Chemical Engineers (NOBCCHE), Society for Advancement of Chicanos/Hispanics and Native Americans in Science (SACNAS). The National Congress of American Indians (NCAI), Black Indians and Intertribal Native American Associations, and others. There are excellent resources that can be used to recruit minority students into STEM departments.

Offering of Cultural Awareness Competency Challenges

Transforming STEM departments for inclusion cannot be achieved without addressing cultural awareness incompetency issues. Cultural insensitivity often leads to low productivity, hostility, intolerance, anxiety, fear, underdeveloped self-confidence, and unfriendliness across cultures and races. On the contrary, cultural awareness competency will promote awareness of one's own culture, understanding the dynamics of difference, awareness, and acceptance of differences, development, and application of cultural knowledge, and celebration of diversity. It is recommended that a cultural awareness competency workshop and/or symposium be conducted for all the educational stakeholders, alliances, and partners and participants. At the symposium, strategies to promote understanding of cultural differences among educational stakeholders should be discussed. The benefit of demonstrating culturally competent attitudes and behaviors among alliances and educational stakeholders should also be discussed at the cultural competency awareness workshops. Finally, strategies for enhancing cultural competence to improve relationships and better performance among alliances and educational stakeholders and the use of cultural competence continuum as a personal and organizational evaluation tool among all educational stakeholders should be discussed. The outcome of such cultural awareness competency training and discussion is envisioned to (1) increase respect and mutual understanding among all educational stakeholders that will promote STEM education across cultures, (2) increase creativity in problem-solving through new perspectives, ideas, and strategies in the formal and informal teaching instructions, (3) increase participation and involvement of different cultural groups on campuses, alliances, and community, (4) increase trust and cooperation between and among all educational stakeholders and community that will promote inclusive STEM education across cultures, races, and genders, (5) help overcome fear of mistakes, competition, or conflict between and among all educational stakeholders, and (6) promote inclusion, diversity, equality, and broaden participation in STEM.

Creating Community Engagement and Educational Outreach Through Partnerships with Non-profit Organizations, Use of Town Hall Meetings, and Social Media

Millions of persons are patrons of local Goodwill, YMCA, Salvation Army, Habitat for Humanity, Boys and Girls Clubs, Girl Scouts, and Boys Scouts. For instance, YMCA provides programs and services that primarily focus on youth development, healthy living, and social responsibility according to the unique needs of the community it engages. The YMCA has over 600,000 volunteers and over 20,000 national and local partnerships, serving persons of all ages and backgrounds to bridge the gaps in community needs. In addition, the YMCA has formed alliances with several schools, government agencies, churches, hospitals, youth agencies, parks and recreation organizations, and other stakeholders to address health issues, including adult and childhood obesity, across the country to improve the lives of children, families, and persons in need. The Salvation Army also operates hundreds of after-school programs and offers sports, arts, music programs, and summer camp opportunity for thousands of children in low- and moderate-income neighborhoods across the country. The Girl Scouts is pushing a science, technology, engineering, arts, and mathematics (STEAM) agenda that integrates arts and opens enormous doors of opportunities for students to explore and is ripe with prospects for STEM in higher education. Likewise, the Salvation Army has formed an excellent partnership with the LGBT community. Together, these organizations already have excellent facilities, infrastructure, and education outreach programs such as Achievement Gap programs platforms, including Early Learning Readiness, Summer Learning Loss Prevention, Afterschool, and Power Scholars Academy in place.

These programs are excellent and often underutilized resources that can be coordinated and used to promote inclusive and broad participation in STEM across all ages and cultures. Formation of alliances with these organizations, local school districts and higher education, and policymakers is recommended to promote community outreach, inclusive, and broad participation in STEM. For instance, increasing awareness through booklets, pamphlets, and flyers containing educational resources, college application procedures, available financial aid and scholarship opportunities for students can be distributed at the Goodwill, YMCA, and Salvation Army, Habitats for Humanity and Boys and Girls Club members, and at similar service providers. To further create awareness among various youth entities, it is recommended that STEM faculty from colleges and universities organize more seminar presentations and joint field trips and summer research opportunities on their campuses for local Goodwill,

YMCA, Salvation Army, and Habitats for Humanity and Boys and Girls Club members.

The educational needs and challenges of each community, region, and state differ. It is therefore imperative for the appropriate entities within each community work together in unity to resolve those challenges. The use of grassroots town-hall meetings and dialogue to stimulate discussion and to address the social, economic, and cultural issues that may promote inclusive and broader participation in STEM education in their communities is highly recommended. Educators from 2- and 4-year colleges and universities, local stakeholders, K–12 personnel (superintendents, principals, counselors, teachers), representatives from local Parent Teachers Associations (PTA), parents, Department of Human Services, local police departments, Department of Corrections, Mayors, Chamber of Commerce members, local elected officials, and members of district councils should be invited to participate in the town-hall meetings and discussions. Officials from non-governmental alliances (Goodwill, YMCA, Salvation Army, Habitat for Humanity, Boys and Girls Club, etc.,) and local businesses should also be invited to the town-hall discussions. Challenges, social issues, hindrances, and obstacles contributing to low participation and achievement gaps in STEM should be discussed in an interactive and constructive format. Strategies to resolve those challenges in each community should be discussed. Research-based best practices should be shared most importantly, the town-hall discussions can be used as an avenue to educate the attendants about educational resources, including regional college application procedures, available financial aid opportunities, and scholarship opportunities for students. Each stakeholder should be encouraged to discuss the available educational resources and community outreach programs and opportunities that students can take advantage of at these town-hall meetings.

The use of social media, including Facebook, Twitter, Instagram, and others, has become crucial for social networking and rapid dissemination of information, globally. For instance, Facebook has over 1.79 billion monthly active users (and 229 million monthly active users in the North American in 2016), and Twitter and Instagram also have 317 million and 400 million monthly active users, respectively, in the third quarter of 2016 (The Statistics Portal, 2016). These are valuable resources that can be used for community engagement and outreach to create educational awareness and to provide information about available opportunities, resources, scholarship, and internship in STEM departments for all students.

Finally, the policymakers and educational stakeholders in the community should recognize the importance of inclusive and broad participation in STEM. Sustained efforts and inclusive recruitment and retention of minority students and personnel in STEM disciplines are imperative and must be a top priority for colleges and universities. A significant amount from college foundation and endowment should be dedicated and used where possible to promote sustained inclusive and broad participation in STEM across the nation. Donations from local industries, philanthropists, and non-governmental agencies must be utilized to promote and sustain inclusive and broad participation in STEM across the nation.

REFERENCES

AARP. (1996). *A profile of older Americans: 1995* (Vol. PF3049). Washington, DC. Accessed on May 21, 2018.

Adams, G. M., & Lisy, J. M. (2007). The chemistry merit program: Reaching, teaching, and retaining students in the chemical sciences. *Journal of Chemical Education*, *84*(4), 721.

Amaral, K. E., Shank, J. D., Shibley Jr, I. A., & Shibley, L. R. (2013). Web-enhanced general chemistry increases student completion rates, success, and satisfaction. *Journal of Chemical Education*, *90*(3), 296–302.

Amendt, J., Campobasso, C. P., Gaudry, E., Reiter, C., LeBlanc, H. N., & Hall, M. J. R. (2007). Best practice in forensic entomology – Standards and guidelines. *International Journal of Legal Medicine*, *121*(2), 90–104.

American Civil Liberties Union. (2012). *Written statement of the American Civil Liberties Union for a hearing on "Ending the School-to-Prison Pipeline."* Submitted to the Senate Judiciary Subcommittee on the Constitution, Civil Rights and Human Rights. Retrieved from https://www.aclu.org/files/assets/aclu_statement_for_sjc_subcomm_hearing_on_the_school_to_prison_pipeline_12_2012.pdf

Arcia, E. (2006). Achievement and enrollment status of suspended students' outcomes in a large, multicultural school district. *Education and Urban Society*, *38*(3), 359–369.

Balfanz, R., Byrnes, V., & Fox, J. (2014). Sent home and put off-track: The antecedents, disproportionalities, and consequences of being suspended in the ninth grade. *Journal of Applied Research on Children: Informing Policy for Children at Risk*, *5*(2), 13.

Balfanz, R., Herzog, L., & Mac Iver, D. J. (2007). Preventing student disengagement and keeping students on the graduation path in urban middle-grades schools: Early identification and effective interventions. *Educational Psychologist*, *42*(4), 223–235.

Bitterman, A., Goldring, R., & Gray, L. (2013). *Characteristics of public and private elementary and secondary school teachers in the United States: Results from the 2011–12 schools and staffing survey (NCES 2013–314)*. Washington, DC: National Center for Education Statistics, U.S. Department of Education.

Blasie, C., & Palladino, G. (2005). Implementing the professional development standards: A research department's innovative masters degree program for high school chemistry teachers. *Journal of Chemical Education*, *82*, 567–570.

Bundschuh, J. E., El-Awady, A. A., Hardin, R. L., Hughes, B. G., & Shelton, R. W. (1982). Continuing education program for high school chemistry teachers. *Journal of Chemical Education*, *59*, 407–408.

Bureau, U. C. (2014). The United States Census Bureau. From http://www.census.gov/

Campbell, C. (2015). The socioeconomic consequences of dropping out of high school: Evidence from an analysis of siblings. *Social Science Research*, *51*, 108–118.

Carbonaro, W., & Workman, J. (2013). Dropping out of high school: Effects of close and distant friendships. *Social Science Research*, *42*, 1254–1268.

Chen, X. (2009). *Students who study science, technology, engineering, and mathematics (STEM) in postsecondary education. Stats in Brief. NCES 2009-161*. Washington, DC: National Center for Education Statistics.

Claudio, L. (1997). Making more minority scientists. *Environmental Health Perspectives*, *105*(2), 174.

Costa, G., & Di Milia, L. (2008). Aging and shift work: A complex problem to face. *Chronobiology International*, *25*(2–3), 165–181.

Craig, N. (1999). The joys and trials of doing research with undergraduates. *Journal of Chemical Education*, *76*, 595–598.

Dahm, D. J., & Nelson, E. A. (2013). *Calculation in chemistry an introduction*. New York, NY: W. W. Norton & Company, Inc.

Dilworth, M. E., &Coleman, M. J. (2014). *National education association report, time for a change: Diversity in teaching revisited a background paper*.

English, D., Lambert, S. F., & Ialongo, N. S. (2016). Adding to the education debt: Depressive symptoms mediate the association between racial discrimination and academic performance in African Americans. *Journal of School Psychology*, *57*, 29–40.

Ewing, B. A., D'Amico, E. J., & Humor, A. N. (2014). Loneliness and acceptance: Predictors of university drop-out intentions. *Procedia – Social and Behavioral Sciences, 152*, 1079–1086.

Fakayode, S. O. (2014). Guided inquiry laboratory experiments in analytical chemistry laboratory curriculum. *Analytical and Bioanalytical Chemistry, 406*, 1267–1271.

Fakayode, S. O. (2015). Purity analysis of the pharmaceuticals naproxen and propranolol: A guided-inquiry laboratory experiment in the analytical chemistry laboratory. *Journal of Chemical Education, 92*(1), 157–162.

Fakayode, S. O., Abel, C., & Pollard, D. A. (2014). Determination of moisture content of over the counter pharmaceuticals and dried powdered food products using coulometric Karl Fischer titration: A summer undergraduate research experience at Historically Black College and University. *The Chemical Educator, 19*, 264–268.

Fakayode, S. O., Mayes, J. O., Kanipes, M. I., Johnson, D., & Cuthbertson, E. L. (2017). Promoting student learning in STEM and criminal justice: Aggie Sleuth Initiative (AggieSI) guided inquiry learning experience. *Journal of Criminal Justice Education, 28*(2), 192–206.

Fakayode, S. O., Pollard, D. A., Mohammed, A. K., Adeyeye, O. M., & Yakubu, M. (2014). Promoting undergraduate STEM education at a Historically Black College and University through research experience. *Journal of Chemical Education, 91*, 662–665.

Fakayode, S. O., Pollard, D. A., & Snipes, V. T. (2014). Offering a geoscience professional development program to promote science education and provide hands-on experiences for K-12 science educators. *Journal of Chemical Education, 91*(11), 1882–1886.

Fakayode, S. O., Snipes, V. T., & Kanipes, M. (2015). Collaboration between Historically Black Colleges and Universities and local school districts to promote K–12 science education. *North Carolina, USA. The Chemical Educator, 20*, 86–94.

Fakayode, S. O., Snipes, V., & Kanipes, M. (2016). Got Caffeine in your beverage and Fe in your diet? Hands-on professional development workshop for K-12 science educators. *The Chemical Educator, 21*, 15–22.

Fakayode, S. O., Snipes, V., Kanipes, M., Mohammed, A. K., & Wilson, Z. S. (2016). Use of innovative pedagogies and creative partnership strategies to promote undergraduate STEM education at an HBCU in setting a new agenda for student engagement and retention at HBCUs Editor Charles Prince (pp. 119–151). Hershey, PA: IGI Global Inc..

Fortin, L., Lessard, A., & Marcotte, D. (2010). Comparison by gender of students with behavior problems who dropped out of school. *Procedia Social and Behavioral Sciences, 2*, 5530–5538.

Foundation, N. S. (2004). *Women, minorities, and persons with disabilities in science and engineering*. Washington, DC: NSF.

Fuller, E. C., Lefler, R. W., Panush, L., & Will, E. L. (1971). Communication between high school and college teachers. *Journal of Chemical Education, 48*, 333–334.

Galama, T., & Hosek, J. R. (2007). *Perspectives on US competitiveness in science and technology* (Vol. 235). Santa Monica, CA: Rand Corporation.

Gladieux, L. E., & Kahlenberg R. D. (2004). *America's untapped resource: Low income students in higher education* New York, NY: Century Foundation Press.

Hemphill, F. C., & Vanneman, A. (2011). *Achievement gaps: How Hispanic and White students in public schools perform in mathematics and reading on the national assessment of educational progress (NCES 2011–459)*. Washington, DC: National Center for Education Statistics, Institute of Education Sciences, U.S. Department of Education.

Herrington, D. G., Luxford, K., & Yezierski, E. J. (2012). Target inquiry: Helping teachers use a research experience to transform their teaching practices. *Journal of Chemical Education, 89*, 442–448.

Hopkins, K. R., Nestleroth, S. L., & Bolick, C. (1991). *Help wanted: How companies can survive and thrive in the coming worker shortage*. New York, NY: McGraw-Hill.

Howell, J. E. (2001). Especially for high school teachers. *Journal of Chemical Education, 78*, 705.

Hrabowski III, F. (2003). Beating the odds: Preparing minorities for research careers in the chemical sciences. *Minorities in the Chemical Workforce: Diversity Models That Work* (pp. 84–89). Washington, DC, USA: THE NATIONAL ACADEMIES PRESS.

Huang, G., Taddese, N., & Walter, E. (2000). Entry and Persistence of women and minorities in college science and engineering education 2000. *National Center for Education Statistics*

(NCES2000-601). Washington, D.C.: U.S. Government Printing Office. https://nces.ed.gov/pubs2000/2000601.pdf Accessed January 2, 2018.

Hutchison, A. R., & Atwood, D. A. (2002). Research with first- and second-year undergraduates: A new model for undergraduate inquiry at research universities. *Journal of Chemical Education*, *79*, 125–126.

Kemmlert, K., & Lundholm, L. (2001). Slips, trips and falls in different work groups – With reference to age and from a preventive perspective. *Applied Ergonomics*, *32*(2), 149–153.

Klug, S. H. (2009). Recruit, respect, and retain: The impact of Baby Boomer nurses on hospital workforce strategy – A Case Study. *Creative Nursing*, *15*(2), 70–74.

Kneeland, P. P., Kneeland, C., & Wachter, R. M. (2010). Bleeding talent: A lesson from industry on embracing physician workforce challenges. *Journal of Hospital Medicine*, *5*(5), 306–310.

Krezmien, M. P., Leone, P. E., & Achilles, G. M. (2006). Suspension, race, and disability: Analysis of statewide practices and reporting. *Journal of Emotional and Behavioral Disorders*, *14*(4), 217–226.

Leventhal-Weiner, R., & Wallace, M. (2011). Racial differences in high school dropout rates: An analysis of U.S. Metropolitan areas. *Research in Social Stratification and Mobility*, *29*, 393–413.

Lindsay, H. A., & Mclntosh, M. C. (2000). Early exposure of undergraduates to the chemistry research environment: A new model for research universities. *Journal of Chemical Education*, *77*, 1174–1175.

Losen, D. J., & Martinez, T. E. (2013). *Out of school & off track: The overuse of suspensions in American middle and high schools*. Los Angeles, CA: The UCLA Centerfor Civil Rights Remedies at The Civil Rights Project. Retrieved from http://civilrightsproject.ucla.edu/resources/projects/center-for-civil-rights-remedies/school-to-prison-folder/federal-reports/out-of-school-and-off-track-the-overuse-of-suspensions-inAmerican-middle-and-high-schools. Accessed on January 2, 2018.

Mabrouk, P. A. (2009). Survey study investigating the significance of conference participation to undergraduate research students. *Journal of Chemical Education*, *86*, 1335–1340.

Mizel, M. L., Miles, J. N. V., Pedersen, E. R., & Tucker, J. S. (2016). To educate or to incarcerate: Factors in disproportionality in school discipline. *Children and Youth Services Review*, *70*, 102–111.

Myers, B. (2016). Where are the minority professors? The Chronicle of higher education February14, 2016. Retrieved from https://www.chronicle.com/interactives/where-are-the-minority-professors. Accessed on January 2, 2018.

National Education Association. (2011). *Transforming teaching: Connecting professional responsibility with student learning*. Washington, DC. Commission on Effective Teachers and Teaching (Cett). http://www.nea.org/assets/docs/Transformingteaching2012.pdf. Accessed December 12, 2017

National Institute of Justice. (2000). *The future of forensic DNA testing: Predictions of the research and development working group*. Washington, DC: National Institute of Justice.

National Science Foundation, N. C. f. S. a. E. S. (2013). *National Science Foundation, National Center for Science and Engineering Statistics. 2013. Science and Engineering Degrees: 1966–2010. Detailed Statistical Tables NSF 13–327*. Retrieved From National Science Foundation, National Center for Science and Engineering Statistics http://www.nsf.gov/statistics/nsf13327/. Accessed on December 12, 2017.

Oyserman, D. (2013). Not just any path: Implications of identity-based motivation for disparities in school outcomes. *Economics of Education Review*, *33*, 179–190.

Poodry, C. A. (2003). Diversity: Why is it important and how can it be achieved. *Minorities in the Chemical Workforce: Diversity Models That Work* (pp. 6–15). Washington, DC, USA: THE NATIONAL ACADEMIES PRESS.

Rausch, C. C. (2015). Interdisciplinary practices in forensics within American law enforcement: The international context. *GSTF Journal of Law and Social Sciences*, *4*(2), 1–11.

Ricard, N. C., & Pelletier, L. G. (2016). Dropping out of high school: The role of parent and teacher self-determination support, reciprocal friendships and academic motivation. *Contemporary Educational Psychology*, *44–45*, 32–40.

Robison, S., Jaggers, J., Rhodes, J., Blackmon, B. J., & Church, W. (2017). Correlates of educational success: Predictors of school dropout and graduation for urban students in the Deep South. *Children and Youth Services Review*, *73*, 37–46.

Rohrbaugh, M. C., & Corces, V. G. (2011). Opening pathways for underrepresented high school students to biomedical research careers: The Emory University RISE program. *Genetics*, *189*(4), 1135–1143.

Sarquis, A. M. (2001). Recommendations for offering successful professional development programs for teachers. *Journal of Chemical Education*, *78*, 820–823.

Science and Engineering Indicators: (2010) NSB 10-01. Retrieved from https://www.nsf.gov/statistics/seind/

Slezak, J. (1999). Student-inspired undergraduate research. *Journal of Chemical Education*, *76*, 1054–1055.

Snow, C. (1982). Forensic anthropology. *Annual Review of Anthropology*, *11*, 97–131.

Sweeney, A. E. (2003). Addressing the professional preparation of future science teachers to teach hands-on science: A pilot study of a laboratory model. *Journal of Chemical Education*, *80*, 171–173.

Taylor, O., Apprey, C. B., Hill, G., McGrann, L., & Wang, J. (2010). *Diversifying the faculty*. Association of American Colleges and University. Retrieved from https://www.aacu.org/publications-research/periodicals/diversifying-faculty. Accessed on December 15, 2017.

The Statistics Portal. (2016). Retrieved from https://www.statista.com/statistics/264810/number-of-monthly-active-facebook-users-worldwide/. Accessed on November 22, 2017.

Widener, A. (2017). Few gains for minority chemistry professors. Number of U.S. chemistry faculty from underrepresented groups remains stagnant, OXIDE survey shows. *Chemical and Engineering New*, *95*(44), 18–20. Retrieved from https://cen.acs.org/content/cen/articles/95/i44/Few-gains-minority-chemistry-professors.html Accessed on January 15, 2017.

Wilson, Z. S., Holmes, L., Sylvain, M. R., Batiste, L., Johnson, M., McGuire, S. Y., & Warner, I. M. (2012). Hierarchical mentoring: A transformative strategy for improving diversity and retention in undergraduate STEM disciplines. *Journal of Science Education and Technology*, *21*(1), 148–156.

Wilson, Z. S., Iyengar, S. S., Pang, S.-S., Warner, I. M., & Luces, C. A. (2012). Increasing access for economically disadvantaged students: The NSF/CSEM & S-STEM programs at Louisiana State University. *Journal of Science Education and Technology*, *21*(5), 581–587.

NANOHU: A SUCCESSFUL COLLABORATIVE STEM MODEL PREPARING AFRICAN AMERICANS FOR ENGAGEMENT IN NANOSCIENCE, LAYING THE FOUNDATION FOR TRANSFORMATIVE, INSTITUTIONAL STEAM ENGAGEMENT

Michelle O. Fletcher Claville, Sainath Babu, Brandon C. Parker, Emorcia V. Hill, Eric W. Claville and Michelle Penn-Marshall

ABSTRACT

The Nanoscience Project at Hampton University (NanoHU) responds to the international call for more workers in the field of science, technology, engineering, and mathematics (STEM) who are nano-savvy and prepared for engagement in the fourth industrial revolution. The project's initial intent to answer statewide and national initiatives was congruent with Hampton University's (HU) desire for increased diversification of research interests across HU and enhanced the preparation of its students for doctoral degrees.

Broadening Participation In STEM: Effective Methods, Practices, And Programs
Diversity in Higher Education, Volume 22, 107–128

ISSN: 1479-3644/doi:10.1108/S1479-364420190000022005

Funded by the National Science Foundation, the five-year project (2012–2017) purposed to develop and systematically implement an integrated, multidisciplinary STEM research and education program in nanoscience at HU. Evidence of NanoHU's success is demonstrated in the following accomplishments at the University: (1) a new Nanoscience Minor, (2) a new "Introduction to Nanoscience" course that has had a total enrollment of 82 students from STEM and non-STEM fields, (3) the NanoHU Scholars Program that has prepared 23 Scholars for entry into graduate programs and 12 NanoHU Fellows for similar pursuits, (4) a Faculty Development Program that has supported a total of 20 STEM and non-STEM faculty members, (5) a NanoHU Seminar Series that has informed the HU community about the science, business, legal, and ethical topics pertaining to nanoscience and nanotechnology, and (6) a viable outreach program that has prepared high school students (NanoHU Pioneers) for successful matriculation as STEM majors at the college level and stimulated STEM interest in the surrounding community. It is worth emphasizing that execution of the project also resulted in engagement between STEM and non-STEM constituents of the University, establishing a platform for a formal science, technology, engineering, arts, and mathematics (STEAM) institutional initiative. Efforts to communicate the importance of nanoscience to the HU community through seminars resulted in an infusion of nanoscience modules in STEM and non-STEM courses including courses in English, Journalism, Ethics, and other pre-law courses. Although NanoHU is specific to the needs of HU, its collaborative construct promises to be an innovative model for STEM and STEAM programs at other institutions with a similar construct.

Keywords: Undergraduate education; HBCU; retention rate; graduation rate; underrepresented groups; collaboration

INTRODUCTION

Background

The culmination of the twentieth century saw the growth of inter- and multidisciplinary approaches to science and technology as researchers realized that accurate responses to scientific questions needed collaboration across the fields in science, technology, engineering, and mathematics (STEM). Accordingly, universities and colleges in the United States (US) revolutionized their existing approach to STEM by growing novel interdisciplinary programs in fields such as nanoscience and sustainability. In fact, institutions such as Arizona State University established a viable sustainability program and now boasts a "School for the Future of Innovation in Society" (The future is for everyone: School for the Future of Innovation in Society at Arizona State University, n.d.). By the beginning of the twenty-first century, advances in fields such as nanotechnology, biotechnology, and materials science are now integral to the fourth industrial revolution.

Nanotechnology in the US

As early as 1992, the nation of Japan began investing significant effort into nanoscience and nanotechnology, promoting global investment in the nano-industry (Schulte, 2005). By the year 2000, the US marked its investment in nanotechnology by establishing the National Nanotechnology Initiative (NNI) (United States Congress, 2009). Since that time, the 20 different federal agencies that collaborated within the NNI have continued to promote the advancement of the nanoscience/nanotechnology research and education. In accordance with the nation's thrust to obtain and maintain a prominent global presence in nanotechnology, many states began taking supportive measures. For instance, the Commonwealth of Virginia became so vested in the enterprise that it was recognized by the 2006 Joint Commission on Technology and Science (JCOTS) Nanotechnology Research & Development and Manufacturing Advisory Committees for possessing notable research expertise in nanotechnology (e.g., self-assembly and nanocharacterization). The JCOTS also reported that Virginia's nanotechnology industry amounted to US$1 billion (JCOTS Nanotechnology Research & Development and Manufacturing Advisory Committees, 2006) and concluded in its report that education must be addressed at all levels in order to prepare the necessary workforce in Virginia.

Efforts to develop nanoscience and nanotechnology curricula became increasingly popular. For instance, Laherto reported on the "theoretical part of a study that employed *the Model of Educational Reconstruction* in order to respond to the perceived need for education in nanoscience and nanotechnology" (Laherto, 2010). The stated model provides a theoretical framework to determine how worthwhile and possible it may be to teach particular areas of science. Using the analytical component of the model, Laherto determined that nanoscience and nanotechnology have certain epistemological and societal features that make them relevant to scientific and technological literacy.

Fazarro, Newberry, Trybula, and Hyder (2012) developed a conceptual framework for a four-year nano-curriculum. The authors proposed a nanotechnology curriculum based on the needs and expectations of the industry. Educators created and acknowledged that the processes should be iterative, with feedback requested from industry upon the review of the educational content, modification, and improvement. Therefore, the conceptual framework was designed with four components (theoretical, applied, societal, and ethical) that address the industry requirements. To implement the nanotechnology curriculum, the authors identify a required multidisciplinary approach that engages faculty with relevant expertise.

The Disparity of African Americans in STEM

Traditionally, the numbers of students from African Americans (Blacks) who pursue degrees in STEM fields have been consistently low. Data from the

National Science Foundation (NSF) (NSF, n.d.) showed that in the year 2015, African Americans who earned doctorates in STEM fields (i.e., Life Sciences, Physical Sciences, Mathematics and Computer Science, Psychology and Social Sciences, and Engineering) represented a mere 0.00035% of the US population (1,360 out of 390.9 million). Moreover, in that year, African Americans represented 5.5% (1,360 out of 24,547) of those who earned doctorate degrees in STEM fields in stark contrast to Whites who represented 72.1% (17,702 out of 24,547). Similarly, in 2016, African Americans represented 5.4% (1,353 out of 25,104) of those who earned doctorate STEM degrees, while Whites earned 71.1% (17,856 out of 25,104). According to the most recent US 2017 census estimates, 13.3% of 325.7 million people are identified as African Americans while 76.9% identified as White (United States Census Bureau, 2017). Since the numbers of Blacks who earn doctorates in STEM fields are disparate when compared to White counterparts and the nation's population, there is an obvious need to implement strategies to improve the numbers.

Historically Black Colleges and Universities (HBCUs) have traditionally played a major role in graduating a significant number of African American students when compared to predominantly White institutions (PWIs). The 101 institutions, formally designated as HBCUs, constitute only 3% of all institutions of higher learning in the nation. However, they graduate 27% of Blacks with bachelor's degrees in STEM fields (Obama, 2016). These historical institutions are also recognized as leaders in educating Blacks who persist to earn doctorate degrees in STEM fields (Gasman, 2013; Perma 2009; The Journal of Black in Higher Education, 2012; Palmer, Davis, & Thompson, 2010; Strayhorn, 2008; Strayhorn, Williams, Tillman-Kelly, & Suddeth, 2012). In fact, many HBCUs are known for their successes in graduating students who are first-generation college students and come from low-income families (Gasman, 2013). Even more notable is the ability of such institutions to accomplish these feats despite dealing with notable challenges that are not as threatening at non-HBCUs. For these reasons, federal agencies like the NSF have elected to invest funding (mostly by way of competition) to ensure the continued success and progression of HBCUs.

STEM at Hampton University

Hampton University (HU) is a private, non-sectarian HBCU that is dedicated to providing a complete education to its students (Hampton University Mission, n.d.). In its mission statement, HU states that its "[…] curriculum emphasis is scientific and professional with a strong liberal arts undergirding." Founded in 1868 after the American Civil War, HU now stands among the top five of all HBCUs (Compass & Historically Black Colleges and Universities Ranking Methodology, n.d.) and is one of the few institutions that graduates Blacks 10% higher than the national average (JBHE, 2012, 2014). According to the Carnegie Foundation for the Advancement of Teaching, HU is currently categorized as Master's Colleges and Universities: Medium Programs (The Carnegie Classification of Institutions of Higher Educations, 2017).

HU, a modest-sized institution, is committed to being a leader among HBCUs in preparing its students for STEM careers. Currently, the University consists of nine schools and colleges (Hampton University Mission, n.d.). While the School of Science and the School of Engineering and Technology are dedicated to preparing students for traditional careers in STEM fields, significant STEM-related activity exists in the School of Pharmacy (SOP) and the School of Liberal Arts and Education (SLAE). More specifically, SOP boasts masterful research in pharmaceutics and related fields, while SLAE produces scholarship from its departments of Sociology, Political Science, and Psychology. Moreover, the University is the first and only HBCU to have 100% control of a NASA mission and four satellites currently in orbit (The Center for Atmospheric Sciences, n.d.). In addition, HU is the only HBCU to have a proton therapy center established for treating cancer and conducting affiliated research. The center, known as the Hampton University Proton Therapy Institute (HUPTI), is the world's largest free-standing proton therapy cancer treatment facility (Simpson, 2010).

HU is purposed to becoming a more research-intensive institution. Accordingly, it has begun to strategically modify its existing construct to increase research activity that is in line with the current global interests. By virtue of the institution's historical undergirding, and its current progressive status, it seeks to establish formal programs that are novel and exemplary. The University seeks to be a number one producer of African Americans who earn diverse baccalaureate through doctorate degrees. The benchmarks above in STEM lay the foundation for novel initiatives in the institution's range of vision as it seeks to pioneer more novel initiatives that are in keeping with the current global trends.

STEAM Efforts across the Nation

When STEM initiatives (e.g., education, research) include aesthetic approaches (art) toward "real-world" problems, science, technology, engineering, arts, and mathematics (STEAM) results. With the advancement in technologies and competency for global education over the past decade, K-12 educators began advocating to integrate the Arts (in its purest sense, art, graphic design) with STEM education to enhance creativity within. In fact, both government and private sectors have been actively trying to incorporate the arts component to existing STEM education models (Bequette & Bequette, 2012; Clark & Burron, 2011; Madden et al., 2013) to form STEAM education. Moreover, in 2013, the US National Endowment for the Arts (NEA) initiated a bipartisan STEAM caucus to promote the addition of arts in STEM education.

A looser definition of the term "arts" – meaning anything not typically considered to be STEM – allows the recent STEAM movement to become more inclusive and permit broader approaches that address world issues. For instance, the World Economic Forum and its network have established a global transformation map (Fig. 1) that explores and provides insight into the interconnectivity and interdependence between industry, economy, and global issues.

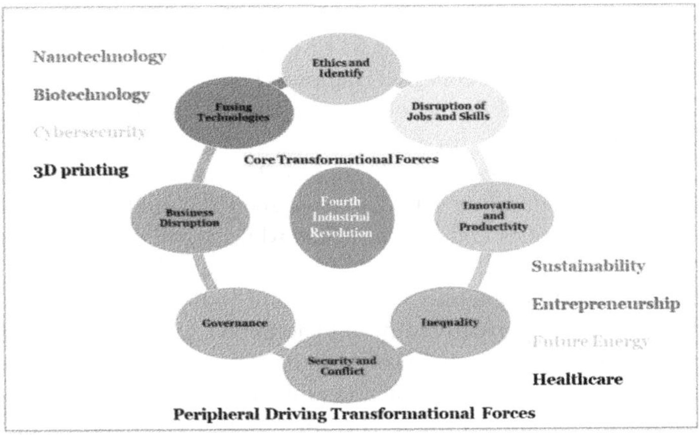

Fig. 1. Global Transformation Map, Identifying Core and Peripheral Transformational Forces for the Fourth Industrial Revolution. An Interactive Transformation Mapping Could Be Seen at the World Economic Forum, 2016. *Source*: https://toplink.weforum.org/page/aboutPublicTM.

The map (Fig. 1) features current research and analysis of respective fields that concern the Fourth Industrial Revolution. Interaction with the active map provides summaries of research findings that lead back to the original sources. As an example, economists (McAfee & Brynjolfsson, 2016) believe that the fourth industrial revolution could yield significant disruption to labor markets since the increased implementation of automation will correspondingly result in the increased displacement of the current workforce. Similarly, ethicists and policymakers query the semantics of creating and utilizing novel materials that defy conventional logic and have no real history about the known deleterious effects.

Efforts to integrate arts expertise within traditional STEM fields are also exemplified by the Society for the Studies of New and Emerging Technologies (S.NET; The Society for the Studies of New and Emerging Technologies, 2017). S.NET describes itself as:

> [...] an international association that promotes intellectual exchange and critical inquiry about the advancement of new and emerging technologies in society. S.NET represents diverse communities, disciplines, viewpoints, and methodologies in the social sciences and humanities. It also welcomes contributions from scientists and engineers that advance the critical reflection of nanoscience and other emerging technologies.

The Society's members discuss how their varied approaches to society's questions may be integrated into the areas of research dynamics and organization, governance and regulation, politics and ethics, roles of public and stakeholders, and so on. Furthermore, the members publish research on emerging technologies and their impact on the global society.

The precedent set by S.NET, and the clarion call by the World Economic Forum, clearly establish the need for more professionals who are skilled practitioners in innovative STEM fields and socially intuitive enough to inform and ethically engage the global public. While large numbers of colleges and universities have long since heeded the call to promote interdisciplinary STEM programs that are relevant to the current industrial revolution, fewer have established true collaborative between STEM and arts fields. Furthermore, fewer HBCUs have begun to embrace initiatives that promote pioneering STEM initiatives that are interdisciplinary, let alone discussions regarding interdisciplinary collaborations that include both STEM and arts.

Motivation for NanoHU

The Nanoscience Project at Hampton University (NanoHU) was birthed in response to a twofold call: (1) the call for more STEM workers across the globe who were skilled in nanotechnology and (2) the nation's need to improve the numbers of underrepresented groups who excel in STEM fields. Intrinsic to the call to prepare the traditional STEM constituents at HU in nanoscience was the need to engage non-STEM (arts) university constituents in the endeavor. Recognizing the need to prepare both the STEM and non-STEM constituents of HU to engage in enterprises about the novel technologies, NanoHU sought to engage all constituents of the institution. The Project was strategically designed to inform university constituents of certain fundamentals of the area so that both the STEM and arts could engage in meaningful conversation regarding advances and associated policies.

In theory, the establishment of NanoHU was supposed to be a logical undertaking. After all, the institution had already disclosed its strategic plan to broaden its research repertoire and become a more research-intensive institution. Inherent in that undertaking was the need to accomplish obvious and not-so-obvious transitions. These would include the establishment of innovative educational and research programs through formal inter- and intradepartmental collaborations. Moreover, such collaborations would be established on the platform of nanoscience/nanotechnology. The institution already boasted researchers across its STEM-related schools who expressed interest and/or conducted research in the field. It seemed inevitable that those individuals should join forces to the benefit of all concerned.

What about HU's non-STEM academic units? Faculty members within those units had already been proficient in status quo scholarship within traditional norms (i.e., within the non-STEM disciplines). Others, however, by way of anecdotal data, had demonstrated an interest in establishing non-traditional collaborations with "hard-STEM" (e.g., chemistry, physics, engineering) academicians. In theory, the novelty of nanoscience and its role in the global economy would be the common ground to establish the desired collaboration. The establishment of NanoHU would provide the community space for interaction between non-STEM and STEM enthusiasts to cooperatively participate in discussions about nanoscience/nanotechnology and its global impact. Such a space would inform the HU community of (1) the fundamentals of such a science, (2) the relevance of the field to society's current mode of operation, (3) the relationship of the current advances to that which is traditionally understood, and (4) the opportunities to explore the future implications of embracing this type of technology and other emerging technologies. Moreover, the interdisciplinary space would prepare undergraduate students who major in business, journalism, graphic arts, music, to engage more competitively in the enterprising global endeavors.

PROGRAM GOALS, STRUCTURE AND COMPONENTS, IMPLEMENTATION, AND ADAPTATION DUE TO INSTITUTIONAL CONTEXT

Program Goals

The NanoHU was established to implement an undergraduate nanoscience program at HU that would recruit, retain, and graduate competitive students for engagement in the STEM enterprise. Funded by the NSF's historically Black colleges and universities undergraduate program (HBCU-UP), NanoHU's focus was the preparation of underrepresented minorities to earn graduate degrees in STEM fields or enter STEM careers. The project's objectives included the following:

- the development of a nanoscience curriculum that would offer STEM undergraduate majors the baccalaureate degree with a nanoscience minor (note: a concentration per STEM subject area was previously proposed);
- the immersion of undergraduate students into nanoscience-related research throughout the academic year and summer session at national laboratories and/or international universities as a condition for earning the minor in nanoscience;
- the establishment of a seminar series that would address both technical and "non-technical" attributes needed for academic and professional success;
- the provision of "next steps" educational and career support that would result in at least 80% of program participants applying to graduate school in STEM and/or nanoscience-related fields;
- the provision of professional development support for faculty to increase their publications record with undergraduates as coauthors by 50%;
- the promotion of the nanoscience project to incoming and current HU STEM students to build their awareness of and interest in pursuing the nanoscience concentration; and
- outreach that would engage a minimum of eight area high school students per year in nanoscience research, and activities that would promote career opportunities in STEM.

The expected outcomes of NanoHU were to simultaneously extend institutional STEM capacity and strengthen the support for undergraduate research across STEM disciplines, in order to further institutionalize a culture of undergraduate research.

Program Construct and Components

The goals and objectives were specifically designed to empower (1) the undergraduate student community, (2) the faculty, (3) area high school students, and (4) the surrounding community. Furthermore, preparing the undergraduate student community for engagement in nanoscience would mandate the creation of a viable curriculum that would serve as a foundation for subsequent study. Similarly, a basic curriculum would be necessary to engage the general public in the Hampton Roads community.

It is worth noting that the design of the project was not based on any existing model or conceptual framework. The initial project design evolved from an intricate study of the needs of the state of Virginia, discussion with research collaborators at NASA Langley Research Centre (Hampton, VA), study of the University's strategic plan, and survey of certain university stakeholders and key players (i.e., students and research-active faculty). The design of the nanoscience curriculum resulted from reviewing the existing nanoscience curriculum of colleges and universities across the country.

Undergraduate Student Community

During the early stages of the project, it became obvious that the student focus would be more extensive than previously recognized. The project initially proposed to track the progression of a group of 20 undergraduate students (Scholars) over four to five years through graduation, to subsequent enrollment in graduate programs. Howbeit, execution of the project revealed the need to establish and designate three tiers of NanoHU students, that is, Scholars, Fellows, and "Affiliate" students. NanoHU Scholars were those students who had expressed intent to earn graduate degrees in STEM fields. They had successfully applied for scholarship funds guaranteed for four years under the conditions that they would (1) earn the newly created nanoscience minor (which required students to take the newly developed introductory course), (2) engage in nanoscience-related research for a minimum of 10 hours per week throughout the academic year and summers, (3) maintain a minimum grade point average (GPA) of 3.0 out of 4.1, and (4) give research presentations at local, regional, and national scientific meetings. NanoHU Fellows were students who were either uncertain about pursuing a graduate degree in a STEM field or/and expressed interest in engaging with NanoHU after the formal selection of the NanoHU Scholars. With the said designation, fellows would receive lower scholarship funds with the condition that they maintain a minimum GPA of 2.7 out of 4.1, take the "Introduction to Nanoscience" course developed by NanoHU, and conduct 10 hours of nano-related research per semester of funding. The third tier of students included all students directly impacted by NanoHU that did not fall in the categories of NanoHU Scholar or Fellow. The latter group primarily comprised students enrolled in the introductory nanoscience course and those who attended NanoHU-sponsored seminars. As seen from the cumulative GPAs (Fig. 2), students who have enrolled in the Scholars program have either maintained or improved their GPA by an average of 0.1 points, since engaging in the program.

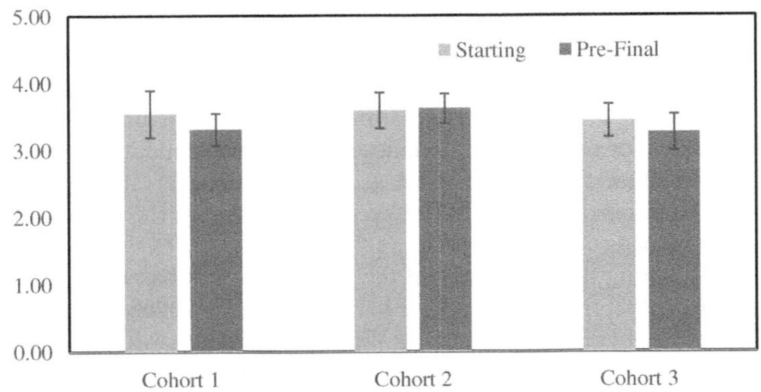

Fig. 2. Comparison of Scholars Cumulative GPA ($n = 5$) at the Start of NanoHU Fellowship to the Last Academic Year (Fall 2017, Except Last Semester) (GPA: grade point average).

To facilitate the desired impact on the undergraduate student community, NanoHU developed a nanoscience minor and an introductory nanoscience course. The nanoscience minor was developed by utilizing existing courses that already contained significant nano-related content and or would be infused with the nanoscience modules. In 2012, NanoHU proposed the nanoscience minor to the University community as the first of its kind. It would be the first HU minor that was NOT directly associated with a particular major. To populate the minor, representatives from certain STEM departments (i.e., Engineering, Physics, Chemistry, Biology, Mathematics, and Computer Science) identified courses and designated sections of certain courses which contained nano-related content. After successfully going through the University's curriculum revision process, the minor was approved at the end of the spring semester of 2013 and offered for the first time in the spring semester of 2014.

The project also developed the introduction above to Nanoscience course that serves as an approved physical science elective course that satisfies the University's general education requirements. The course is required for all Scholars and Fellows but is available to any HU student as long as they fulfill the prerequisite of completing high school mathematics. In the course, students learn the fundamentals of nanophysics, nanochemistry, nanobiology, technology, business enterprise, and federal regulations that pertain to nanoscience and nanotechnology. The course is team-taught by three faculty members representing three different disciplines each semester.

The Faculty

NanoHU invested in faculty development in order to support the University's mission to become more research-intensive and to facilitate the learning of the student community. At first, NanoHU invited junior faculty from STEM fields to apply for mini-grants to conduct nanoscience research over the summer sessions, with the anticipation of funding eight faculty members per year. Grants paid for an average of six weeks of summer salary, research supplies, and research-related travel. Each faculty awardee was required to serve as a research mentor to two high school students during the summer and an average of two undergraduate NanoHU students per fiscal year. Faculty awardees were invited to give technical seminars in the NanoHU seminar series and present their findings at the School of Science's annual research symposium.

By 2015, faculties from non-STEM/Arts units were invited to apply for summer funding with the condition that they partnered with a faculty member from a STEM field. Faculty members from Schools of Education and Journalism took advantage of that allowance

In 2016, a supplemental award provided additional support for faculty that was more robust. The funding supported 25% release time for six faculty members during the spring 2017 semester. Moreover, those selected faculty members received two personalized grant-writing tutelage sessions with an external consultant over the semester. The first grant-writing session was provided at the beginning of the semester while the second was provided at the end of the semester, with the expectation the faculty would submit proposals to funding agencies shortly after that.

The High School Students

In an effort to support the transition of high school students with interest in STEM careers to college programs, NanoHU instituted its NanoHU Pioneers program. Every summer since the summer of 2013, NanoHU has invited applications from rising sophomores and juniors who attend local high schools to engage in a six-week summer research program at HU. As many as 40 students have competed each year for a mere eight slots. Students who were selected to the program have been recognized as NanoHU Pioneers and did not necessarily possess the highest GPAs of the applicants. Students who were selected received stipends on the condition that they remained fully committed to the six-week program.

At the beginning of the summer program, both Pioneers and their parents are oriented over a day and a half. During that time, parents and Pioneers meet faculty mentors; Pioneers receive accelerated training in lab and fire safety, code of conduct, and program expectations. For the remainder of the six weeks, they engage in technical research (STEM and nano-related) for eight hours per day, four days per week. One day per week is reserved for math enrichment and seminars. Seminars are given by academic or industrial "experts" in nanoscience, representatives from HU's recruitment team or NanoHU administrators. Seminar topics include (1) tutorials on the preparation for oral and poster presentations, (2) career opportunities in STEM fields, and (3) practical examples of research in nanoscience and nanotechnology. Throughout the summer, students give oral presentations to a STEM audience. By the end of the summer, students participate in a poster session where campus administrators, peers, family, and invited friends attend.

The Surrounding Community

Although NanoHU originally intended to reach the community that was external to the University by way of the Pioneers, it also reached out to area elementary schools and civic organizations, with an emphasis on impacting underrepresented groups. NanoHU Scholars and Fellows were encouraged to "pay-it-forward" by participating in at least one local or regional event per year. To that end, deliberate efforts were made in the first year of funding to identify local elementary schools and request opportunities to engage students with hands-on activities. By the second year, the NanoHU received at least three invitations per year to participate in events that included Career Days (K5 schools), Mad Scientist Night (K5 schools), and Educational and Health Fairs (civic organizations including local and regional church organizations).

One of NanoHU's goals for participating in outreach activities was to incite excitement and wonderment in youth regarding nanoscience. NanoHU introduced its various audiences to basic nanoscience principles including the role of size on (1) reactivity (i.e., the impact of size on surface area to volume ratio), (2) Newton's Laws of Motion (i.e., non-Newtonian fluids), and (3) thermal energy and elasticity. Moreover, parents, grandparents, and other guardians were encouraged to participate with children in hands-on activities, with the hope of alleviating concerns about the ability to succeed in STEM. It is worth noting that supplies used in demonstrations primarily consisted of regular household items, including but not limited to Alka-Seltzer®, cooking oil, food coloring, and cornstarch. Nitinol memory wire and nano-sized gold were used for demonstrations about the impact of size on thermal energy and elasticity and color, respectively.

In March 2015, NanoHU was instrumental in coordinating and participating in a STEAM Expo at HU's "38th Annual Conference of the Black Family." At the half-day event, area fourth-graders were invited to obtain a "passport" that gave them access to the various STEAM units where they could engage in numerous hands-on activities. For example, students engaged in activities that involved robotics, chemistry, marine biology, physics, and planetary science. They also learned how music might be correlated with biology and mathematics.

In response to an invitation from The National Resource Centre for Materials Technology Education (MatEdU), NanoHU faculty and staff participated in the "2017 Materials in STEM (M-STEM)" event on November 6, 2017, at Thomas Nelson Community College in Hampton, Virginia. NanoHU representatives were invited to conduct a two-hour session for high school teachers. In a session entitled, "When Size Matters," the NanoHU team engaged teachers in simple, hands-on experiments that introduced learners to simple concepts that pertain to nanoscience. Teachers were allowed to interact with memory wire and magic sand, among other simple, everyday items, so that they (1) explore how fundamental properties change at the nanoscale, (2) deliberate on how these ideas might be incorporated into courses, and (3) explore how biomimetics can inform the next generation of materials in nanotechnology.

PROGRAM OUTCOMES AND EVOLUTION – NANOHU AND STEAM AT HU

Impact on Campus Constituents

Students

Since the inception of the project, NanoHU has engaged a total of 23 NanoHU Scholars, 12 NanoHU Fellows, and over 100 "Affiliates" students, including students who have participated in the seminars and outreach programs. To date, one Scholar and two "Affiliate" students have completed the nanoscience minor. Two of those students continued to pursue graduate degrees in STEM fields (one Scholar and one "Affiliate"), while the third is currently serving active duty in the military. By the end of the spring 2018 semester, seven additional Scholars will complete the nanoscience minor.

All Scholars and Fellows engaged in ongoing nanoscience research projects in the departments of biology, chemistry, computer science, engineering, marine and environmental science, and physics. Furthermore, all Scholars and Fellows engaged in summer research in laboratories across the US including Brandeis University, the University of Nebraska, and the University of California, Berkeley. Furthermore, two Scholars and one Fellow serve as coauthors of peer-reviewed articles that discuss research at the named institutions (Xie et al., 2017; Aghvami et al., 2017; Osoba et al., 2018).

Two Scholars who conducted research in HU's Department of Atmospheric and Planetary Sciences have since been awarded the prestigious Ernest F. Hollings Undergraduate Scholarship funded by the National Oceanic and Atmospheric Administration (NOAA) Office of Education. Both Scholars were recruited immediately after leaving high school. One of the Scholars engaged in the summer program for high school students two consecutive years as a NanoHU Pioneer.

To date, a total of 82 students have enrolled in the Introduction to Nanoscience course. Not surprisingly, the majority of students (85.4%) who enrolled in the course were either STEM or pre-professional majors. (Note: For the purposes of this publication, STEM majors include architecture, aviation, biology, chemistry, engineering, marine and environmental science, mathematics, physics, political science, psychology, and social science. Pre-professional majors include communicative sciences and disorders, kinesiology, nursing, physical therapy, and pre-pharmacy.) The distribution of Nanoscience Course concerning STEM and Non-STEM disciplines over the last five years can be seen in Table 1. When contrasted to other physical science elective courses offered over the past five years, we found that the 11.7% of the students (249 out of 394) who took "Introduction to Physical Science" course (a 100-level course in existence since 1988) majored in STEM or pre-professional majors. Also, 14.5% of students (24 out of 166) who elected to take the "Climate Change" physical science elective course (100-level course in existence since 2011) in that same time frame were either STEM or pre-professional majors. In the "Introduction to Nanoscience" course, the 14.6% of students who majored in Arts majors, majored in the following disciplines: Education, Fine and

Table 1. Distribution of STEM vs Non-STEM in the Participants of Introduction to Nanoscience Course for the Years 2014 to 2018 of NanoHU Program.

Majors	2014	2015	2016	2017	2018
Percentage of STEM majors	78.6	87.5	80.0	84.6	100.0
Percentage of non-STEM majors	21.4	12.5	20.0	15.4	0.0

Note: STEM: science, technology, engineering, and mathematics; NanoHU, Nanoscience Project at Hampton University.

Table 2. Gender Distribution of Students Who Have Participated in the Introduction to Nanoscience Course for the Year 2014 to 2018 of NanoHU Program.

Gender	2014	2015	2016	2017	2018
Percentage of male students	50.0	52.2	44.0	46.2	28.6
Percentage of female students	50.0	47.8	56.0	53.8	71.4

Note: NanoHU: Nanoscience Project at Hampton University.

Performing Arts, Music Education, Accounting, Economics, Business Management, International Studies, and Journalism. Most of the STEM majors came from Biology, Chemistry, Computer Science, and Physics fields.

Interestingly, the distribution of female students to male students was higher in the "Introduction to Nanoscience" course (Table 2). The higher female to male ratio is consistent with the other two introductory physical science elective courses. More specifically, 53.7% (44 out of 82) of students in the "Introduction to Nanoscience" course, 63.2% (249 out of 394) of students in the "Introduction to Physical Science" course, and 54.2% (90 out of 166) of students in the "Climate Change" course identified as female.

We have begun to see evidence that the "introduction to nanoscience" course has stimulated students' interest in learning more about nanoscience. Since 2012, at least ten students elected to apply for either scholar or fellow status while taking the course. All ten have stated that their interest was stimulated by the course in part. An additional four students, who are neither scholar nor fellow, elected to earn the nanoscience minor since taking the course. Two of those students have already graduated, and one of those two is currently pursuing a graduate degree in a STEM field.

By way of seminars and special events, NanoHU has impacted over 100 students who were neither Scholars, Fellows nor enrolled in the introductory nanoscience course. Since the fall semester of 2014, NanoHU has hosted or co-hosted seminar series that have showcased numerous STEM professionals. While Scholars and Fellows were required to attend, the audience often consisted of other students who were not affiliated with NanoHU. For instance,

NanoHU hosted a number of sessions at HU's 38th Annual Conference on the Black Family. In a session entitled "The STEM Void," at least 200 students from STEM and non-STEM majors had the opportunity to participate in a combined Town Hall/Breakout session that featured a senior program officer from the NSF, a senior administrator from the Association of American Colleges and Universities, and a renowned STEM Scholar. There students learned of the need for more practitioners in STEM from underrepresented groups. They were also encouraged to explore interdisciplinary approaches to solve societal problems. Post-session surveys showed that 95% of participants were extremely satisfied with the content (including topic and break-out groups) of the session. Surveys also revealed that the main source of discontent was the lack of seating.

In other sessions, between 50 and 100 students (STEM and non-STEM) per event heard from a diverse group of professionals (STEM and non-STEM practitioners) who demonstrated the importance of collaboration across disciplines. Featured speakers included a physicist on the 2017 Nobel Prize-winning Laser Interferometer Gravitational-Wave Observatory Project, Presidential awardee for Mentoring, a Professor of Law and Ethics who is also a scientist, an entrepreneur who specializes in nanoscience education at institutions across the globe, chemists, geoscientists, and researchers in nanomedicine, among others. The variety of speakers drew diverse student audiences from pre-law, journalism, education, business, and fine arts as well as from the traditional STEM fields.

Faculty

NanoHU has contributed to the professional development of HU faculty across numerous STEM and non-STEM disciplines. The initial intent of the project was to aid the faculty from STEM fields develop their research careers by providing modest support that might be utilized for one summer. Howbeit, as the project progressed, the faculty development component grew out of definite necessity and probable serendipity. Now, in the final stages of the project, it has supported faculty by providing faculty start-up awards, release time, research supplies, research-related travel awards, seminars, and grant-writing workshops.

To date, 20 faculty members representing seven STEM departments and two non-STEM departments have received support from NanoHU after responding to requests for proposals (RFPs). These faculties have also mentored all tiers of students (Scholars, Fellows, and Pioneers). Faculty reported that they had leveraged the results of their projects for career advancement in several ways. Scholarly activities include submitting manuscripts/publications, seeking external grants from federal and non-federal sources, applying for early career awards, and establishing research collaborations. In fact, 100% of those faculties presented at professional conferences along with students, 50% submitted manuscripts to peer-reviewed journals, 45% published manuscripts in peer-reviewed journals, 40% submitted proposals to federal agencies, 20% received federal grants, and 60% established research collaborations.

Regarding the *objective to provide professional development support for (STEM) faculty to increase their publications record with undergraduates as coauthors by 50%,* the publication/scholarship record across the STEM departments has generally increased, and to date, all faculties have given a research presentation at local, regional, and/or national meetings. All faculties are in varying stages of preparing chapters for technical peer-reviewed publications. As required, all faculties have been engaged with NanoHU undergraduate and high school students per year, have given technical seminars in the seminar series, have collaborated with NanoHU institutional partners, and have consulted with technical consultants.

NanoHU has also provided professional development opportunities for non-STEM faculty in the form of a grant-writing workshop. As a result of supplemental funding from the NSF, NanoHU hosted a grant-writing workshop in the 2017 spring semester facilitated by an external consultant. Faculty from diverse departments (STEM and non-STEM), many of whom held to a teaching focus, learned the fundamentals of grant-writing in a four-hour session. Analysis of the post-session survey showed that 100% of workshop attendees were extremely satisfied with the workshop, 98% of attendees felt more empowered to write a grant proposal for funding, and 89% of attendees stated that they intended to write and submit a grant proposal within one year of the grant-writing session.

High School Students

The NanoHU Pioneers program has hosted 65 high school students representing 20 high schools since the beginning of the project. In the first year, students with GPAs above 3.8 were selected for participating in the six-week program. After participating in pre- and post-summer program surveys, it was evident that the summer program activities did not have a significant impact on the students. Responses to survey questions suggest that those students already had the interest, support, and preparation to enter undergraduate STEM programs with the intent to earn corresponding graduate degrees. In fact, 80% of those students have indicated that they entered undergraduate programs at colleges and universities across the country and are pursuing the baccalaureate degree in various STEM fields.

In subsequent years, deliberate efforts were made to select students with lower GPAs and/or deficiencies with regard to pursuing STEM degrees in college or universities. Surveys showed that all of those students were engaged in research for the first time when they engaged in the NanoHU summer program. Moreover, by the end of the summer, 95% of those students indicated that they felt more prepared to enter undergraduate STEM programs.

As a result of their participation in the summer programs, NanoHU Pioneers have made notable achievements. All students presented the results of their summer research at poster sessions at conferences at HU. Even more noteworthy, in 2015, is that three of those students received travel awards to participate in the Emerging Researchers Network Annual Conference in Washington, DC. In that same year, three students coauthored a laboratory manual that was pilot tested with preservice STEM education teachers. The results of this effort were published by the faculty mentors at the American Educational Research Association 2016 spring conference (Berube & Dash, 2016).

All but two students from the first 2013 cohort have indicated that they have successfully entered either a college or a university and are pursuing baccalaureate degrees in various STEM fields. Harvard University, University of Alabama, HU, and Thomas Nelson Community College are just a few of the institutions that those students currently attend. Data are currently being gathered from latter cohorts.

Surrounding Community

NanoHU has exceeded its goal to engage in at least one community event per year. To date, NanoHU has participated in, and/or hosted, 13 STEM exposition-type activities over five years. While no formal count was taken of all K5 students impacted, we estimate that NanoHU has directly impacted over 200 students and 100 adults, including parents, grandparents, other guardians, and K-12 teachers (Claville et.al, 2016). From anecdotal data and surveys administered at all events, NanoHU's demonstrations were perceived to be engaging and informative.

NanoHU's intent to quell the apprehension of targeted communities regarding the engagement in and pursuit of STEM fields proved successful. In fact, after one outreach activity that was conducted in an urban setting, at least three adults over the age of 50 and with "blue-collar" professions stated:

"If science were like this when I was in school, I would have enjoyed school much more."

That repeated statement, combined with the popularity of our table at the various events, informed us that we were accomplishing our intent of demystifying the combined STEM fields. In another instance, elderly women who succumbed to our invitation to "play with Oobleck" (cornstarch and water mixture) while we discussed fundamental differences between Newtonian and non-Newtonian movement discussed their historical experience with the substance. By the end of their interaction with NanoHU's presentations, they indicated that they had a more informed (i.e., scientific) understanding of their formative experiences with cornstarch. Similarly, surveys from the M-STEM event indicated that NanoHU had provided an engaging presentation as well as useful information that would be incorporated into their respective high school science classes.

Program Design

The multifaceted design of NanoHU was derived from the personal and professional expertise that was embedded in its primary architects – seasoned faculty across HU's STEM Schools and departments – and not existing theoretical models. Thus, NanoHU was not designed to test existing theories but rather to accumulate data and evidence based on the real-life empirical observations made during NanoHU's implementation. It was anticipated that, as an innovative endeavor, new empirically derived NanoHU learnings and insights would be aggregated that could be used to explicate the NanoHU model. Over the project's five years, we have become increasingly aware of and attuned to the theoretical perspectives and frameworks – scientific identify, social cognitive career theory, institutional change, and curricula development/reform – that are relevant to the work as previously described. These provide a context within which to situate NanoHU innovations, as well as to begin to specify the constructs, as indicators of success that have been observed, described, measured, and aligned with existing theories that have been determined. We acknowledge that essential to establishing the NanoHU model is an examination of the relationship between how constructs are described and classified and the anticipated outcomes as specified in each objective. As NanoHU continues to emerge as an iterative empirical, descriptive model-building and theory-generating process, we are guided by multiple sources, which is the innate expertise, as well as theoretical perspectives and frameworks, which when coupled will potentially expand the knowledge base and the constellation of constructs that theories and perspectives ultimately embody (Carlile & Christensen, 2004).

LESSONS LEARNED, KEY SUCCESSES, AND FUTURE ASPIRATIONS

Lessons Learned

Throughout the execution of the project, we made a number of discoveries that altered our approach from the original design. For instance, when writing the grant proposal, we assumed that larger numbers of STEM students at HU would naturally embrace the opportunity to engage in preparation that would lead to graduate degrees in STEM fields. A large number of the initial scholarship applicants sought support for preparation for entry into medical professions and exhibited very little interest or recognition of viable careers in STEM fields. Accordingly, creative marketing strategies were developed to reach large audiences of students. In fact, this initial lesson informed the subsequent execution of the project such that all events and opportunities affiliated with the project were marketed widely across the institution.

Annual evaluations and advisory boards were integral to informing the direction of the project. For instance, evaluation of the first-year activities revealed the need to alter the selection process of participants in order to effect the desired change in high school participants. Subsequent evaluations led to (1) multi-tiered designation of undergraduate student participants (Scholars, Fellows, and Affiliates), (2) the adoption and execution of contracts for all students participants (Scholar, Fellows, and Pioneers), (3) the incorporation of weekly update meetings for Scholars and Fellows, (4) specialized professional development for faculty, and (5) other miscellaneous management practices.

It was important to obtain support for the project of NanoHU from various sectors across the University. During the proposal preparation, it became evident that the funder's charge to accomplish institutional transformation would only succeed if the project design aligned with the University's strategic plan.

The proposed NanoHU was a plausible response to the University's intent to increase and diversify its research repertoire and existing innovation trends across the state of Virginia, the nation, and the globe.

Key institutional support came from the University's Executive Leadership, the Deans and Chairs of STEM units, key personnel in non-STEM units, and personnel in non-academic units. Support was first evident when the Executive Leadership invited the newly funded PI to summarize the intent of the funded proposal to the University faculty and the Board of Trustees. In the presentation to the faculty, the PI was certain to inform the faculty of interdisciplinary of nanoscience education and execution. By doing so, non-STEM faculty members (e.g., Business, Pre-Law, Journalism, Public Policy, English Literature, Fine Arts) were invited to learn about the science and explore innovative collaborations.

We learned that active engagement of STEM leaders in the project would result in desired STEM interdisciplinary engagement. For instance, NanoHU's Internal Advisory Board consisted of all STEM leaders (i.e., department chairs and School deans). This construct was instrumental in ensuring nanoscience infusion in corresponding curricula, the design of the nanoscience minor, evaluation of students' (Scholars, Fellows, and Pioneers) performance, and subsequent actions in response to students' performance. This collaborative approach proved to be vital in ensuring compliance to scholarship requirements and promotion of departmental research.

We also understand that the successful execution of this project necessitated the avoidance of certain pitfalls. For instance, it was always important to remain true to the mission of the project. Consistent evaluation and response to the same ensured that the project's goal remained central to all activities. In addition, we recognized the importance of understanding the policies of the University and evaluating congruence with the funder's policies, before executing activities. This point was especially crucial when spending funds (e.g., for student stipends, release time) and scheduling events.

Key Successes

The NanoHU boasts multiple successes. These include the genesis of formal nanoscience program that is evident from the development of the nanoscience minor, the new nanoscience course, the engagement of students in innovative research, the professional development of faculty, and community engagement. Here, we focus on the establishment of functional collaboratives.

From Silos to Collaboratives

The efforts to establish a more cohesive STEM atmosphere across, particularly in the area of nanoscience, have experienced significant success. For instance, more nanoscience researchers on campus have embraced the opportunity to exchange research results in the NanoHU Seminar Series. Moreover, faculties who received NanoHU awards have embraced opportunities to collaborate across STEM and Liberal Arts fields.

On a larger scale, constituents across the University have begun to engage in efforts that lessen the "divide" between the STEM and Arts fields. For instance, in political science classrooms, pre-law professors have begun to discuss the ethical implications of advances in STEM. Moreover, Journalism students have been tasked with writing about aligning the efforts of NanoHU with the recent 2016 Nobel prizes in chemistry (LaTorre, 2016), and English professors have begun to collaborate with STEM faculty about exploring the influence of science in science fiction literature. In fact, in the 38th Annual Conference on the Black Family, the aforementioned session entitled, "The STEM Void" was moderated by a law and policy professor who charged the audience (comprising STEM and pre-law students) to collaboratively effect policy changes that would result in increased numbers of underrepresented groups in STEM fields.

Future Aspirations

With the marked successes, there is still more work to do. Most notable is the need to formalize a nanoscience community that continues to support the University's transition to a more research-intensive institution. Accordingly, plans are currently underway to host a town hall that will allow institutional stakeholders to discuss next steps. In April 2018, STEM units across the HU campus will meet and develop plans that will result in the growth of a materials science/nanoscience community on campus. By the end of the session, it is anticipated that an enhanced interdisciplinary science center will be established at the university. Furthermore, it is the intent of these stakeholders to continue discussions across disciplines, solidifying a formidable STEAM enterprise that will prove to be innovative and pioneering.

REFERENCES

Aghvami, S. A., Opathalage, A., Zhang, Z. K., Ludwig, M., Heymann, M., Norton, M., ... Fraden, S. (2017). Rapid prototyping of cyclic olefin copolymer (COC) microfluidic devices. *Sensors and Actuators B: Chemical, 247*, 940–949.

Bequette, J. W., & Bequette, M. B. (2012). A place for art and design education in the STEM conversation. *Art Education, 65*(2), 40–47. doi:10.1080/00043125.2012.11519167

Berube, C., & Dash, S. (2016). *Nanoscale and nanoparticle research modules for African-American pre-service STEM teachers; core Nanoscience concepts as a vehicle in STEM education*. Paper presented at the 2016 annual meeting of the *American Educational Research Association*. Retrieved from the AERA Online Paper Repository. Accessed on February 16, 2018.

Carlile, P. R., & Christensen, C. M. (2004). *The cycles of theory building in management research*. Bingley, UK: Emerald Group Publishing Limited 2011. Retrieved from http://citeseerx.ist.psu.edu/viewdoc/summary?doi=10.1.1.136.8158

Clark, B., & Button, C. (2011). Sustainability transdisciplinary education model: Interface of arts, science, and community (STEM). *International Journal of Sustainability in Higher Education, 12*(1), 41–54. doi:10.1108/14676371111098294

Claville, M. O., Babu, S., Parker, B. C., Hill, E. W., Ramdon, R., & Lowe, C. W. (2016). *NanoHU: Pioneering the preparation of under-represented groups for engagement and leadership in the global STEM enterprise* (7th ed., pp. 105–123). Berlin: AKA Press.

Compass, U. N., & See the Historically Black Colleges and Universities Ranking Methodology. (n.d.). *The 10 best Historically Black Colleges and Universities*. Retrieved from https://www.usnews.com/best-colleges/rankings/hbcu. Accessed on February 26, 2018.

Fazarro, D., Newberry, D., Trybula, W., & Hyder, J. (2012). Introducing a nanotechnology curriculum and considerations for bridging academic/industry relationships for applied technology-oriented programs: The new challenge for ATMAE. *Journal of Technology, Management, and Applied Engineering*, *28*(1), 1–8.

Gasman, M. (2013). *The changing face of historically Black colleges and universities*. Philadelphia, PA: Center for Minority Serving Institutions, University of Pennsylvania.

Hampton University Mission. (n.d.). Mission. Retrieved from http://www.hamptonu.edu/about/mission.cfm. Accessed on February 26, 2018.

The Journal of Blacks in Higher Education. (2012), "Tracking Graduation Rates at HBCUs" 2012. Retrieved from https://www.jbhe com/2012/01/tracking-graduation-rates-at-hbcus/ Accessed on Dec 19, 2018.

Laherto, A. (2010). An analysis of the educational significance of nanoscience and nanotechnology in scientific and technological literacy. *Science Education International*, *21*(3), 160–175.

LaTorre, D. (2016, December 11). *Machines thinner than hair, studied at Hampton U* (Web log post). Retrieved from http://blog.shsjc.hamptonu.edu/index.cfm/2016/11/22/Machines-thinner-than-hair-studied-at-Hampton-U#. Accessed on February 16, 2018.

Madden, M. E., Baxter, M., Beauchamp, H., Bouchard, K., Habermas, D., Huff, M., … Plague, G. (2013). Rethinking STEM education: An interdisciplinary STEAM curriculum. *Procedia Computer Science*,*20*, 541–546. doi:10.1016/j.procs.2013.09.316

McAfee, A., & Brynjolfsson, E. (2016). Human work in the robotic future: Policy for the age of automation. *Foreign Affairs*, *95*, 139.

Obama, B. (2016). *White House initiative on educational excellence for African Americans*. Information Resource Center, U.S. Department of Education, Washington, DC.

Osoba, B., Saha, B., Almeida, S. F., Patil, J., Brandt, L. E., Roots, M. E., … Liu, T. J. K. (2018). Variability study for low-voltage microelectromechanical relay operation. *IEEE Transactions on Electron Devices*, *65*(4), 1529–1534.

Palmer, R. T., Davis, R. J., & Thompson, T. (2010). Theory meets practice: HBCU initiatives that promote academic success among African Americans in STEM. *Journal of College Student Development*, *51*(4), 440–443. doi:10.1353/csd.0.0146

Perna, L., Lundy-Wagner, V., Drezner, N. D., Gasman, M., Yoon, S., Bose, E., & Gary, S. (2009). The contribution of HBCUs to the preparation of African American women for STEM careers: A case study. *Research in Higher Education*, *50*(1), 1–23.

S.NET - The Society for the Studies of New and Emerging Technologies (SNET). (2017). About S. Net. Retrieved from http://www.thesnet.net/. Accessed on February 26, 2018.

Schulte, J. (2005). *Nanotechnology: Global strategies, industry trends and applications*. Chichester: Wiley (Scripps Howard School of Journalism and Communications Blog).

Simpson, E. (2010). Hampton U. celebrates opening of proton therapy center. *The Virginian Pilot*. Retrieved from http://pilotonline.com/news/local/health/hampton-u-celebrates-opening-of-proton-therapy-center/article_2ef67349-b62a-5b5d-a4b7-c888147d915d.html

Strayhorn, T. L. (2008). Influences on labor market outcomes of African American college graduates: A national study. *The Journal of Higher Education*,*79*(1), 28–57. doi:10.1353/jhe.2008.0003

Strayhorn, T. L., Williams, M. S., Tillman-Kelly, D., & Suddeth, T. (2012). Sex differences in graduate school choice for Black HBCU bachelor's degree recipients: A national analysis. *Journal of African American Studies*,*17*(2), 174–188. doi:10.1007/s12111-012-9226-1

NSF. (n.d). Demographics of Doctorates. (Table 23. U.S. citizen and permanent resident doctorate recipients, by broad field of study, ethnicity, and race: Selected years, 1995–2015.) Retrieved February 26, 2018, from https://www.nsf.gov/statistics/2017/nsf17306/datatables/tab-23.htm

The 2006 JCOTS Nanotechnology Research & Development and Manufacturing Advisory Committees. (2006). *Nanotechnology in Virginia*. Retrieved from http://dls.virginia.gov/commission/pdf/JCOTS%20NanoDay%20Whitepaper%20.pdf. Accessed on 2 February, 2018.

The Carnegie Classification of Institutions of Higher educations. (2017). Institution Lookup: Hampton University. Retrieved from http://carnegieclassifications.iu.edu/lookup/lookup.ph

The Center for Atmospheric Sciences. (n.d.). The Center for Atmospheric Sciences. (Research. Outreach. Education.) Retrieved from http://cas.hamptonu.edu/. Accessed on February 26, 2018.

The future is for everyone: School for the Future of Innovation in Society at Arizona State University. (n.d.). Welcome to the School for the Future of Innovation in Society. Retrieved from https://sfis.asu.edu/. Accessed on February 26, 2018.

United States Census Bureau. (2017). *Population estimates*, July 1, 2017. Retrieved from https://www.census.gov/quickfacts/fact/table/US/PS`T045217#. Accessed on February 2, 2018.

United States Congress. (2009). *The national nanotechnology initiative: Overview, reauthorization, and appropriations issues* (J. F. Sargent, Author) [Cong.]. Washington, DC: Congressional Research Service, Library of Congress.

Xie, Y., Yu, F., Tang, W., Alade, B. O., Peng, Z., Wang, Y., Li, J., & Oupický, D. (2017). Synthesis and evaluation of chloroquine-containing DMAEMA copolymers as efficient anti-miRNA delivery vectors with improved endosomal escape and antimigratory activity in cancer cells. *Macromolecular Bioscience*, *18*(1), 1700194. doi:10.1002/mabi.201700194

ALL FOR ONE AND ONE FOR ALL: COORDINATING THE RESOURCES OF INDIVIDUAL STUDENT RESEARCH TRAINING INITIATIVES IN BIOMEDICAL SCIENCES AT XAVIER UNIVERSITY OF LOUISIANA

Maryam Foroozesh, Marguerite Giguette, Teresa Birdwhistell, Kathleen Morgan, Kelly Johanson, Tiera S. Coston and Clair Wilkins-Green

ABSTRACT

Xavier University of Louisiana has a national reputation for producing science, technology, engineering, and mathematics (STEM) graduates who go on to obtain MD and PhD degrees. According to a 2013 National Science Foundation report, Xavier is ranked first in producing African American graduates who go on to receive life sciences PhD degrees, fifth in the nation in producing African American graduates who go on to receive science and engineering PhD degrees, and seventh in producing African American graduates who go on to receive physical sciences PhD degrees. Xavier is currently third among the nation's colleges and universities in the number of African American graduates enrolled in medical school, according to data compiled by the Association of American Medical Colleges, and ranked first in the number of African American alumni who successfully complete their medical

Broadening Participation in STEM: Effective Methods, Practices, And Programs
Diversity in Higher Education, Volume 22, 129–149

ISSN: 1479-3644/doi:10.1108/S1479-364420190000022006

*degrees. The success of Xavier's graduates is due to a combination of university-based student support initiatives and externally funded programs, in particular, the Building Infrastructure Leading to Diversity (BUILD), Maximizing Access to Biomedical Research Careers (MARC) U*STAR, and Research Initiative in Scientific Enhancement (RISE) programs. These three programs, funded by the Training, Workforce Development, and Diversity (TWD) Division at the National Institutes of Health (NIH), offer select trainees undergraduate research opportunities, support mechanisms, and a variety of activities designed to improve their potential for success in graduate school. The BUILD, MARC U*STAR, and RISE programs work closely together and with the University to leverage the resources provided by each in order to provide the best experience possible for their students with a minimum of redundancy of effort. This chapter focuses on the program components and how the programs work together.*

Keywords: Student research training; underrepresented; biomedical workforce; supportive environment; mentoring; cultural responsiveness

INTRODUCTION AND BACKGROUND

National statistics point to the decreasing numbers of US students, especially minorities, enrolling in science, technology, engineering, and mathematics (STEM)-related fields of study at both the undergraduate and the graduate levels, resulting in a decreasing number entering the related research workforce. As a result, the country is increasingly falling behind other nations in growing its STEM workforce. This represents a serious challenge to our economic productivity as the advanced technology and science sectors, requiring large numbers of STEM workers, have been the fastest growing sectors of the world's economy (U.S. Department of Education, 2016).

Given their percentage of the US population, African Americans obtain far fewer science degrees than expected. Based on the most recent US census estimates (2016), African Americans constitute about 13.3% of the United States' population; however, they earn less than 9% of the bachelor's degrees, 8% of the master's degrees, and 3% of the doctoral degrees conferred in STEM disciplines in the country (National Science Foundation, National Center for Science, & Engineering Statistics, 2015). This disparity is due to many socioeconomic factors, including a lack of access to high-performing school systems and thus receiving substandard preparation in STEM disciplines, and unfamiliarity with STEM careers other than the ones related to medicine.

Xavier University of Louisiana (Xavier) has a strong commitment to enrolling students who have not had access to the educational resources required for a solid foundation in the STEM disciplines. Along with serving the nation's best and brightest students attracted to Xavier by its reputation in STEM, the University aims to provide all of its students with exceptional educational opportunities inside and outside of the classroom, including hands-on research training. Xavier works to overcome unequal educational backgrounds and socioeconomic challenges to develop student talent through expanding STEM and biomedical training opportunities and building on an established reputation in science education. A diverse scientific workforce taps into unique perspectives, broadening the scope of inquiry into unexplored or neglected areas, challenging disparities and promoting equality

THEORETICAL UNDERPINNINGS AND ASSUMPTIONS

While over the years, many individual, innovative courses, workshops, student support systems, and faculty development programs have been developed across the country, most have been limited in their focus and reach, and implemented to function in isolation from others. This chapter reports attempts at Xavier to foster cooperation between programs.

Xavier recognizes the value of hands-on research training and direct engagement of students in research projects and has a long history of providing research training opportunities to its students through programs funded by various federal agencies or private foundations. The University's success in securing funding for these programs, and the institution-wide participation in them, demonstrates faculty and staff's grant-writing and management skills, their knowledge in designing and implementing innovative interventions, and their deep commitment to improving student training and educational opportunities.

Xavier's suite of student training programs in biomedical disciplines has been designed to respond to (1) extensive reviews of the national literature regarding factors that lead to, or become barriers to, African Americans achieving terminal degrees in STEM/biomedical fields, (2) surveys of Xavier graduates who went on to graduate schools in such fields, to determine which Xavier educational and/or support programs especially helped or hindered them in being successful in their graduate school programs, and (3) surveys of their graduate school faculty to determine perceived strengths and weaknesses. Based on the data obtained from these reviews and surveys, Xavier redesigned often unrelated programs (e.g., career advising, academic advising, tutoring, graduate placement, faculty and student development, academic courses) and integrated and focused them on the major goal of moving students along the STEM/biomedical research pathway.

INSTITUTIONAL CONTEXT AND STRATEGY OF CHANGE

Xavier University of Louisiana is a historically Black and Catholic university (HBCU) that is nationally recognized for its STEM curricula. A Xavier education also features a solid liberal arts foundation and broad learning experiences that extend beyond the classroom and campus (Hannah-Jones, 2015). Of the 3,044 students enrolled at Xavier in Fall 2017, approximately 72% were African Americans, and about 72% of the 2,345 undergraduates majored in the biomedical sciences. Xavier is a national leader in the number of STEM majors who go on to receive MD and PhD degrees in science and engineering. The University is best known for its reputation in the health professions, ranking first in the number of African Americans who earn bachelor's degrees in the physical sciences and second in the number earning bachelor's degrees in the biological and biomedical sciences (Diverse Issues in Higher Education, 2016). A report released by the National Science Foundation in 2013 confirms Xavier's success in educating science graduates. According to the report, Xavier is ranked first in producing African American graduates who go on to receive life sciences PhD degrees, fifth in the nation in producing African American graduates who go on to receive science and engineering PhD degrees, and seventh in producing African American graduates who go on to receive physical sciences PhD degrees (Fiegener & Proudfoot, 2013). Additionally, Xavier is currently third among the nation's colleges and universities in the number of African American graduates enrolled in medical school, according to data compiled by the Association of American Medical Colleges, and ranked first in the number of African American alumni who successfully complete their medical degrees (American Association of American Medical Colleges, Applicants, & Matriculants Data, 2016). Xavier alumni contribute significantly to the number of underrepresented minority physicians, pharmacists, dentists, and other professionals in medically related fields who practice throughout the United States. The success of Xavier's pre-medical program was also recently highlighted in *The New York Times* (Hannah-Jones, 2015).

Although many well-prepared, highly motivated students are attracted by Xavier's reputation in the sciences, other students, though bright and capable, come from underperforming public school systems and thus have received substandard preparation in STEM disciplines. Xavier has a mission and history of inclusion in the education of underserved populations. Grounded in the Christian principles exemplified by its founders, Saint Katharine Drexel and the Sisters of the Blessed Sacrament, Xavier has the distinction of being the only Catholic university founded by an American-born saint (Griffin & Hurtado, 2010). True to its HBCU roots, Xavier's mission continues to be to create a more just and humane society by preparing its students to assume roles of leadership and service in a global society. Xavier is dedicated to a curriculum that focuses on the liberal arts, ethical and moral values, critical thinking, and career preparation for further graduate or professional school education.

Approximately 55.1% of Xavier's students are from Louisiana, primarily from the New Orleans area. The remaining 44.9% come from other states, the District of Columbia, the Virgin Islands and abroad. While Xavier's student body is predominantly African American, the University is open to all and strongly supports diversity among its students, faculty, and staff.

As previously mentioned, due to its exemplary research and education training programs, Xavier has been successful in graduating large numbers of Black scientists and health professionals, thereby addressing a national need. The Building Infrastructure Leading to Diversity (BUILD), Maximizing Access to Biomedical Research Careers (MARC U*STAR), and Research Initiative in

Scientific Enhancement (RISE) programs are three student training programs funded by the National Institutes of Health's (NIH) National Institute of General Medical Sciences (NIGMS) under the Training, Workforce Development, and Diversity (TWD) umbrella. These programs have significantly contributed to Xavier's success by providing cohorts of students with knowledge of, and preparation for, biomedical research careers. Xavier's TWD programs have focused on several initiatives proven to increase student success in graduate school, including research skills training and enhancement of scientific communication skills (Burelli & Rapoport, 2008; Doyle, 2000; Hakim, 2000; Lopatto, 2004; Rothman & Narum, 1999). It is important to note that there are a number of other successful student training programs at Xavier; however, we are focusing only on the TWD programs in this document.

PROGRAM GOALS, COMPONENTS, AND IMPLEMENTATION

The overarching goal of the TWD programs at Xavier University of Louisiana is to increase the number of individuals from underrepresented groups who choose careers in biomedical research and pursue terminal degrees in these disciplines. Toward this end, Xavier's TWD programs were designed with the following goals:

(1) stimulate interest of Xavier students in biomedical research careers;
(2) increase retention in the biomedical disciplines at Xavier; and
(3) increase the number of Xavier students entering graduate programs in biomedical research areas.

By implementing isolated initiatives, rather than looking at the whole inclusive picture, educational institutions run the risk of fragmenting the student experience into smaller and smaller pieces and further isolating students from the greater scientific community (Longden, 2006). The TWD programs at Xavier work together to offer the successful programming that they have been providing over their past project cycles, improving the academic and research competitiveness of Xavier's undergraduates, while putting emphasis on effectively integrating and evaluating initiatives. In order to achieve a greater impact on a larger cohort of students, leverage the available funding, and prevent duplication of efforts, the programs intersect and closely collaborate with each other. In this chapter, rather than focusing on the differences between the three aforementioned programs, their integrated best practices and how the programs are informing and improving each other are discussed in detail.

The TWD programs at Xavier are highly innovative programs designed to broaden the career interests of students early on and engage them in activities that entice them to continue their education toward biomedical terminal degrees and research careers. Strategies used involve transformation of Xavier's academic and non-academic programs through the redesign, supplementation, and integration of academic advising (BUILD, MARC, RISE), tutoring and supplemental instruction (SI) (BUILD), career services (BUILD, MARC, RISE), research skills training (BUILD, MARC, RISE), on- and off-campus hands-on research opportunities (BUILD, MARC, RISE), mentor training (BUILD, MARC, RISE), mentee training (BUILD, MARC, RISE), and development of new biomedical and research skills courses (BUILD, RISE). The programs also focus on providing faculty members with opportunities to improve their teaching skills (BUILD, RISE) and increase collaborations and communication among faculty engaged in STEM course/curriculum development (BUILD, RISE), as well as their research competitiveness (BUILD). In addition to the wide range of activities supported by these programs within the institution, Xavier is partnering with a number of major research universities across the nation to provide the students with summer research opportunities on their campuses and Xavier faculty with more research collaboration prospects.

Addressing a National Need

Minorities currently represent an expanding portion of the US population, and yet many groups are underrepresented in the STEM and biomedical research workforce. As reported in the 2016 *US News/Raytheon STEM Index*, the number of White students who earned STEM degrees grew 15% in the prior five-year period, while the number of Black students fell by about the same percentage (Neuhauser & Cook, 2016). In general, the Index showed that women, Blacks, and Hispanics still lag far behind White and Asian men in earning degrees and securing jobs in STEM fields. Thus, unless scientific education becomes more inclusive, the US will be denied the talents of a large segment of its population.

Xavier is already a leader in placing African American students in professional and graduate programs despite the facts that it is a relatively small school and that the majority of its students and their families are socioeconomically challenged. Xavier does not require its students to report their financial status to the University. However, institutional data from the past five years supplied by Xavier's Office of Planning, Institutional Research and Assessment, indicate that an average of 43% of its students receives Pell grants. Notably, in 2017, Xavier was highlighted in a *New York Times* article for its sixth-place ranking among all colleges and universities nationwide for the upward economic mobility of its students from the bottom fifth of the income distribution to the top three-fifths (Leonhardt, 2017). Even with this positive economic trend for Xavier's graduates, there is always room for improvement.

Consistently, the Higher Education Research Institute (HERI) Cooperative Institutional Research Program (CIRP) Freshman Survey indicates that approximately 50% of incoming freshmen enter Xavier with a "strong sense of belonging to a community of scientists" and approximately 65% of incoming students indicate that they will definitely or probably pursue a STEM-focused career. Further, less than 5% of Xavier's incoming freshmen say that the highest degree they plan to attain is a bachelor's degree; of the remainder, 10% intend to go on to a master's, 34% to a medical degree, 26% to a PhD and a further 19% to a professional doctorate. Seventy-one percent of Xavier's current undergraduates are majoring in a STEM discipline, and if we include all biomedical disciplines, this number is closer to 83%. Therefore, Xavier, with its unique culture and brand in STEM and biomedical fields and the large proportion of its students with interest in these disciplines, has the potential to lead the way in making a difference.

The literature identifies unique barriers faced by African Americans and other individuals from underrepresented groups entering into biomedical research careers, including the lack of (1) *Early Awareness of and Deepening Exposure to* biomedical research careers and the type of rewards associated with them, (2) *Supportive Relationships*, particularly those related to faculty advising and mentoring, (3) *Suitable Educational Infrastructure*, namely (a) innovative STEM courses that engross students in activities that promote a scientific mindset and (b) the supplemental help needed when they face educational challenges, and (4) *Active Engagement* in meaningful biomedical research experiences and the presence of faculty and institutional resources needed to do this (Butts et al., 2012; Tabak & Collins, 2011; Baxter, Botelho, & O'Donnell, 2015; Pender, Marcotte, Domingo, & Maton, 2010). Surveys of Xavier STEM alumni and the graduate faculty at other institutions who teach them (made possible by an NIH-funded BUILD planning grant) echoed many of these findings and pointed to areas in Xavier's curricula and support services that needed to be revised or supplemented.

Three Distinctive Programs, One Common Goal

Within the confines of each funding opportunity announcement, Xavier's TWD programs have been designed to maximize the impact of the resources on the Institution and its students. The three programs use a multi-pronged approach with many concurrent and sequential activities and initiatives that aim to enhance the academic and research environment at Xavier.

The youngest and most extensive of these programs is the BUILD Program. Xavier is one of only 10 institutions across the country receiving support from this experimental funding mechanism at the NIH (2014–2019, and now with a competitive renewal for 2019–2024). The Xavier BUILD Program, *Project Pathways*, provides resources for key offices and centers across the campus that assists students with academic support, professional development, and undergraduate research activities. This program has contributed to the expansion of both personnel and support of *the Student Academic Success Office* (SASO), *the Office of Career Services* (OCS), *the Center for Undergraduate Research and Graduate Opportunity* (CURGC), and *the Center for the Advancement of Teaching and Faculty Development* (CAT + FD). These expanded and significantly improved offices and centers have greatly enhanced the quality of student and faculty support services at Xavier and have positively impacted the entire institution.

The BUILD Program coordinates a number of activities designed to educate all interested freshman and sophomore students about the variety of possible biomedical research careers and helps them identify careers that match their interests. Beginning in the freshman year, through a series of broad-based workshops, CURGO and OCS assist Xavier students in gaining a better understanding of the nature of biomedical research careers. Such activities allow students to gradually and voluntarily increase their involvement, based on their level of interest in biomedical research. Early exposure to biomedical research career paths allows students the opportunity to build the skills necessary for success. Also, when motivated to pursue a specific career, students are more likely to engage in their courses, which leads to enhanced retention and improved graduation rates (Jimerson & Ferguson, 2007). These offices guide students through a process that results in their ability to identify a specific biomedical research interest, obtain experiences that advance their research skills, and develop the tools needed to be accepted to and succeed in graduate programs. They also offer assistance to students in creating Individual Development Plans (IDPs) as well as assistance in preparing applications, *curriculum vitae*, and personal statements. CURGO hosts the annual *Festival of Scholars* and *Summer Research Symposium*, which encourage Xavier students to present their research and creative work to the broader campus community.

Both freshmen and sophomores are given opportunities through BUILD to explore research labs at Xavier through the *Freshman Open House* held during the *Biomedical Week* at Xavier and also through *Research Shadowing Experiences*, during which they shadow a current research student on campus and interact with a research group. The BUILD Program also organizes a series of peer-led discussion groups where interested students, primarily sophomores, discuss current scientific topics in the news and the various careers of people involved in the studies. They also hear research-active students discuss their specific research projects and talk about their experiences with the graduate school application process. This peer approach is critical to building and sustaining interest in science careers (Snyder, Sloane, Dunk, & Wiles, 2016).

Each year, the Program chooses 12 sophomore students from among the participants in these different activities who apply to become BUILD Scholars starting the summer before their junior year. The student's grade point average (GPA) is just one factor considered during the selection, and the process is holistic in nature. The Scholars receive two years of mentored research, including a summer at a research-intensive partner institution. In order to increase the reach of the Program beyond that already included in the funded proposal, each year, several additional students are chosen as BUILD Research Students. Like the Scholars, the BUILD Research Students are matched with mentors and participate in hands-on research; however, due to budget limitations, a BUILD Research Student's appointment is for a period of 12 months at a lower funding level but with potential for competitive continuation for an additional year.

The oldest of the TWD programs at Xavier, the MARC Program, supports 10 students each year in their junior and senior years and provides them with hands-on research experiences, including a summer at a research-intensive institution. The students are selected at the end of their sophomore year and enter the program in the fall of their junior year. The MARC Program is an honors program with a GPA requirement of 3.0 or above.

The RISE Program accepts students in their freshman, sophomore, and occasionally junior years (while MARC and BUILD programs only accept rising juniors as Scholars, BUILD accepts students at any level as BUILD Research Students who participate in most but not all program activities) and has a minimum GPA requirement of 2.75. This program supports 12 students each year. Earlier involvement with support programming through RISE increases the reach of the TWD student training programs and provides earlier access to hands-on research and longer periods of research training to the selected students. This program also requires one summer at a research-intensive university.

To improve efficiency, the three TWD programs at Xavier use a common online application process coordinated through the Center for Undergraduate Research and Graduate Opportunity. All program applicants are required to work with the Office of Career Services to complete a basic IDP and submit it with their application materials. The qualified applicants are interviewed by a selection committee comprised of faculty members from the biomedical departments and the Program Directors of the three programs. The selected students are placed in the different programs for which they are the best fit.

It is important to note that all three programs have submitted renewal proposals in 2018. Due to the timing of this publication, the descriptions in this document are based on the current programs and do not include changes proposed in the new cycles. If funded and implemented, information and results from the new funding cycles will be shared with the greater academic community through future publications.

Programs' Components and Implementation

Student interest in biomedical research careers, science identity, self-efficacy, and motivation is measured on entry to Xavier by the HERI CIRP Freshman Survey (as a baseline) and again upon entry into the programs, with a survey that measures interest in biomedical research careers, motivation, science identification, and self-efficacy. Graduating students also complete a final post-test, which captures these same measures. The objective is a significant increase in the different measures from pre- to post-test for the program students.

Entering Research Workshop (Currently BUILD; Going Forward, also RISE)

Program students complete the "Entering Research at Xavier University of Louisiana" (ER XULA) Program modeled after the program developed at the University of Wisconsin-Madison by Drs Branchaw, Pfund, and Rediske (Handelsman, Pfund, Miller Lauffer, & Maidl Pribbenow, 2005; Branchaw, Pfund, & Rediske, 2010). These faculty members served as consultants for Xavier's BUILD Program in 2015, assisting in the implementation of the workshops at Xavier, which have subsequently been adapted to better fit the needs of Xavier students. ER XULA consists of 10 workshops that focus on preparing students to be effective mentees and researchers. The curriculum involves helping students become aware of and acquire a number of soft skills necessary for becoming contributing members of a research group. Topics covered include but are not limited to choosing a mentor (based on students' interests), mentor–mentee communication, searching the literature, reading scientific articles, establishing goals and expectations, defining hypotheses or research questions, and designing scientific experiments. As part of these workshops, the students also learn how to effectively update and follow their IDPs. To increase effectiveness, the first three introductory ER XULA sessions are offered in the spring semester, soon after program students are selected. The other seven modules are covered over seven consecutive weeks during the students' first summer in the programs. Each session is 1.5 hours long.

Grant-writing Workshops (Currently BUILD; Going Forward, also RISE)

At the end of their first summer, program students complete a series of three grant-writing workshops and are provided an opportunity to write at least one small proposal for CURGO funding. These workshops are offered following the ER XULA Program and are delivered by CURGO Staff. The combination of these workshops, the *Science and Technical Writing* course, and the *Science Communication* course (described below) prepares the students for writing small proposals. Each fall semester, CURGO, releases a request for proposals specifically to students. These proposals are to secure funds for research supplies or travel to scientific meetings. To practice and improve their proposal writing skills, program students are expected to submit at least one small proposal while in the programs. Mentors review the proposals and provide feedback before submission.

Intramural Research (BUILD, MARC, RISE)

All program students participate in an intramural research project with an active investigator at Xavier (or another local university as appropriate). It is very important for students to work for an extended period of time with the same research mentor which results in the students becoming truly engaged in the different aspects of their research project and developing a sense of ownership and commitment to it. Students can then have the opportunity to follow changes in the research direction, learn several research protocols, troubleshoot problems, collect their own data, and evaluate the significance of their results. In addition, this engagement results in the development of a true mentor–mentee relationship. An article in the *Council on Undergraduate Research Quarterly* identifies the relationship between students and mentors as most important, stating, "Most minority students who end up being scientists have typically had outstanding mentors. We suggest that establishing strong mentoring programs will be the key to any successful effort in recruiting and retaining minority students in the sciences" (Wubah et al., 2000). Mentors are required to complete the Preparing for Mentoring and Advising at Xavier (P-MAX) Program, as described in detail later, and register with the National Research Mentoring Network (NRMN). All research staff directly working with the program students are also required to complete P-MAX. Program students are also required to register with the NRMN and use the resources provided by that national network.

Progress Reports (BUILD, MARC, RISE)
All program students write annual progress reports on their intramural research projects. These progress reports are technical in nature and include experiments attempted and completed, as well as a complete record of collected data. The program students are required to complete the *Science and Technical Writing* course (described below), which includes instruction on writing technical reports.

Research Poster (BUILD, MARC, RISE)
Each program student presents at least one research poster at a national scientific meeting. The students prepare and submit a research abstract following the conference guidelines. Before submission, the research mentor reviews the abstract and provides input. Poster presentations provide students with experience explaining their research projects to scientists and other students in a more relaxed atmosphere than a formal seminar. Program students also are expected to make poster presentations at Xavier's annual *Festival of Scholars, Summer Research Symposium*, or other local meetings.

Library Research Techniques Workshop (BUILD, MARC, RISE)
All program students receive instruction in library research techniques. Students learn about the organization of the library, basic library search processes, hard copy indices available, various science citation manuals, and select print and online primary and secondary scientific resources. They also learn how to search databases, including American Chemical Society Publications (ACS Web Editions), Medline, Scifinder Scholar, Chemica, Science Direct, and Biological Abstracts online. Xavier librarians provide this instruction. At the conclusion of this training, students complete a bibliographic exercise to measure their ability to search the scientific literature competently. To reinforce these skills, the students are asked to search for background information about their research topics and obtain research papers related to upcoming seminars for the *journal club* meetings. This activity is now wholly supported by University funds and is fully sustainable. Xavier librarians are also available to meet with the students one-on-one as needed to assist them in performing literature searches for their specific research projects.

Extramural Research Experience (BUILD, MARC, RISE)
All program students spend at least one summer at a research-intensive summer program. This experience provides the students with a first-hand perspective of performing research at a large majority institution. An article in the *Council of Undergraduate Research Quarterly* stated, "SROP (summer research opportunity programs) are designed to not only give minority undergraduates valuable research experience; they also allow them to see what life as a graduate student at a CIC (Committee on Institutional Cooperation) institution would be like"

(Foertsch, Alexander, & Penberthy, 1997). These experiences allow participants to be better informed in their decisions regarding graduate school selection and better prepared for graduate work once they enroll. In addition, Xavier's Center for Undergraduate Research and Graduate Opportunity provides students with summer research experience information. Xavier's BUILD Program currently has 17 research-intensive partner institutions. MARC and RISE Programs have also formed relationships with a number of research-intensive universities over the years and are also leveraging the relationship between Xavier and BUILD Partner Institutions to provide more summer research program placement opportunities to their students. By participating in both a long-term intramural research experience and an extramural summer program, the students are provided with a more complete perspective on biomedical research. Note that BUILD students are placed in an external lab where the research is in some way related to their research project at Xavier; the student ideally can carry out experiments that cannot be done at Xavier or can learn a new technique to bring back to their Xavier lab.

Science and Technical Writing Course (ENGL 3001) (BUILD, MARC, RISE)
All program students complete the *Science and Technical Writing* course offered by the English Department at Xavier. Although many undergraduate science majors are technically prepared to pursue advanced graduate degrees, they too often have less than adequate technical writing skills. ENGL 3001 is a one-credit-hour course that meets once each week for one hour and 15 minutes. The course was recently fully revised. Writing topics covered include but are not limited to letters of application, *curriculum vitae*, research summaries or abstracts, article reviews, lab reports, research documentation, and proposals. Grammar and syntax are taught on an "as needed" basis. All writing assignments relate to the students' majors and problems of scientific interest. The course also explores the use of graphics for communicating information. Pre- and post-tests are administered by the Instructor to evaluate the progress of the students. If the pre-test shows a significant deficiency in writing skills, the student is also required to use the services of the Xavier Writing Resource Center. This course is now taught on regular basis and is open to all Xavier students. Writing skills are further reinforced throughout the students' period of participation in the programs through the preparation of research progress reports, research abstracts, a senior thesis, and a short research proposal.

Science Communication Course (CMST 3125) (Currently BUILD; Going Forward, also RISE)
Program students complete the *Science Communication* course offered by the Communication Studies Department. An understanding of and ability to engage in science communication, where scientists present scientific information to expert and non-expert audiences, is key to success in scientific careers, graduate and professional schools and in generating financial and policy support. In this course, students gain the skills to (1) demonstrate the ability to dispense

scientific information in face-to-face, group, and interpersonal settings in long and short formats, (2) deliver in-person and electronically mediated scientific presentations with a measure of poise and confidence, (3) give scientific presentations and respond to questions extemporaneously, (4) understand how language use and style are powerful tools in effectively delivering ideas and that this power comes with ethical, professional, and social responsibility, (5) utilize effective language and style to deliver understandable and persuasive scientific messages and arguments, (6) make use of clearly designed and meaningful visual aids to inform or persuade during presentations and as stand-alone representations of work, (7) analyze and speak to various audiences, and (8) transfer scientific knowledge for the purposes of generating support and informing decision-making. Course assignments include "Basic Scientific Presentation," "Elevator Pitch," "Narrative Research Presentation," "Video News Research Presentation," and "Poster Presentation." The Course is now taught on regular basis and is open to all Xavier students.

Ethical Conduct in Scientific Research (PHIL 3400) (BUILD, MARC, RISE)

All program students complete the *Ethical Conduct in Scientific Research* course offered by the Philosophy Department. The students practice recognizing ethical problems in research and resolving them in a well-reasoned manner. Upon completion of the course, students should be able to (1) identify and explain the prevailing legal and professional norms relevant to resolving ethical problems in research, (2) relate these norms to the broader values they reflect, (3) recognize when an ethical problem in research has arisen and identify the legal and professional norms relevant for resolving it, (4) generate ethically sound, well-reasoned answers to ethical problems in research, both orally and in writing, (5) explain the roles virtue and integrity play in research ethics, and (6) identify and explain practical strategies they can use to develop the character traits and habits of mind characteristic of virtuous researchers. The Course is now taught on regular basis and is open to all Xavier students.

Career Roundtable (BUILD, MARC, RISE)

An annual career roundtable is organized at which students can compare and contrast biomedical research careers in academia, industry, and government. This has been a highly successful approach and one that is easily sustainable. Three biomedical researchers from underrepresented groups are invited to participate in each roundtable. These researchers are asked to give a short presentation on the type of research they perform, their job responsibilities, and how their profession affects their personal life. They are asked to talk to the students about the journey to their current positions, the ups and downs, and the pros and cons. The students are then provided the opportunity to ask questions and have a dialogue with the researchers. The roundtables are widely publicized and open to all students. All program students are expected to participate. At the end of the meeting, the students in attendance are asked to evaluate the roundtable and the individual speakers.

Journal Club (RISE)

Journal club meetings are organized in advance of certain program-sponsored seminars. Students are asked to review and discuss two or more publications by the incoming seminar speaker in order to familiarize themselves with the speaker's research topics and techniques (technical scientific reading comprehension training). Students are informed of the seminar dates and titles in advance and are asked to conduct a literature search for recent articles by the speaker, reinforcing their literature search skills. They are asked to read the articles and be ready to discuss the topic and the scientific techniques used at the journal club meeting (scheduled one week before the seminar). During the meeting, a faculty member with expertise in the field leads the discussion. All program students are expected to participate in the discussion. At the end of the academic year, students are asked to evaluate the journal club meetings and the associated seminars.

Seminars (BUILD, MARC, RISE)

Established biomedical researchers are invited to present research seminars and spend time with students discussing graduate school opportunities and programs. This has been another successful activity which is easily sustainable, especially in that many speakers are eager to recruit Xavier students and pay their own travel expenses. Speakers from academia also have the opportunity to recruit students for the graduate programs at their institutions. Speakers are asked in advance to direct their talks toward undergraduate students and to emphasize the use and importance of various research techniques. They also meet with students to discuss career goals, issues faced by underrepresented groups, and graduate school expectations and environment. While attempting to expose students to a variety of biomedical fields, all efforts are made to choose speakers whose research interests are similar to those expressed by the students. Researchers are invited from a variety of research environments. Many of these researchers are members of underrepresented groups who serve as excellent role models for Xavier students. The seminars are advertised widely. Program students are expected to participate.

Scientific Conferences (BUILD, MARC, RISE)

Program students are required to attend scientific meetings and present their research results while networking with other scientists. In general, many undergraduates are unaware of the larger scientific community and, even when involved in research, are still not aware of how various research projects at different institutions are parts of a puzzle to solve a specific scientific problem. They often have a very myopic view of science. Attending a national scientific meeting is an enlightening experience for any undergraduate student. Attending national meetings such as the Annual Biomedical Research Conference for Minority Students (ABRCMS) is even more impactful, since for many underrepresented minority students, seeing a large number of scientists from underrepresented groups in one location provides a sense of community and belonging.

Research Seminar (Currently BUILD; Going Forward, also MARC and RISE)
All program students are required to present one formal seminar to their departments or at Xavier's annual *Festival of Scholars* in their senior year. *Festival of Scholars* is an on-campus research symposium held in April of each year, where Xavier students present their research outcomes in the form of a poster or oral presentations. Mentors actively help the students prepare for their presentations. The students are evaluated on their performance by their peers and faculty members.

Senior Thesis (BUILD, MARC, RISE)
All program students write a senior thesis. The intention is that these theses are technically complete and well-written. Mentors review the theses and provide feedback, giving the students an opportunity to revise before submitting to the programs for evaluation. All theses include an introduction describing the background and significance of the research problem, an experimental methods section, the data collected, a statement of results, a discussion of the significance of the results, a conclusion, and a bibliography.

Curriculum Enhancement (BUILD and RISE)
The BUILD and RISE programs also focus on strengthening the supportive environment needed for Xavier students to overcome barriers to success through curricular enhancement. Several new courses have been developed with support from these programs including the above-mentioned *Science and Technical Writing*, *Scientific Communication*, and *Ethical Conduct in Scientific Research* courses. These courses are required for the program students, but open to all.

In addition, to increase the retention and success of Xavier students in the biomedical sciences and to assure courses taken by students are providing effective learning experiences, the programs provide funding and support for evaluation and revision/development of science and mathematics courses. Results from curricular assessment instruments are used by the various departments to identify the areas of weakness in the biomedical curricula at Xavier. The programs provide curriculum development mini-grants for faculty to develop new courses or modify existing courses in order to fill gaps in the biomedical curricula at Xavier.

Academic Support (BUILD)
The mission of Xavier's SASO is to improve retention and graduation rates of Xavier students by offering academic support through tutoring, SI, and academic success workshops. These services were originally offered only to students taking introductory courses in biology, chemistry, physics, mathematics, writing, and reading. Using BUILD Program support, SASO has expanded its tutoring services and SI to include sophomore- and junior-level courses. SASO also provides academic monitoring of all students using an alert system through which faculty report any student academic, attendance, or behavioral issues. The alerts are then provided to academic advisors who, in turn, meet with the students to

assist them in overcoming the underlying issues, ideally, before their grades suffer.

Post-baccalaureate Technician Program (BUILD)

An aspect of *Project Pathways* showing early success, the BUILD Technician Program, is a post-baccalaureate research training program which provides recent Xavier graduates with the additional training and research experience needed to successfully enter and complete graduate work. It is not uncommon for senior-level undergraduates to express the first-time interest in graduate school, and this program gives such students the time to become competitive applicants for admission to and completion of biomedical graduate programs. Others joining the Program as BUILD Technicians are students who have applied unsuccessfully to graduate school or who graduate mid-year. BUILD Technicians participate in the Program for approximately one year, during which time they are considered full-time staff employees receiving all staff benefits. Additionally, they participate in BUILD enrichment activities such as mentor training and GRE workshops and attend and present at scientific meetings (Fig. 1).

Preparing Mentors and Advisors at Xavier (BUILD, MARC, RISE)

To improve the quality of undergraduate research mentoring at Xavier, the programs provide training activities for faculty and staff, which educate Xavier research mentors on the challenges faced by their mentees and equip them to provide the necessary support both in and out of the laboratory. Xavier's CAT + FD works closely with the NRMN to present workshops on topics such as cultural awareness, stereotype threat, setting reasonable goals and expectations, and using technology to foster the mentor–mentee relationship. P-MAX is based on the Entering Mentoring seminar developed at the University of Wisconsin Madison by Drs Handelsman, Pfund, Miller Lauffer, and Pribbenow and is designed to provide participating faculty and staff with the knowledge and skills needed to mentor and advise undergraduate students, particularly those engaged in research (Handelsman et al., 2005). The Program addresses topics such as mentor–mentee communication, goal- and expectation-setting, cultural competence, issue identification and resolution, and best practices for good mentoring and advising. Faculty are also provided a repository of resources to support their growth as mentors. P-MAX begins with an intensive, day-long, workshop followed by at least three additional one-hour workshops during each of the subsequent fall and spring semesters. Case studies (found in literature or based on faculty experiences) are routinely used to stimulate discussion on the topic to be addressed, and faculty are encouraged to bring their own real-world experiences for discussion. P-MAX programming is made accessible, relevant, and inviting and is offered in varying formats (panel discussions, hands-on workshops, seminars), at different locations (program locations vary across campus and online) and by expert presenters/facilitators (outside speakers and consultants are frequently brought in to conduct workshops and seminars

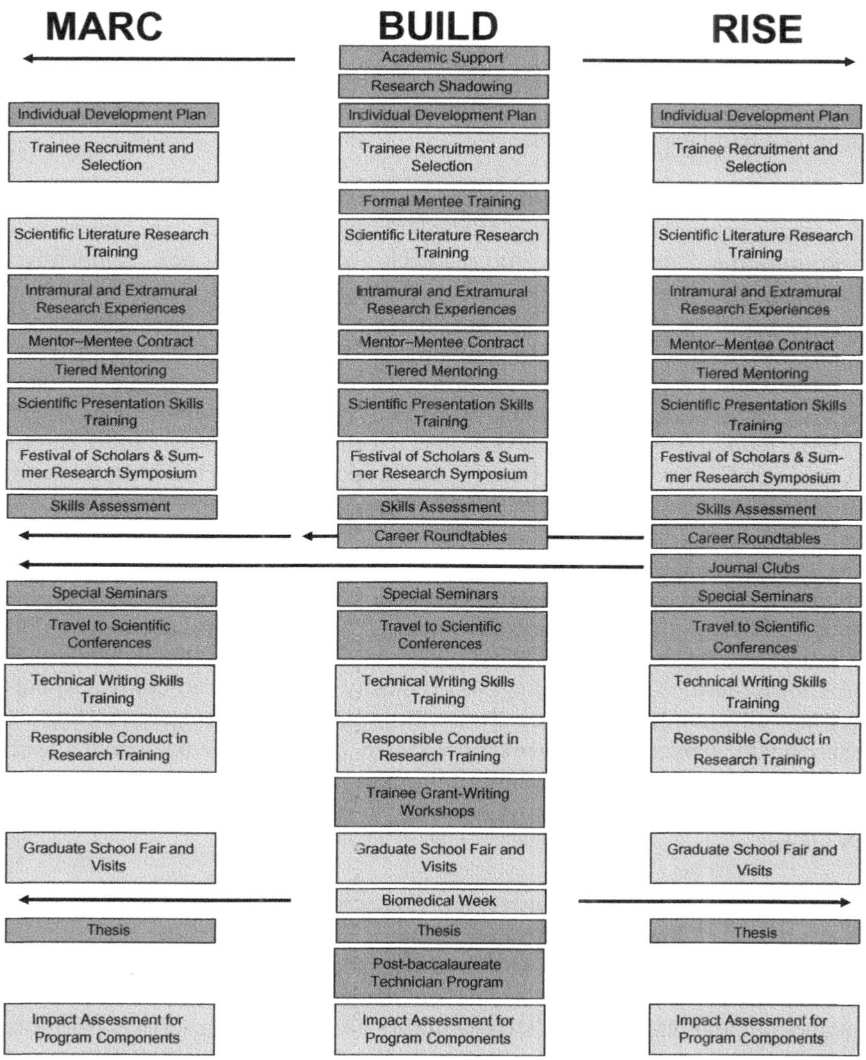

Fig. 1. Leveraging Institutional, BUILD, MARC, and RISE Resources for Trainees. Shown are Selected Trainee Activities from the BUILD, MARC, and RISE Programs. Green Boxes Indicate that the Activity is Provided by that Program, while Gold Boxes Indicate that the Activity is Provided by Shared University and Program Funding through a University Office or Center, such as the Center for the Advancement of Teaching and Faculty Development (CAT + FD), Center for Undergraduate Research and Graduate Opportunity (CURGO) or the Student Academic Success Office (SASO), or an Academic Course. Arrows Indicate Activities that are Utilized by the Program, but Provided by the TWD Program from which the Arrow Originates.

and advice CAT + FD staff). All research staff working directly with undergraduate research students – including the post-baccalaureate technicians – are also expected to participate in the P-MAX workshops. Efforts are underway to take P-MAX online, so that faculty from other institutions who mentor Xavier students can participate in training.

Faculty Research Funding (BUILD)

To assure engagement of Xavier students in quality research experiences, it is also important to have faculty members who are research competitive. Consequently, it is necessary to address the need for junior faculty mentoring and release time, to expand Xavier's collaborative research network, and to expose Xavier's undergraduates to a broader range of research opportunities mentored and guided by faculty at Xavier and at other institutions. Pilot project funding is provided through the BUILD Program to address these needs. *Project Pathways* partners with two other faculty research programs at Xavier (Louisiana Cancer Research Consortium (LCRC) and NIH-funded Research Centers at Minority Institutions (RCMI)) to leverage available funds and provide funding to a larger group of faculty members than would be possible without the partnership. Applicants/awardees fall into one of the following three categories: (1) recently hired faculty who need start-up funding, (2) more experienced investigators with current or recently terminated external grant support who are seeking bridge funds to keep their laboratories functioning while they seek longer-term external funding, or (3) faculty who need seed funding to pursue new directions in their research. Faculty with BUILD Program funding are expected to expand the network of research opportunities by working with Xavier undergraduates seeking research experiences. They are also expected to develop a well-defined student research experience and complete the P-MAX Program.

Research Opportunities for Non-program Students (BUILD)

The BUILD Program provided funding to expand the Center for Undergraduate Research and Graduate Opportunity (CURGO) at Xavier. To provide Xavier students with information about undergraduate research opportunities on- and off-campus, CURGO utilizes an online career services platform, as well as one-on-one support in identifying and applying to research opportunities. CURGO also sponsors student workshops on applying to summer programs, preparing for summer research experiences, and proposal writing. In addition to workshops and seminars, CURGO provides funding to students and faculty for research and travel to conferences.

The Science Education Research Group (BUILD, RISE)

Science Education Research Group (SERG) is an open forum for faculty to meet bi-weekly to bring questions, concerns, or suggestions related to teaching and learning for discussion with other faculty. SERG meetings are informal and multidisciplinary to encourage and support collaboration and communication

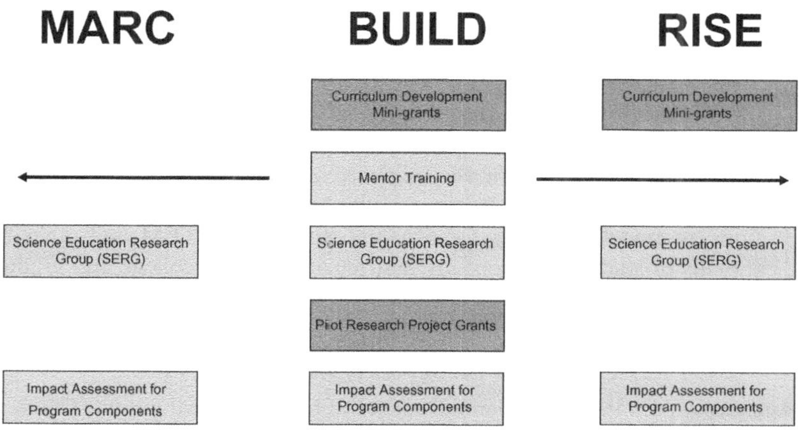

Fig. 2. Leveraging Institutional, BUILD, MARC, and RISE Resources for Faculty. Shown are Selected Faculty Activities from the BUILD, MARC, and RISE Programs. Green Boxes Indicate that the Activity is Provided by that Program, while Gold Boxes Indicate that the Activity is Provided by Shared University and Program Funding through a University Office or Center, such as the Center for the Advancement of Teaching and Faculty Development (CAT + FD), Office of Research and Sponsored Programs (ORSP) or Academic Departments. Arrows Indicate Activities that are Utilized by the Program but Provided by the TWD Program from which the Arrow Originates.

among and between science and non-science faculty. Topics discussed are suggested by the faculty participants, and discussions are facilitated by the *Project Pathways*' Education Improvement Specialist (EIS), who is housed in CAT + FD. The EIS also provides support in the form of topic research and presentation of pedagogical trends. SERG is fully supported by the institution (Fig. 2).

PROGRAM OUTCOMES AND EVOLUTION

Looking at the programs holistically, it is evident that their various initiatives touch every aspect of a student's journey at Xavier from freshman year to graduation and beyond. The strong focus on evaluation makes these projects data-driven experiments, which are expected to lead to the determination of effective practices, which could be shared with the greater academic community with the ultimate goal of effectively increasing diversity in the biomedical workforce. Data on the successfulness of the various initiatives are being collected and analyzed, and will be reported at a future time.

LESSONS LEARNED AND RECOMMENDATIONS

The strategies developed by the TWD programs at Xavier are designed to address the challenges and barriers Xavier students face as they work toward graduate studies and entering the biomedical workforce. Xavier University of Louisiana has a long history of providing high-quality, rigorous education to African American students in a very supportive environment with highly dedicated faculty and staff. The initiatives highlighted here could be used by other institutions as model initiatives for assisting students in STEM and other biomedical fields of study to successfully matriculate through college and graduate school and develop their research careers.

FUNDING ACKNOWLEDGMENT

Research reported in this publication was supported by the National Institute Of General Medical Sciences of the National Institutes of Health under Award Numbers UL1GM118967 8TL4GM118968, RL5GM118966, 5T34GM07716, and 2R25GM060926. The content is solely the responsibility of the authors and does not necessarily represent the official views of the National Institutes of Health.

REFERENCES

American Association of American Medical Colleges, Applicants and Matriculants Data. (2016). *Undergraduate institutions supplying 15 or more Black or African-American applicants to US Medical Schools, 2016–2017* [Data File]. Retrived from https://www.aamc.org/download/321446/data/factstablea2-1.pdf

Baxter, S., Botelho, J., & O'Donnell, K. (2015). *Increasing student success in STEM: A guide to systemic institutional change*. Retrieved from https://www.aacu.org/sites/default/files/files/pkalkeck/casestudies.pdf

Branchaw, J., Pfund, C., & Rediske, R. (2010). Entering research: A facilitator's manual. *Workshop for Students Beginning Research in Science*. New York, NY: W.H. Freeman and Company.

Burrelli, J., & Rapoport, A. (2008, August). Role of HBCUs as baccalaureate-origin institutions of Black S&E doctorate recipients. *National Science Foundation Info Brief*, (NSF 08–319). Retrieved from https://files.eric.ed.gov/fulltext/ED502482.pdf

Butts, G. C., Hurd, Y., Palermo, A.-G. S., Delbrune, D., Saran, S., Zony, C., & Krulwich, T. A. (2012). Role of institutional climate in fostering diversity in biomedical research workforce: A case study. *Mount Sinai Journal of Medicine, New York*, *79*(4), 498–511.

Diverse Issues in Higher Education. (2016). *Top 100 degree producers 2016: baccalaureate* [Data File]. Retrieved from http://diverseeducation.com/top100/BachelorsDegreeProducers2016.php?AppKey=38d31000g2h3a9a0b3f1b9i5j0e0&ComparisonType1_1=%3D&MatchNull1_1=N&school=Xavier+University+of+Louisiana&ComparisonType2_1=%3D&MatchNull2_1=N&state=zip&ComparisonType3_1=%3D&MatchNull3_1=N&major=zip&ComparisonType4_1=%3D&MatchNull4_1=N&race=African+American

Doyle, M. P. (2000). *Academic excellence: The role of research in the physical sciences at undergraduate institutions*. Tucson, AZ: Research Corporation.

Fiegener, M. K., & Proudfoot, S. L. (2013, April). Baccalaureate origins of US-trained S&E doctorate recipients. *National Science Foundation, National Center for Science and Engineering Statistics* (NSF 13–323). Retrieved from https://www.nsf.gov/statistics/infbrief/nsf13323/nsf13323.pdf

Foertsch, J., Alexander, B. B., & Penberthy, D. (1997). *Summer research opportunity programs (SROPs) for minority undergraduates – A longitudinal study of program outcomes, 1986–1996*. Retrieved from http://citeseerx.ist.psu.edu/viewdoc/download?doi=10.1.1.583.5609&rep=rep1&type=pdf

Griffin, K. A., & Hurtado, S. (2010). Institutional diversity in American higher education. In J. H. Schuh, S. R. Jones, & S. R. Harper (Eds.), *Student services: A handbook for the profession* (pp. 24–42). San Francisco, CA: Jossey-Bass.

Hakim, T. M. (2000). *How to develop and administer institutional undergraduate research programs*. Washington, DC: Council on Undergraduate Research.

Handelsman, J., Pfund, C., Miller Lauffer, S., & Maidl Pribbenow, C. (2005). *Entering mentoring: A seminar to train a new generation of scientists*. Madison, WI: University of Wisconsin Press.

Hannah-Jones, N. (2015, September 9). A prescription for more Black doctors: How does Tiny Xavier University in New Orleans manage to send more African-American students to medical school than any other college in the country? *The New York Times Magazine*. Retrieved from https://www.nytimes.com

Jimerson, S. R., & Ferguson, P. (2007). A longitudinal study of grade retention: Academic and behavioral outcomes of retained students through adolescence. *School Psychology Quarterly*, *22*(3), 314–339.

Leonhardt, D. (2017, January 18). America's great working-class colleges. *The New York Times*. Retrieved from https://www.nytimes.com

Longden, B. (2006). An institutional response to changing student expectations and their impact on retention rates. *Journal of Higher Education Policy and Management*, *28*(2), 173–187.

Lopatto, D. (2004). Survey of undergraduate research experiences (SURE): First findings. *Cell Biology Education*, *3*(4), 270–277.

National Science Foundation, National Center for Science and Engineering Statistics. (2015, June 30). *Science and engineering degrees: 1966–2012. Detailed statistical tables NSF 15–326*. Retrieved from https://www.nsf.gov/statistics/2015/nsf15326/

Neuhauser, A., & Cook, L. (2016, May 17). 2016 US News/Raytheon STEM Index shows uptick in hiring, education. *US News and World Report*. Retrieved from https://www.usnews.com/

Pender, M., Marcotte, D. E., Domingo, M. R. S., & Maton, K. I. (2010). The STEM pipeline: The role of summer research experience in minority students' PhD aspirations. *Education Policy Analysis Archives*, *18*(30), 1–36.

Rothman, F. G., & Narum, J. L. (1999). *Then, now, and in the next decade: A commentary on strengthening undergraduate science, mathematics, engineering and technology education*. Washington, DC: Project Kaleidoscope.

Snyder, J. J., Sloane, J. D., Dunk, R. D. P., & Wiles, J. R. (2016). Peer led team learning helps minority students succeed. *PLoS Biology*, *14*(3), 1–7. e1002398. doi:10.1371/journal.pbio.1002398

Tabak, L. A., & Collins, F. S. (2011). Weaving a Richer Tapestry in biomedical science. *Science*, *333*(6045), 940–941.

U.S. Department of Education. (2016, March 16). *Fact sheet: Spurring African American STEM degree completion*. Retrieved from https://www.ed.gov/news/press-releases/fact-sheet-spurring-african-american-stem-degree-completion

Wubah, D., Schaefer, D., Gasparich, G., Brakke, D., McDonald, G., & Downey, D. (2000). Retention of minority students through research. *Council on Undergraduate Research Quarterly*, *20*(3), 120–125.

PART II

CULTIVATING AGENCY THROUGH THE CHEMISTRY AND BIOCHEMISTRY CURRICULUM AT SPELMAN COLLEGE

Leyte L. Winfield, Lisa B. Hibbard,
Kimberly M. Jackson and Shanina Sanders Johnson

ABSTRACT

The racial and ethnic representation of individuals in the workforce is not comparable to that in the general population. In 2010, African Americans constituted 12.6% of the US population. However, African Americans represented less than 5% of PhD recipients in 2010; African American women comprised less than 1% of the degrees awarded in that same year. These disappointing statistics have sparked conversations regarding the retention of underrepresented groups with a focus on what helps to ensure these individuals will transition through the science, technology, engineering, and mathematics (STEM) pipeline. This chapter provides insight into the elements of the Spelman College learning environment that empower women of African descent to become agents of their success while facilitating their movement through the STEM pipeline. The chapter focuses on interventions and resources developed in the Chemistry and Biochemistry Department to foster student-centered learning. Described herein are cocurricular strategies and course-based interventions are used synergistically to enhance student outcomes. The approach to curricular innovation is framed by theories related to community of inquiry (CoI), metacognition, agency, and self-regulated learning. Strategic institutional investments have underpinned these efforts. In addition to providing a snapshot of student outcomes, the authors discuss

Broadening Participation in STEM: Effective Methods, Practices, and Programs
Diversity in Higher Education, Volume 22, 153–181

ISSN: 1479-3644/doi:10.1108/S1479-364420190000022007

lessons learned along with the realities of engaging in this type of intellectual work to elucidate the feasibility of adopting similar strategies at other institutions.

Keywords: Spelman college; HBCU; active learning; flipped classroom; chemistry; women

SPELMAN'S ROLE IN THE BROADER NARRATIVE ON DIVERSITY

Documented achievement discrepancies in the field of science, technology, engineering, and mathematics (STEM) have steered relevant discussions on the retention of underrepresented groups in these disciplines (Harkavy, Cantor, & Burnett, 2015; National Academy of Sciences, 2011). While a growing body of work has characterized the deficits of students of color who struggle in STEM learning environments, narratives of their accomplishments remain scarce. Such narratives are essential to advancing the understanding of minority students' STEM persistence and realizing the national goal of broadening participation in the STEM workforce (Augustine et al., 2010). As scholars continue to investigate the academic performance of individuals from underrepresented groups who excel in STEM and persist in related careers, they often look to minority-serving institutions (MSIs) for strategies leading to diversity, equity, and inclusion in science. MSIs and historically black colleges and universities (HBCUs) have routinely addressed the challenge of educating a diverse society with a degree of impact that has not been fully realized by primarily white institutions. According to Conrad and Gasman, "mainstream institutional models are often indifferent to the needs of a diverse society that includes students from an array of backgrounds and cultures" (2015). Therefore, all institutions can learn how to better support their diverse student populations by looking at models from MSIs and HBCUs.

Institutions such as Spelman College have long been providing culturally relevant learning environments that build social capital, create avenues for academic development, and increase students' sense of agency (Jackson & Winfield, 2014). As the nation's largest producer of African American women who receive STEM doctorates, Spelman is often regarded as the standard for training empowered women of African descent (Estrada et al., 2016; Leggon & Gaines, 2017; Malcom & Malcom, 2011; Palmer, Maramba, & Gasman, 2015). This chapter summarizes efforts in the Chemistry and Biochemistry Department to build upon and expand this legacy of success through the synergy of curricular and cocurricular activities.

SPELMAN COLLEGE INDICATORS OF SUCCESS

Spelman College is located in southwest Atlanta and is a fully accredited, private liberal arts institution dedicated to educating women of African descent.

Founded in 1881, Spelman is the oldest institution established for this purpose. The student body represents 43 states, territories, or military posts and 13 countries. With an annual enrollment of 2,100 students, Spelman College is one of only two HBCUs classified by the Carnegie Foundation for the Advancement of Teaching and Learning as a highly selective and highly competitive Baccalaureate I institution. The college is also one of only four HBCUs in the nation to hold a chapter of Phi Beta Kappa Honor Society. Further, Spelman is the sole HBCU ranked among the Top 75 Best Liberal Arts Colleges by *US News & World Report*, which listed the institution as the best HBCU in America.

Spelman offers a robust, challenging liberal arts curriculum with 31 academic majors, nine of which are in STEM. The college employs approximately 173 full-time faculty members who are dedicated to providing an outstanding interdisciplinary experience that allows students to reach the highest levels of academic, community, and professional achievement (Office of Institutional Effectiveness, 2017). The college has continued to expand its national visibility and increase the scope and vitality of its educational curriculum, particularly in the sciences. The college conferred 463 degrees in 2016, only 99 of which were awarded in STEM disciplines. Spelman has also been recognized as one of six institutions in the country designated by the National Science Foundation (NSF) as a *Model Institutions for Excellence* in undergraduate science and mathematics education. The award was based on the college's significant track record of recruiting, retaining, and graduating African American women in science and for catalyzing their subsequent graduate study and pursuit of professional careers (Lederman, 2008; Thompson & Scriven, 2008). Since receiving the award, the college has sustained its record of accomplishments and currently ranks second among all institutions as the bachelor's degree-granting institution of African Americans who have obtained doctorate degrees in STEM fields (Fiegener & Proudfoot, 2013). Notably, Spelman has one-fourth the student enrollment of Howard University, which currently holds the top spot. As a testament to the college's sustained success in this area, Spelman was also the first HBCU to be awarded the 2014 Carnegie Corporation award for efforts in broadening Spelman's liberal arts education to enhance STEM initiatives. Beverly Tatum (President Emerita of Spelman College) stated:

> Consider the fact that while the 105 public and private HBCUs make up only 3% of today's colleges and universities, more than 20% of all African-American college graduates attended an HBCU. Particularly in the fields of science, technology, engineering and mathematics (STEM), where Black students are woefully under-represented in most predominantly white institutions, HBCUs have demonstrated great effectiveness in fostering academic success. [...] Spelman College is leading the way, having sent more African-Americans (150 women) on to earn PhD degrees in the STEM fields in the ten years between 1997 and 2006 than Georgia Tech (32), Emory (24), Duke (34), and the University of North Carolina Chapel Hill (54) combined. (2010)

MSIs such as Spelman have been recognized for fostering bachelor's degree attainment in general and in STEM specifically (Figueroa, Hurtado, & Eagan, 2013; Ladeji-Osias et al., 2015; Nocnan, Lindong, & Jaitley, 2013; Palmer et al., 2015). The college's record of achievement illustrates the critical role this unique institution plays in preparing students of color for future careers, offering hope for a STEM workforce reflective of the diversity found in the US population.

INSTITUTIONAL CONTEXT

Strategic Plans: Past and Present

Spelman College has a well-articulated mission to sustain the pipeline of talented black women in all fields and engages stakeholders who are committed to ensuring this aim is realized. The college has achieved remarkable success in this regard owing to the strategic utilization of federally funded programmatic efforts. These achievements underscore the impact of an institution seeking to advance a rigorous intellectual culture. In its 2015 strategic plan, the college defined student development using the acronym GOALS: Global experiences for all students, Opportunities for career-related internships and undergraduate research experiences, Alumnae engagement and mentoring, Leadership development, and Service learning through community involvement (Office of the President, 2010). These activities anchor the Spelman My Integrated Learning Experience (MILE), which embeds applied and service learning, teamwork, leadership development, and global experiences into a rigorous liberal arts curriculum (Office of the President, 2010; Office of the Provost, 2010). By coordinating these and other academic initiatives, Spelman is adapting high-impact practices that provide students with essential experiences that promote critical thinking, effective communication, and quantitative reasoning. Opportunities such as research experiences and active, authentic engagement in the learning environment are especially beneficial to STEM majors' development (Kuh & Schneider, 2008; Olson & Riordan, 2012). The current plan (Spelman 2022: Imagine, Invent, Ascend) aims to expand previous efforts and is summarized under the following themes: (1) deliver the Spelman promise of ensuring that every student graduates with a competitive edge, (2) elevate the Spelman difference by anchoring the Spelman narrative in educational excellence and teaching and learning expertise, (3) enhance operational excellence, and (4) promote academic innovation (Office of the President, 2017).

Spelman College has also established the Office of Institutional Research, Planning, and Effectiveness to support its efforts to sustain innovative, data-driven curricula. The office is responsible for, among other activities, providing performance data related to accreditation. In doing so, the office supports the development of measurable student learning objectives. Individuals from the office work directly with departments to develop methods of data collection and guide the use of outcomes to inform curricular practices.

To further support the faculty's ability to respond to the strategic plan, the college established the Teaching Research and Resources Center (TRRC) which provides support for faculty in developing courses and implementing research-based teaching strategies. Assistance is afforded through interactive workshops and technical training. The TRRC also hosts faculty learning communities to encourage collaboration around interdisciplinary projects. As the college has become more focused on the scholarship of teaching and learning as well as discipline-based education research, the TRRC has supported faculty endeavors in these areas by making available experts for consultation on the development of proposals and publications. Another venue that supports curricular innovation is the Faculty Institute. In 2007, the Office of the Provost implemented two-day institutes at the beginning of each semester as a means of sharing faculty initiatives. In addition, the event provided updates related to progress on the strategic plan and interactive breakout sessions that established an agenda for realizing the next stage of the plan.

In 2015, the college began scaffolding metacognition throughout the curriculum by training students on metacognitive skillfulness. These measures have allowed students to integrate the awareness of how one is thinking into their approach to learning. Additionally, the college is assisting faculty to develop and adopt pedagogies that promote sound metacognitive practices. Originally funded through the Department of Education's First in the World Program, the initiative has since been institutionalized in part through the newly established Student Success Center, providing additional support to the department for securing faculty buy-in related to teaching innovations.

The Division of Natural Science and Mathematics

Over the past 10 years, 33% of entering first-year, first-time students have declared majors in STEM disciplines available at Spelman (biochemistry, biology, chemistry, computer and information science, dual degrees in engineering, environmental science, health sciences, mathematics, and physics) with 25% of graduates having received degrees in these disciplines during that same time frame (Office of Communication, 2018; Office of Institutional Effectiveness, 2017). Spelman's STEM departments and programs include 54 full-time faculty, 83% of whom are racial/ethnic minorities and 52% of whom are women. Of the 28 female faculty, 64% are African American. These demographics represent exceptional diversity while providing supportive role models for Spelman students.

Infrastructure developments in the past decade reflect the institution's strong commitment to building an active research environment to sustain innovative science curricular and training resources. The Science Center offers a venue for teaching in modern laboratories with an emphasis on hands-on instruction, exploration, and infusion of technology into every aspect of the curriculum. Classroom spaces have been remodeled to support active learning; rooms now include modular workstations to support group work along with updated technology (e.g., short-throw projectors, computers, and technology ports) at each workstation. More than 50% of the STEM faculty use web-enhanced technology or active learning applications.

With its expanded strategic plan, support of faculty-led activities, and investment in infrastructure, Spelman College has positioned itself to strengthen its capacity for preparing graduates who contribute to a science workforce representative of the national population.

THE DEPARTMENT'S SUCCESS AND EVOLUTION

Until 1977, Spelman and Morehouse Colleges shared a chemistry department, which offered service courses and produced only two Spelman chemistry graduates. Since branching off from its Atlanta University Center neighbor, the Chemistry Department (renamed the Chemistry and Biochemistry Department in 2015) at Spelman College has offered a rigorous, comprehensive, and multifaceted curriculum that prepares students to become future scientists and health professionals. The department is composed of 12 full-time teaching faculty who serve an average of 105 majors annually (Office of Institutional Effectiveness, 2017). From 2006 to 2016, 212 students completed degrees in chemistry and biochemistry (an average of 19.2 graduates per year). Typically, three majors graduate as Phi Beta Kappa, and five graduate with Latin honors. More than half of all graduates have gone on to pursue academic training beyond a bachelor's degree (Table 1), further characterizing the competitiveness of department alumnae.

To ensure the continued production of competitive graduates, departmental faculty have implemented a more progressive chemistry and biochemistry curriculum based on modern pedagogical approaches and student interest. The current chemistry major is certified by the Committee on Professional Training of the American Chemical Society (ACS), which requires the completion of five foundation courses (organic chemistry, analytical chemistry, biochemistry, physical chemistry, and inorganic chemistry), advanced coursework in foundation areas, and 400 hours of lab-based instruction. The biochemistry major is being modified to reflect these recommendations as well. The department also offers a dual-degree option in chemistry and chemical engineering; this course of study allows students to earn a bachelor's degree in chemistry from Spelman College while earning a bachelor's degree in engineering from a partnering institution. Finally, students interested in pursuing careers in the K–12 sector can complete coursework in education and chemistry to earn a degree and be certified in

Table 1. Academic Activities Completed by Students Who Earned a Degree in the Department (1978–2015).

Post-baccalaureate Activities	Total Number of Alumnae (%)
Professional masters (nursing and public health)	9 (3.0%)
Secondary education certification	11 (3.7%)
Professional doctorate (medical, dental, and pharmacy)	29 (9.6%)
STEM masters	34 (11.3%)
STEM doctorate	81 (26.9%)
Other (including JD, MBA, finance, and industry)	137 (45.5%)
Total majors in the department's history	301

Source: Data obtained from department archival information, verified by graduate and professional programs and the National Student Clearing House (2018).

secondary science education. As students navigate these degree options, they engage in the curricular and cocurricular activities illustrated in Fig. 1.

Structured by the aforementioned degree options, the department has broadened its curriculum within the context of a liberal arts education to provide students with the tools needed to become problem-solvers and critical thinkers. Several pedagogical innovations have been combined to promote learning and knowledge retention throughout courses offered in the department. In addition, the department has established broad learning outcomes that are assessed throughout curriculum throughout as follows:

- Students will develop and apply critical thinking and scientific reasoning skills in problem-solving.
- Students will develop and apply chemical techniques while engaged in scientific investigations.
- Students will demonstrate how to search for, analyze, and apply information from scientific literature.
- Students will demonstrate scientific knowledge through effective oral and written communication.

These targeted outcomes motivate all departmental decisions regarding teaching strategies and learning activities. The enhancements described here align with Spelman's agenda for strengthening the academic core and contributes to the college's sustained preeminence as an academic training ground for STEM talent. In a press release, Joyce Davis (Director of Marketing & Communications at Spelman College) wrote:

> As the highest ranked institution from which Black women science and engineering doctorate recipients earn bachelor's degrees, Spelman is an exceptionally strong training ground for women who go on to careers in which they lead and excel in the traditionally male-dominated fields of STEM. This excellence was reflected in the College's ability to attract a broad range of funding, awards and recognitions during the year [2016]. Spelman faculty have built and sustained a strong record of instilling intellectual excellence in Black women, while using the unique formula of a Spelman education – undergraduate research, global studies, creativity, leadership and sisterhood – to transform students into solution-oriented global change agents. (2017)

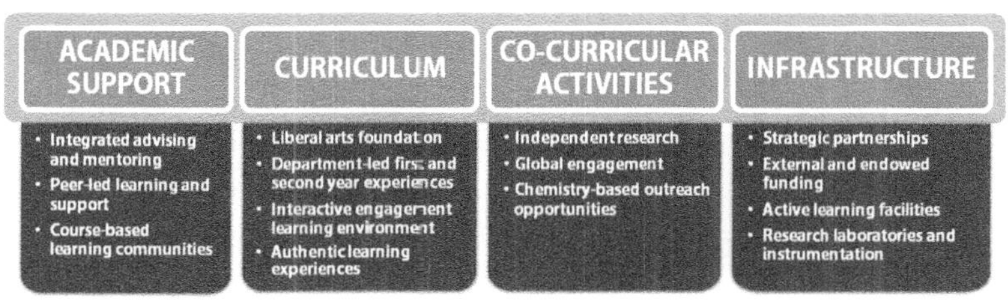

Fig. 1. An Overview of the Chemistry and Biochemistry Department's Learning Environment.

THEORETICAL FRAMEWORK FOR CHANGES IN THE DEPARTMENT OF CHEMISTRY AND BIOCHEMISTRY

Departmental efforts are anchored in student-centered learning strategies that are employed thoughtfully and systematically to enhance learning. Within the science education community, these strategies can encompass various activities including inquiry-, problem-, case-, and team-based learning. Each of these strategies focuses on deeper learning, with students being actively involved and having control over their learning process while encouraging collaboration and building a community of learners. Self-guided control of learning in the department is supported by the community of inquiry (CoI) framework. In addition, success in this environment is thought to be influenced by students' agency and metacognitive regulation. Beyond the classroom, students' science identities and career development are further nurtured through peer-led learning support, independent research, global experiences, and STEM outreach. To this end, work in the department has been informed by the triangulation of theories that define effective learning communities, affective traits, and metacognition (Fig. 2). Additional theories are interwoven as needed to enhance the transportability of the framework (i.e., theories on socially responsible teaching and authentic learning, discussed later in this chapter) from introductory to advanced courses and from lecture to laboratory settings (Kober, 2015).

Community of Inquiry

The CoI model posits that the student learning experience is influenced by teaching, cognitive, and social interactions and is rooted in a constructivist approach to learning (Garrison & Vaughan, 2008; Rahmawati, Ridwan, & Nurbaity, 2017). The components of this model are as follows: (1) *cognitive presence*, defined as the extent to which the participants in any particular CoI configuration can construct meaning through sustained communication, (2) *social* presence, learners' ability to project their personal characteristics into the CoI and thus present themselves as holistic beings, and (3) *teaching* presence, encompassing the design, facilitation, and direction of cognitive and social processes for realizing personally significant and educationally worthwhile learning outcomes. The educational experience occurs at the intersection of each domain. The social

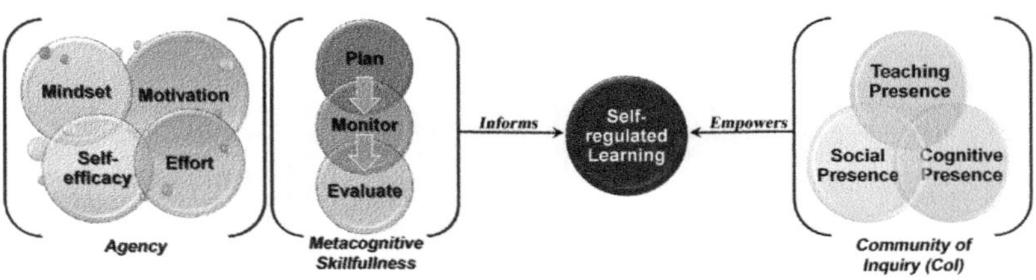

Fig. 2. Framework for Curricular Changes in Chemistry and Biochemistry Department.

and teaching components overlap to establish a climate for learning, the social and cognitive components overlap to support meaningful communication, and regulation of learning occurs at the interface of teaching and cognitive presence. Although this framework was originally designed for online learning, Swan, Garrison, and Richardson (2009) proposed that the framework is a dynamic model in which "constructivist approaches and community are necessary for creating and confirming meaning and are essential for achieving effective critical thinking" (p. 4). Based on this and reports from other scholars (Smiley & Anderson, 2011), CoI was used to guide the Chemistry and Biochemistry Department's efforts to empower self-regulated learning.

Metacognition

Where cognitive engagement involves psychological effort toward mastering knowledge and skills, metacognition is the awareness of and ability to critically assess and modify one's efforts to ensure that mastery (Metcalfe & Shimamura, 1994). The process of metacognition has been critically linked to solving complex problems and understanding concepts such as those in STEM disciplines (Shin, Jonassen, & McGee, 2003; Vidal-Abarca, Mañá, & Gil, 2010; J. R. Wang & Chen, 2014). Metacognition is also believed to be instrumental to the mastery of chemical concepts and in knowledge transfer (Francisco, Nicoll, & Trautmann, 1998; Rickey & Stacy, 2000; Schraw, Crippen, & Hartley, 2006; Tsai, 2001). Since the introduction of metacognition in the psychological and educational literature, many terms have been devised to describe this robust construct. As reviewed in an article by Veenman, Van Hout-Wolters, and Afflerbach (2006) and supported by others, metacognition is often conceptualized as metacognitive knowledge/awareness and metacognitive regulation/skillfulness (Cooper & Sandi-Urena, 2009; Schraw & Dennison, 1994; Tanner, 2012; Veenman et al., 2006). Metacognitive knowledge refers to when, how, and why learners demonstrate particular learning behaviors, whereas metacognitive regulation is considered the application phase of metacognition—namely, the ability to plan, monitor, and evaluate one's learning strategies. The third component of metacognition is the metacognitive experience, which entails the intellectual occurrences that lead to metacognitive knowledge and skillfulness (i.e., failing or passing an exam).

Work in the department centers on metacognitive skillfulness, which is more highly correlated with academic performance than other components of metacognition (Scherer & Tiemann, 2012; J. R. Wang & Chen, 2014; Zhao, Wardeska, McGuire, & Cook, 2014). The level of confidence in one's abilities is key to judging the effectiveness of cognitive strategies, which serve as a reflection of metacognitive skillfulness (Hudlicka, 2005; Mathabathe & Potgieter, 2014; Potgieter, Ackermann, & Fletcher, 2010). Researchers believe that the overestimation of one's impending performance does not hinder successful outcomes and can be properly adjusted with exposure to the material (Mathabathe & Potgieter, 2014; Potgieter et al., 2010). However, negative bias about impending performance and conceptual understanding has been shown to predict failure.

Monitoring, a component of metacognitive skillfulness, has been used to explain differences in students' mental conceptions, visual interpretation of chemical information, and learning gains (Kelly, 2014; Mathabathe & Potgieter, 2014). Intentional instruction on metacognitive regulatory strategies is recommended for those who are less skilled at monitoring learning metacognitively (Mathabathe & Potgieter, 2014). Regarding evaluation, another component of metacognitive skillfulness, researchers have confirmed that a learner's ability to accurately reflect on tasks and related scientific explanation is positively associated with better performance (C. Y. Wang, 2015). Given its correlation with course performance and its actionable nature, metacognitive skillfulness is believed to be a key component of students' success in the Spelman learning environment, particularly when applying the types of learning strategies employed in the department.

Agency

Affective traits, especially a sense of agency, dictate how learners engage in course activities and activate metacognition. Agency captures the extent to which a learner is willing to engage in the learning environment and become an active participant in the process of gaining a specific outcome (Fancsali, Jaffe-Walter, & Dessein, 2013). Being the capacity to act, agency is shaped by the following: (1) a willingness to ***act*** (motivation), (2) a belief in one's ability to ***act*** (efficacy), (3) a belief that one's ***act***ions will lead to intelligence (mindset), and (4) the degree to which one takes ***act***ion (effort). The collective traits that inform agency are associated with higher levels of engagement and commitment to learning (Lawson & Lawson, 2013; Reeve & Tseng, 2011; Toohey & Norton, 2003). Affective traits such as agency influence whether students are able and willing to regulate their approach to learning (J. Martin, 2004; Zimmerman, 1989).

Self-regulated Learning Theory (SLT)

Self-regulation refers to a learner's ability to understand and control learning behaviors to achieve desired academic goals (Schraw et al., 2006). Thus, in self-regulated learning, students can construct circumstances that produce their intended learning goals. Learners who are highly self-regulated are effective at setting goals, implementing appropriate learning strategies, monitoring the success of those strategies, and modifying goals accordingly (Schunk & Zimmerman, 2012). These learners also seek help as needed, an attribute that is reflective of metacognitive skillfulness and activated by the agency. Despite the promise of self-regulation, instructors must realize that although active learning environments promote self-regulation, the self-regulation necessary to be successful is not automatic (Mercer, 2011). The teaching presence component of the CoI model underscores the role of the instructor in this respect.

CREATING A COMMUNITY IN FIRST- AND SECOND-YEAR CHEMISTRY COURSES

The Gateway and Bottleneck

Introductory general chemistry courses, traditionally taken by chemistry and biochemistry majors in their first year and other science and non-science majors in their second year, mark students' first experience grappling with their potential to navigate the STEM curriculum. With the belief that mastery of chemical knowledge shapes one's ability to comprehend other scientific subjects, the challenges encountered in introductory courses often compel students to leave the major. These observations imply that general chemistry courses aim to separate those who are prepared for the rigors of a STEM course of study from those who are not. General chemistry has long been considered the gatekeeper designed to challenge students' readiness for what comes next academically. Likewise, organic chemistry is regarded as the course designed to determine whether a student can continue as a STEM major and persist in a STEM career, particularly those related to health care. The high demand for this course combined with the inherent difficulty and abstract nature of its concepts results in low success rates, creating a bottleneck in the curriculum. Gateway and bottleneck courses have typically served to reduce the number of STEM majors, not just chemistry majors, at many institutions. Given the centrality of chemistry to other sciences and the need to improve diversity in STEM, faculty in the department are reshaping the curriculum to assist students in overcoming the hurdles of the first- and second-year curriculum. Targeting students' emerging scientific thinking at this level should translate into a skill set that will serve students well in subsequent science courses (Smith, Chungh, Sadouq, & Kandiah, 2015; Tai, Sadler, & Loehr, 2005). This section details strategies to address students' developmental needs at these critical junctures.

Connecting with Majors through Structured Cohorts

Traditionally, most students enrolled in general chemistry courses include sophomore biology majors and others interested in health-related careers. In the 1990s and early 2000s, during which the number of STEM majors at Spelman increased substantially, the department noticed that first-year chemistry and biochemistry majors were not connecting with each other and often did not participate in class to the same extent as the more mature and college-savvy upperclass students. Their limited engagement was reflected in the fact that attrition in the major was typically between 30% and 40% after the first year. The observation that majors were being "lost in the crowd" emphasized the need to establish a chemistry/biochemistry learning community.

To support the retention of majors in the general chemistry course, the department one of its three General Chemistry I sections for majors. Beginning in 2009, chemistry and biochemistry majors were required to enroll in this section while other majors were permitted to take the course as space allowed. The initial intent of the majors-only general chemistry sequence was to ensure that first-year majors would move through the sequence as a cohort while receiving targeted curricular interventions. Based on the curricular and pedagogical revisions in the general chemistry sequence, an organic chemistry majors-only course was created in 2010.

PEDAGOGICAL INNOVATIONS IN FIRST-AND SECOND-YEAR LECTURE COURSES

In 2012, the general and organic chemistry majors-only courses adopted interactive engagement pedagogies with the goal of promoting subject mastery in the learning community created by the structured cohorts. Revisions were based on the CoI framework, with learning occurring at the intersection of online content delivery, face-to-face engagement, and enhanced peer−peer/peer−instructor social interactions (Garrison & Vaughan, 2008). Within this framework, the faculty facilitates discourse and course engagement that encourage students to take responsibility for their learning. Student preparation for face-to-face class meetings is an essential aspect of self-regulated learning; hence, the majors-only section was reduced from four to three days per week of face-to-face engagement, providing students additional time to engage with resources outside of the class. Activities completed during and outside of the formal lecture period support self-guided learning, Fig. 3. Class time is largely devoted to team-based, skill-building activities. The instructor sets the tone for discussions and serves as a resource to answer individual or team questions rather than simply lecturing, although instructor-facilitated discussions take place as needed. Due to the level of active learning in the classroom, participation is considered a necessary feature of the learning process and is evaluated daily via individual or team-based assessments, fulfilling the social component of CoI.

	General Chemistry	Organic Chemistry
Pre-class Activity **Deliver Content**	• Textbook reading assignments • Narrated lectures online	• Moodle lessons • Instructors notes • Narrated PowerPoint lectures • Online quizzes
In-class Activity **Build Skills**	• Process oriented guided inquiry learning (POGIL) • iPad simulations • In-class puzzlers and problem-solving Demonstrations • Team-based case studies • Discussion and Q&A	• Discussion and Q&A • ChemDrills© – Computational modeling – Analogy – Problem drills – Guided inquiry
Post-class Activity **Reinforce & Assess**	• Gated exams (Gates) • Quizzes • Rubric graded assignments • Concept maps • Standardized final exam • Reflections	• Online discussion and Q&A • Traditional Exams & Quizzes • Progressive Knowledge Assessments • Concept Maps • Reflections

Fig. 3. Overview of Learning Activities in General and Organic Chemistry Lectures.

General Chemistry-specific Innovations

A collection of novel learning resources was created to support this teaching methodology in general chemistry courses, many of which are technology-enhanced (i.e., case studies, mini-projects, and online lectures). In addition, more sophisticated just-in-time assessment techniques have been infused throughout the general chemistry sequence. To reinforce learning, students use an online adaptive learning platform tied to their textbook. Daily low-stakes assignments (either from the textbook publisher, such as McGraw-Hill's adaptive learning platform LearnSmart, or from instructor-developed assessments) based on previously covered concepts allow students to identify which concepts they understand and which require additional review. Higher-stakes assessments measuring concept mastery include individual and team-based activities inside or outside of class. The assessment cycle of preparation–participation–performance is meant to promote student engagement throughout the learning process.

General chemistry follows a semi-self-paced, subject-mastery model governed by individual chapter exams, "gates" (Hibbard, Sung, & Wells, 2015; Wells, 2014). In this iterative and sequential assessment strategy, students are allowed up to four attempts to earn a minimum score of 85 on a chapter's gate before proceeding to the next gate. Gated tests are made available once approximately half of the chapter material is covered. The assessment closes one week after associated chapter content has been covered. This method provides students who have mastered chapter concepts the opportunity to move ahead by taking tests at self-selected times. Students who are struggling with the material have more time to correct their misconceptions and learn from their mistakes, whether through their efforts or with the assistance of the instructor and/or tutors (Chemistry Learning Apprentices described later in this chapter). Students will then be better prepared to reattempt the gate. Questions change for each test iteration, but the concepts covered remain the same.

Organic Chemistry-specific Innovations

In organic chemistry, peer-led, inquiry-based, and computational exercises are used to guide classroom discussions (Fullilove, Hibbard, & Winfield, 2017; Winfield, 2015). These multifaceted exercises, copyrighted as ChemDrills, were developed at Spelman to add structure to group and class engagement. The activities are supported by iPad applications such as iSpartan (Winfield, McCormack, & Shaw, in press), ChemOffice for iPad, and Explain Everything. Mediated learning strategies are extended beyond the classroom through online resources (e.g., video lectures, structured reading assignments, and virtual office hours).

Skill-building workshops were developed for the Organic Chemistry II course to support self-guided learning outside of the classroom. This intervention mirrors a structured study group. Workshops sections (limited to 10 students), led by a student who previously completed the course, are assigned by instructors, and the assignments are maintained throughout the semester. Students enrolled in the Organic Chemistry II course are required to attend 9 of the 11 workshops offered during the semester. During the workshop, the leader provides a brief overview of the topic, and then students work in small groups to complete a worksheet-based activity. Students are given pre- and post-assessments during each session to assess their levels of understanding and reveal self-reported gains in knowledge. In 2015 and 2016, students' knowledge increased by 24% on average following the completion of each workshop; students' confidence increased by 31% on average. These data suggest that the workshops are successful in providing students with immediate feedback on their level of content comprehension while serving as a resource to promote skill mastery.

Outcomes of Innovations Introduced in First- and Second-year Lecture Courses

Since establishing the majors-only general chemistry cohort, a five-year evaluation indicated that students in this class perform better than those taught using a traditional lecture format as measured by improved performance on the ACS standardized exams (Hibbard et al., 2015; Wells, 2014). The majors-only course sequence has also led to improved retention in the major compared to past years, although this trend appears directly related to data such as incoming math placement and preparation. Similar to the results observed for general chemistry, those remaining in the cohort through organic chemistry performed comparably to those in the non-majors' course, but demonstrated a higher sense of agency and peer engagement compared to those in the non-majors' course (Winfield, 2015; Winfield & Hibbard, 2015). These strategies characterize what has come to be known as *community-based learning*.

The majors-only courses have consistently revealed that students involved in these cohorts demonstrate a high degree of peer interaction within and outside the classroom. From the instructor's vantage point, majors interact more with their peers by learning essential team skills and have more one-on-one interaction with the instructor; these features promote student confidence and motivation to succeed. Students' efforts in these course arrangements allow them to demonstrate more responsibility and commitment to completing coursework. In addition, students demonstrate concept mastery as evidenced by improved performance on comprehensive standardized exams and assessments in upper-level courses (i.e., physical chemistry). Despite their success, the strategies described here tend to run counter to students' expectations in general and organic chemistry courses. Students have interpreted the method of content delivery (i.e., online lectures) as "teaching oneself." It is believed that this incorrect perception can be overcome by thoughtfully acclimating students to the required out-of-class activities.

MERGING CULTURAL RELEVANCE WITH AUTHENTIC LEARNING

The term "culturally relevant teaching" was coined by Gloria Ladson-Billings to describe the meaningful infusion of students' cultural references into learning experiences (1995). The model has been framed as a way to raise engagement and scholastic accomplishments of minority students and bridge achievement gaps (Gay, 2010; Malter, 2016; Stroder, 2008). It may also encourage the formation of professional identity and has been cited as an underlying motivation for learning, specifically for organic chemists (Bhattacharyya, 2008). Even at a historically black women's college such as Spelman, STEM teaching materials are largely constructed without regard for students' culture, background, or language. Socially responsible curricula complement culturally relevant theories by providing students with the tools to sustain themselves and their communities. This philosophy emphasizes the need for scientists to consider how their work benefits or harms the community at large. Most students are taught the ethics of conducting science, but the larger implications of science outside the classroom (i.e., macroethics) may not be explained. Advocates of socially responsible science suggest it should be a part of science literacy and student training (Bird, 2014). Therefore, the Chemistry and Biochemistry Department is leveraging theories of cultural relevance and social responsibility to reinforce students' individuality and nurture their emerging science identity.

To support the implementation of a curriculum that emphasizes cultural relevance, traditional verification or cookbook labs are being replaced by authentic learning experiences. Such activities engage students more deeply in the scientific process. The three most prevalent strategies are guided inquiry (discovery), inquiry, and project-based lab formats (DIPs as referred to by Kimberly Jackson).

- *Discovery or guided inquiry-based labs* incorporate critical thinking into the course-based laboratory experience because instructors do not inform students of the expected outcome of an experiment. In this method, the instructor guides students through experiences to meet selected learning objectives. Students are granted a certain degree of independence in conducting experiment, so ideas can be explored and evaluated. The level of inquiry can also be adjusted to suit student and instructor expertise. This approach has been found to foster student engagement and interest (Chatterjee, Williamson, McCann, & Peck, 2009; Coppola, 2015; Dockery & Potts, 2014; Dunlap & Martin, 2012). The discovery model is also credited with encouraging millennial students to apply the scientific method and develop critical thinking skills. As the experimental outcome is unknown, students must analyze the corresponding procedure and methods to develop a conclusion.
- *Inquiry-based labs* require students to design all or part of a procedure with little instructor guidance. Students formulate the problem, connect the investigation to previous work, state the purpose, predict the result, and perform the investigation. Students take genuine ownership through this inductive approach (Chatterjee et al., 2009), which cultivates critical thinking and a positive attitude toward chemistry.
- *Project-based labs* involve multi-week activities that closely mimic an independent research experience; thus, these labs are common in Organic Chemistry II and higher-level courses (Demare & Regla, 2012; Flynn & Biggs, 2012;

Woods, 2014). In the purest form of this approach, students choose the project and methods. Modified projects may involve an instructor-chosen topic that is either derived from the literature or aligns with the instructor's research interests. The topic typically has real-world applications while being appropriate for the curriculum.

These three approaches offer structure through which contemporary societal problems can be explored in the chemistry curriculum. They give students greater ownership over laboratory activities and expand the open-ended nature of experimental tasks. More importantly, these methods expose students to advanced techniques as they implement authentic research in laboratory classes. In recent years, the department has sought to ensure that the chemistry we teach reflects the chemistry we do. The shift in the academy that recognizes how students learn and that understands teaching strategies that promote the learning of science substantiates the transition from traditional cookbook or validation laboratories to DIPs. These approaches have been integrated into departmental laboratory courses to provide spaces to conduct authentic research and, in some cases, offer alternative research experience to the conventional internship model.

The DIP-based Biochemistry Laboratory

In 2013, the first DIPs were introduced into the biochemistry laboratory curriculum. The course, intended for biochemistry majors and biology minors, allows students to experience the excitement and challenges of authentic research while increasing their mastery of biochemical techniques. The laboratory is a one-credit, four-contact-hour course that enrolls a maximum of 16 students each spring semester. Approximately two-thirds of students taking the course are juniors, the remaining third are seniors, and all have completed the four-semester sequence of general and organic chemistry along with one semester each of introductory cell biology and biochemistry. Because the laboratory covers a broad range of topics from mammalian cell cultures to molecular genetics, the laboratory proceeds independently of the corresponding lecture course. The instructor selects fundamental assays (e.g., cell signaling, and enzymology) to frame course activities for the semester. The 15-week laboratory experience allows students to (1) develop skills for critical analysis of research results and scientific literature, (2) provide training in technical writing, and (3) engage in the learning of laboratory skills while collecting data related to real-world issues.

Due to the rigor of research projects, limiting the class size to 16 students support impactful learning. A research question and several learning objectives are established for each weekly experiment. Students incorporate a culturally relevant *New York Times* bestseller as the primary text for the semester (e.g., *The Immortal Life of Henrietta Lacks*; *Wit*, a play about a college professor dying of cancer; or *How Cancer Crossed the Color Line)*. Core biochemistry topics (e.g., buffers, statistics, spectroscopy, and protein chemistry) are covered regardless of the project. One example course project was related to cancer therapeutics;

students examined the treatment of triple-negative breast cancer cells, prostate cancer, and lymphoma cells using natural products. Students chose from three natural products to treat various human carcinoma cells. They learned cell-free and cell-based techniques such as mammalian cell culture, cell proliferation assays, viability studies, drug dilutions, fluorescence and light microscopy, cell morphology, cell lysate preparation, protein quantification using a standard curve (statistics), spectroscopy, flow cytometry, electrophoresis, and Western blotting. Although the laboratory officially meets once a week for four hours, student pairs often conduct supplemental experiments such as cell culture media changes outside of scheduled laboratory periods. The additional laboratory work is considered homework and is monitored by a faculty member or graduate teaching assistant. Culminating course activities include the generation of a technical chapter and group presentations at Spelman College's Annual Research Day, a five-minute oral defense, and a class presentation on StoryCorps® (located at the Atlanta History Center). The presentations, in particular, provided opportunities for students to discuss their projects in various professional settings.

The implementation of DIPs into the biochemistry laboratory course has been well received by students. Over the years, free responses from student course evaluations have included the following: *The lab helped me understand the materials we learned about in class"* and "*I really enjoyed the 'different' lab. Usually biochemistry labs are merely validation labs; however, I feel that this lab helped me understand more about the physiology of the body.* Several students have reported receiving offers from summer research programs, graduate schools, and post-baccalaureate programs at top-tier universities thanks to their experience in this course.

EXTENDING THE CURRICULUM BEYOND THE CLASSROOM

Independent Research

Independent research is the traditional form of authentic learning in the undergraduate environment. The educational benefits of undergraduate research are essential to student development and predict the persistence of underrepresented minorities in STEM majors (Merisotis & Kee, 2006; Schultz et al., 2011). Spelman College believes undergraduate research is key to connecting knowledge across disciplines and developing lifelong learning habits. In addition, it is believed that conducting authentic research develops students' confidence by affording them the ability to contribute to solutions for complex, real-world problems.

In 2012, the department established a plan to scaffold research-based skills throughout the curriculum. The first- and second-year seminar courses offer formal and informal opportunities to explore research and careers related to the chemical sciences. In the first year, majors are introduced to research methods and chemistry-specific lab skills through the general chemistry laboratory course. Students receive an overview of research expectations through the First-year Seminar in Chemistry, which is followed and complemented by the discipline-specific Sophomore-year Experience. In the second-year course, students are introduced to scientific writing, public speaking, and research methods; tasks include oral presentations, deconstruction of journal articles, and the development of a one-page project proposal. Students gain lab skills through various laboratory and independent research projects. In the case of the latter, students are required to maintain laboratory notebooks, provide biweekly progress reports, and develop an extended abstract of their work. Students who make significant progress have presented at professional meetings, developed and published manuscripts with their supervising faculty member, and developed an honors thesis based on the work.

Student research activities at Spelman have been supported through various mechanisms (e.g., federal programs, industry partners, and private donors). These opportunities are extended to all students who meet the criteria of the program of interest. Funded by an anonymous donor and supplemented by support from Corning Incorporated, the department established the Chemistry and Biochemistry Scholars Program which expands the slate of research training opportunities available at the college. The program provides support for students to conduct research and present their findings at professional meetings. Students apply annually but can participate in the program for multiple years if accepted. To apply for the program, interested students submit a proposal drafted with the assistance of their research advisor. Success in the program is determined based on students' progress reports and academic standing along with the research advisor's evaluation of the student. Since 2013, this program has supported 18 students and 8 faculty advisors. Students receive a stipend and travel support and faculty receive funding for supplies or a modest stipend and travel funds. Of the 18 Chemistry and Biochemistry scholars, 14 have graduated and are now either pursuing graduate education in STEM or have joined the STEM workforce.

Curricular Support

The primary focus of traditional tutoring or supplemental instruction is remediation (D. C. Martin & Arendale, 1992; Topping, 1996) in that such programs often focus on acute problems with short-term goals in mind (i.e., an upcoming test or assignment). The Chemistry Learning Center (CLC) was established in 2012 to go beyond remediation by providing students the following:

(1) an introduction to fundamental concepts before the class discussion;
(2) reinforcement of strategies for approaching traditionally challenging topics and those requiring more in-depth exploration; and
(3) tools to empower students to take ownership of their learning success.

The CLC supports learning outside of the classroom by providing resources that are interactive and promote "learning by doing." This strategy mirrors the active learning environment described for first- and second-year chemistry courses.

When students come to the CLC, they receive focused reinforcement of skills identified by the instructor as necessary for mastery of a specific concept. In addition, the CLC offers strategies that enhance students' preparedness for upcoming lectures and summarizes key points covered in blended classroom activities.

Undergraduate students serving as *Chemistry Learning Apprentices* (*CLAs*) lead sessions in the CLC, assisting peers in navigating the content of general and organic chemistry courses. CLAs provide feedback to instructors regarding students' general course concerns and the most commonly reported challenges. Additional CLA duties include moderating an online Q&A database and assisting with the development of study resources to provide additional practice. Administrative responsibilities of student apprentices include maintaining student visitation logs and submitting biweekly reports detailing activities completed in the CLC. The latter provides instructors with an additional source of information regarding students' problem-solving approaches and their attitudes toward strategies employed in the course. This wide range of tasks provides CLAs a developmental opportunity that supports their growth as scientists.

The establishment of the CLC was preceded by the ASPIRE Peer-facilitated Learning Program (APFL, established in 2007 as a component of the NSF's Historically Black Colleges and Universities-Undergraduate Programs, HBCU-UP, Implementation Award #0714553) and supplemental instruction, established under the NSF's Model Institutions of Excellence effort (Thompson & Scriven, 2008). Each was developed to enhance critical thinking and lifelong learning through peer-led instruction. The APFL program included an additional goal of incorporating informatics-based technologies and concepts into the curriculum. Like these programs, the CLC supports an interactive and collaborative academic environment that guides students through pre-designed activities that enhance their mastery of a particular concept. Ultimately, the focus is on empowering students to be proactive and responsible for their success.

Global Experiences

The challenges facing our nation are influenced by forces beyond the borders of the contiguous United States. A multicultural perspective is necessary for the identification of solutions that will shape future societies. Therefore, the college has implemented a curriculum that expands the classroom to locations abroad through its Quality Improvement Plan (QEP), "Spelman Going Global." Established as a component of accreditation from the Southern Association of Colleges and Schools Commission on Colleges, Spelman's QEP calls for a curriculum that extends the learning environment to locations abroad. To sustain this effort, the college created the Gordon-Zeto Center for Global Education, which provides international opportunities to faculty and students. The center establishes and manages partnerships that facilitate global student and faculty activities. In addition, the Center oversees key pre-departure activities including support for obtaining passports and visas, coordinating vaccinations and travel insurance, and providing required pre-departure courses and orientations.

In providing developmental opportunities to faculty and students, the center offers a series of short-term study travel trips that emphasize the interdisciplinary exploration of academic environments and cultures. Led by two faculty, each trip hosts 30 rising juniors and seniors across all majors. The two-week trips begin the Sunday following the spring semester final exams and conclude by Labor Day, leaving students the full summer to complete internships or summer study. The study-travel experiences were developed primarily through a partnership with Spelman College and the Council on Internal Educational Exchange (CIEE). Funding for faculty participation is provided by the college, and students' costs are largely subsidized by the same. While CIEE-sponsored trips are non-discipline-specific, other trips have been sponsored by the Halle Foundation and the Spelman Health Careers Program that focus on science and technology and global health issues, respectively.

The short-term international travel experiences described above are believed to help students develop skills that will promote their evolution as professionals who (1) have a respect for other cultures and languages, (2) demonstrate diverse scientific perspectives when engaging in societal narratives, and (3) are knowledgeable about international scientific capabilities and the global need for scientists (Jackson, Jeilani, Sanders, Thompson, & Winfield, 2015). Historically, STEM majors have found participating in a full-semester study abroad program to be difficult due to the demands of the curriculum and the limited number of courses they are allowed to take off-campus. Therefore, study travel trips provide a chance for students to fulfill global engagement requirements while staying on track with their studies. Since 2011, 64 chemistry and biochemistry majors have participated in the courses (Gordon-Zeto Center for Global Education, 2018), and four department faculty have led courses during this time. Beyond the short-term experiences, faculty in the department have collaborated with the Spelman College Global STEM initiative (a STEM-specific arm of the Gordon-Zeto Center) to establish partnerships in France, Belgium, and England (McCormack et al., 2017). The initiative allows students to complete an 8- to 10-week research experience in laboratories abroad while being co-mentored by a Spelman faculty and a researcher at the host facility. Students who participate in college-sponsored opportunities are required to complete journal reflections, surveys, and other assessment activities unique to the activity (i.e., presenting outcomes at a professional meeting or Research Day).

Outreach and Professional Development

The Spelman College Chemistry Club is a student chapter affiliate of the ACS. The club, whose mission is to advocate for science by promoting education and awareness of chemistry, has amassed several national awards from the society over the last ten years. Club members are involved in professional development and outreach events in the Atlanta area. Students often earn community service hours through these activities, and through service, are able to reinforce core chemistry and biochemistry concepts. Some of the significant activities in which the club has participated include the Super STEM Expo (hosted by the Girl Scouts of Greater Atlanta) and the Atlanta Science Festival, both of which host thousands of participants. Club members operate booths at these events, performing interactive chemistry experiments and distributing STEM reading materials to K–12 participants. Graduate School Prep is a component of the chemistry club that provides internship opportunities, graduate lab tours, and workshops for students interested in graduate and professional schools. Graduate students from Emory University and the Georgia Institute of Technology lead these sessions at Spelman College. Through faculty partnership, the chemistry club is also affiliated with the Center for C-H Functionalization, which provides summer research experiences for undergraduates through its 18 member institutions located in the United States. Since partnering with CCHF in 2014, four Spelman students have been offered internships.

LESSONS LEARNED AND THE REALITIES OF IMPLEMENTATION

Time, Resources, and Expertise

Out of a desire to see students take ownership of their learning experience, a small group of faculty spearheaded the efforts that are reported in this chapter. The process was a grand undertaking, with some faculty requesting sabbatical leave to revise entire courses and others taking steps to modify courses incrementally while teaching a full load. Increasing faculty advocacy for novel approaches was essential for ensuring the expansion of these innovations. An important means of convincing faculty colleagues of the benefits of self-regulated learning involved communicating a detailed analysis of the positive effects the efforts have had on student progress.

Ensuring that individuals involved in this type of work possess the requisite expertise is another aspect that is critical when implementing novel strategies. To this end, training opportunities related to active learning strategies have been offered within the department and on campus via the TRRC. Additionally, faculty have traveled to conferences and workshops to stay abreast of new pedagogical developments. Through rigorous training and practice, some faculty have become highly competent in areas and routinely operate workshops and training sessions for faculty nationally.

Classroom structures can act to either impede or facilitate active learning classroom environments. As such, facilities that support pedagogical advancements must also be considered. Faculty have been intentional in collaborating to obtain external funding and have leveraged institutional resources to secure necessary resources (i.e., state-of-the-art laboratory instrumentation), as many of the innovations described here have been resource intensive. In recognition of this, Spelman has transformed several classrooms into spaces that promote collaboration and teamwork.

Select the Right Data

Longitudinal programmatic assessments should not be utilized to measure the efficacy of an intervention such as that described in this chapter. Although metrics such as retention data may be useful for determining the impact of a four-year curriculum, such assessments are influenced by factors beyond the control of an individual course. For instance, retention trends for students majoring in chemistry or biochemistry exhibited little to no change (Fig. 4). Utilizing such data to determine the success of a course modification would imply that the change does not harm student retention; trends in student retention before the formation of major cohorts, 2008 and 2009, were comparable with post-implementation, 2010–2016 (Office of Institutional Effectiveness, 2018). In addition, overall course success may not provide the type of meaningful information indicative of an intervention's success. Like data on retention of majors, course pass rates can be analyzed to ensure that implemented interventions are not adversely impacting student progress. These data do, however, demonstrate the role of general chemistry (the first-semester course in particular) as a gatekeeper course and that of organic chemistry as a bottleneck course, Fig. 5 (Office of the Registrar, 2018). As a juxtaposition to retention and course outcomes data, the department is defining the quality of student performance (e.g., meta-analysis of students' problem-solving, approaches, and pre- and post-knowledge assessments) to characterize the overall success of the intervention (Fullilove et al., 2017; Hibbard et al., 2015; Winfield & Hibbard, 2015; Winfield & McCormack, 2018). Students' affective traits and attitudes toward learning are also monitored. Such data have shown a 9.4% increase in students'

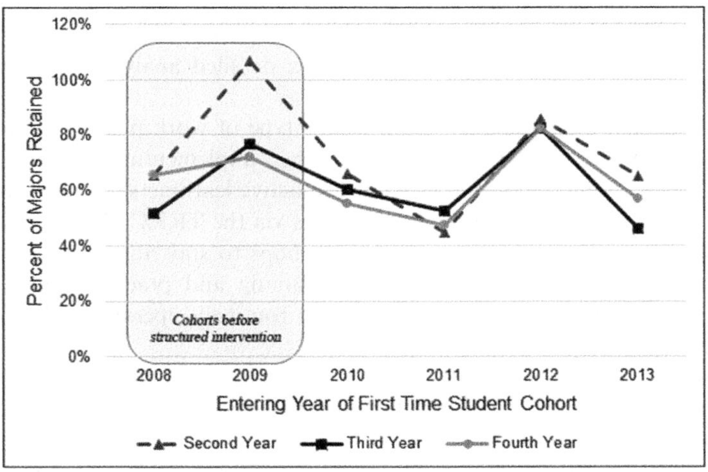

Fig. 4. Retention Trends for First-year, First-time Biochemistry and Chemistry Majors (2008–2013). *Notes*: Change in the composition of majors entering class was determined based on the number of entering first-year, first-time majors that remained as majors in the second, third, and fourth years of study.

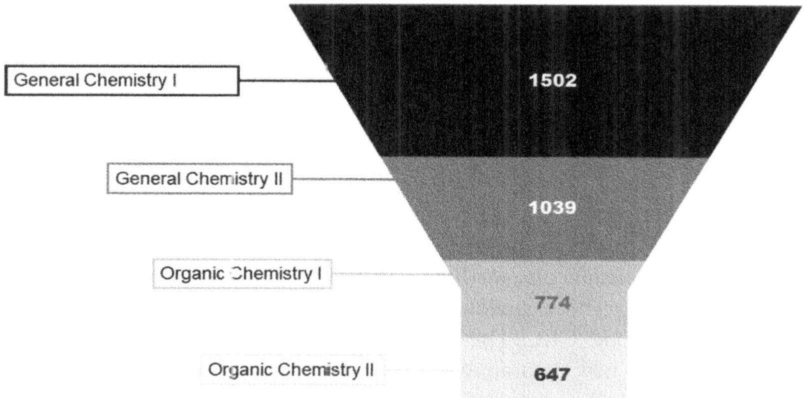

Fig. 5. Visualizing General Chemistry as the Gatekeeper an Organic Chemistry as the Bottleneck. Average Enrollment across All Sections of a Given Course Fall 2008–Spring 2018.

self-determination for students matriculating through three semesters of the structured cohort (completing the majors-only sections of General Chemistry I and II and Organic Chemistry I).

Consider the Student Population

Student push-back was identified as a challenge when implementing the interactive engagement in general and organic chemistry, as students expected a more traditional style of instruction. In response to this commentary, the TRRC has hosted several forums and workshops for faculty to discuss strategies for altering students' understanding of how learning occurs. Overcoming student resistance has also been facilitated by institutional support for these efforts.

While Spelman has successfully implemented strategies into courses with enrollments that are considered relatively small in comparison with that at other institutions, faculty have collaborated with colleagues at larger institutions (e.g., organic chemistry faculty at Georgia State University, general chemistry faculty at Georgia Institute of Technology, and organic chemistry and biochemistry faculty at New York University) to examine how chemistry classes enrolling large groups of students can successfully implement these strategies (Mooring, Mitchell, & Burrows, 2016). Although the scale of a given intervention or strategy may differ, methods are available to actively engage students whether in a class of 30 or 300 (e.g., online polling question responses, nearest-neighbor pairing in problem-solving, and online adaptive learning/and homework). When considering interventions, faculty should carefully determine the amount, extent, and frequency of implementation.

FUTURE PERSPECTIVES

The Chemistry and Biochemistry Department at Spelman College is continually revising and improving its curricular offerings to ensure that each student is holistically prepared for life as a scientist. Students continue to enter college with diverse academic backgrounds, yet are faced with the same expectation of excellence. It is critical that student cohorts matriculating through these courses be monitored and assessed to measure knowledge retention over time and to maintain students in the major. The department is, therefore, vigilant in nurturing student accomplishments, working closely with the college's Student Success Program to identify those students needing additional academic support and guidance outside of the CLC. The college is also focused on enhancing academic advising by using a new data analytics platform that allows advisors to monitor student progress in all of their courses and to request intervention meetings as needed. The Arch system provides long-term analytics to identify the points at which students are most likely to be challenged in moving forward in the curriculum. To bring these initiatives together, we must continue to train and mentor new faculty as they enter the department to sustain a culture that intentionally trains competitive black women in STEM.

The department is dedicated to sustaining a relevant and rigorous curriculum to fit the needs of Spelman's students. Our core pursuit is the cultivation of a community of student-scholars in an academic environment that is culturally relevant and socially responsive. Departmental activities are dynamic, and the curricular modifications are ongoing.

CONCLUSION

Through forums such as the Faculty Institute and TRRC, Spelman faculty have been encouraged to adopt student-centered course engagement and research-based pedagogies. In some cases, the college has offered incentives (i.e., stipends) for relevant professional development and commitment to implementation. Although the initial impetus behind these approaches was related to accreditation, faculty have since self-selected to utilize proven methods with an understanding of the implications for Spelman's ability to sustain academic relevance. Additionally, faculty engaged in these activities have served as advocates of proven teaching strategies.

In the Chemistry and Biochemistry Department, multiple theories are being leveraged to create academic strategies that nurture agency, model self-regulated learning behaviors, and allow students to manage their progress toward becoming independent learners. Departmental faculty have increased the use of instructor-developed resources to guide students through comprehending and applying chemical concepts inside and outside of the classroom. Many now use online polling software for quizzing purposes in order to provide students with immediate feedback on their level of comprehension. While the adaptation of these strategies may depend on the specific instructor, curricular changes have been substantiated due to their positive impacts on student performance and learning.

ACKNOWLEDGMENTS

Lecture activities reported herein have been funded in part by grants from the NSF: the HBCU-UP Targeted Infusion Project Award No. HRD-1332575 and the Improving Undergraduate STEM Education: Education and Human Resources Award No. 1626002. Funding was also provided, in part, by the Howard Hughes Medical Institute Award No. 52007559. Lab activities have been funded, in part, through the Associated Colleges of the South Grant Program and the Proctor & Gamble Higher Education Grant Program. Corporate funding was also provided by Corning, Incorporated. Additional funding and support were provided by Spelman College through the Teaching Research and Resources Center and the Gordon-Zeto Center for Global Education.

The authors are appreciative to Dr Myra Burnett, Vice President of Institutional Effectiveness, for reviewing the initial chapter draft. The authors would also like to thank Lakesha Stevenson (Internal Evaluation Services), Dimeji Togunde and Rokhaya Falls (Gordon-Zeto Center for Global Education), John Brown (Registrar), and James Sanders (Office of Institutional Effectiveness) for providing a portion of the data reported here.

The department is grateful to the anonymous donor whose generous support produced the endowment for Chemistry and Biochemistry Scholars.

REFERENCES

Augustine, N. R., Barrett, C., Cassell, G., Grasmick, N., Holliday, C., Hackson, S., & Murray, C. (2010). *Rising above the gathering storm, revisited: Rapidly approaching category 5* (N. A. Press Ed.). Washington, DC: National Academy of Sciences, National Academy of Engineering, Institute of Medicine.

Bhattacharyya, G. (2008). Who am I? What am I doing here? Professional identity and the epistemic development of organic chemists. *Chemical Education Research and Practice*, *9*(2), 84–92. doi:10.1039/b806222f

Bird, S. J. (2014). Socially responsible science is more than "good science". *Journal of Microbiology and Biology Education*, *15*(2), 169–172. doi:10.1128/jmbe.v15i2.870

Chatterjee, S., Williamson, V. M., McCann, K., & Peck, M. L. (2009). Surveying students' attitudes and perceptions toward guided-inquiry and open-inquiry laboratories. *Journal of Chemical Education*, *86*(12), 1427–1432. doi:10.1021/ED086P1427

Conrad, C., & Gasman, M. (2015). *Educating a diverse nation: Lessons from minority-serving institutions*. Boston, MA: Harvard University Press.

Cooper, M. M., & Sandi-Urena, S. (2009). Design and validation of an instrument to assess metacognitive skillfulness in chemistry problem solving. *Journal of Chemical Education*, *86*(2), 240–245.

Coppola, B. P. (2015). Do real work, not homework. In J. García-Martínez & E. Serrano-Torregrosa (Eds.), *Chemistry education: Best practices, opportunities and trends* (pp. 203–258). Weinheim: Wiley-VCH Verlag GmbH & Co. KGaA.

Davis, J. (2017). *Robust 2016 funding fuels Spelman's rigorous research, programs, and initiatives focused on STEM* [Press release]. Retrieved from http://www.spelman.edu/about-us/news-and-events/news-releases/2017/01/04/robust-2016-funding-fuels-spelman-s-rigorous-research-programs-and-initiatives-focused-on-stem

Demare, P., & Regla, I. (2012). Synthesis of two local anesthetics from toluene: An organic multistep synthesis in a project-oriented laboratory course. *Journal of Chemical Education*, *89*(1), 147–149. doi:10.1021/ed100838a

Dockery, C. R., & Potts, G. E. (2014). Millennial students and undergraduate chemistry. *Addressing the Millennial Student in Undergraduate Chemistry*, *1180*, 1–10. doi:10.1021/bk-2014-1180.ch001

Dunlap, N., & Martin, L. J. (2012). Discovery-based labs for organic chemistry: Overview and effectiveness. *Advances in Teaching Organic Chemistry*, *1108*, 1–11. doi:10.1021/bk-2012-1108.ch001

Estrada, M., Burnett, M., Campbell, A. G., Campbell, P. B., Denetclaw, W. F., Gutiérrez, C. G., ... McGee, R. (2016). Improving underrepresented minority student persistence in STEM. *CBE – Life Sciences Education*, *15*(3), es5. doi:10.1187/cbe.16-01-0038

Fancsali, C., Jaffe-Walter, R., & Dessein, L. (2013). *Student agency practices in the middle shift learning networks, 3*. Retrieved from https://www.impaqint.com/sites/default/files/files/Student%20Agency%20Practices%20in%20the%20Middle%20Shift%20Learning%20Networks_0.pdf

Fiegener, M. K., & Proudfoot, S. L. (2013). *Baccalaureate origins of U.S.-trained S&E doctorate recipients (NSF-13–123)*. Retrieved from http://www.nsf.gov/statistics/infbrief/nsf13323/

Figueroa, T., Hurtado, S., & Eagan, K. (2013). *Making it! [...] Or not: Institutional contexts and biomedical degree attainment*. Retrieved from http://www.heri.ucla.edu/nih/downloads/AIR-2013-Biomed-Completion.pdf

Flynn, A. B., & Biggs, R. (2012). The development and implementation of a problem-based learning format in a fourth-year undergraduate synthetic organic and medicinal chemistry laboratory course. *Journal of Chemical Education*, *89*(1), 52–57. doi:10.1021/ed101041n

Francisco, J. S., Nicoll, G., & Trautmann, M. (1998). Integrating multiple teaching methods into a general chemistry classroom. *Journal of Chemical Education*, *75*(2), 210–213. doi:10.1021/ed075p210

Fullilove, F., Hibbard, L., & Winfield, L. L. (2017). *Engineering course success through interactive engagement*. Paper presented at the Teaching a New Generation of Students: A National Symposium, New York, NY.

Garrison, D. R., & Vaughan, N. D. (2008). *Blended learning in higher education: Framework, principles, and guidelines*. Hoboken, NJ: John Wiley & Sons.

Gay, G. (2010). *Culturally responsive teaching: Theory, research, and practice*. New York, NY: Teachers College Press.

Gordon-Zeto Center for Global Education. (2018). *Institutional data on global travel*. Atlanta, GA: Spelman College.

Harkavy, I., Cantor, N., & Burnett, M. (2015). Realizing STEM equity and diversity through higher education-community engagement. *Netter Center for Community Partnerships Supported White Paper*, 1–52. Retrieved from https://www.nettercenter.upenn.edu/sites/default/files/Realizing_STEM_Equity_Through_Higher_Education_Community_Engagement_Final_Report_2015.pdf. doi:10.1073/pnas.1211286109

Hibbard, L., Sung, S., & Wells, B. (2015). Examining the effectiveness of a semi-self-paced flipped learning format in a college general chemistry sequence. *Journal of Chemical Education*, *93*(1), 24–30. doi:10.1021/acs.jchemed.5b00592

Hudlicka, E. (2005). *Modeling interactions between metacognition and emotion in a cognitive architecture*. Paper presented at the AAAI Spring Symposium: Metacognition in Computation, Menlo Park, CA.

Jackson, K., Jeilani, Y., Sanders, S., Thompson, A., & Winfield, L. (2015). *Promoting global engagement in chemistry*. Paper presented at The Global Imperative for Higher Education: A National Symposium, San Juan, PR.

Jackson, K., & Winfield, L. L. (2014). Realigning the crooked room: Spelman claims a space for African American women in STEM. *Peer Review*, *16*(2), 9–12.

Kelly, R. M. (2014). Using variation theory with metacognitive monitoring to develop insights into how students learn from molecular visualizations. *Journal of Chemical Education*, *91*(8), 1152–1161. doi:10.1021/ed500182g

Kober, N. (2015). *Reaching students: What research says about effective instruction in undergraduate science and engineering* (978-0-309–30043-8). Retrieved from https://www.nap.edu/catalog/18687/reaching-students-what-research-says-about-effective-instruction-in-undergraduate

Kuh, G. D., & Schneider, C. G. (2008). *High-impact educational practices: What they are, who has access to them, and why they matter*. Retrieved from https://keycenter.unca.edu/sites/default/files/aacu_high_impact_2008_final.pdf

Ladeji-Osias, J., Hohmann, C., Hargett, S., Brown, L., Hughes-Darden, C., & Reece, M. (2015). *The impact of undergraduate research in STEM at Morgan State University on the production of doctoral degrees in engineering and the sciences* (Vol. 17). West Yorkshire: Emerald Group Publishing.

Ladson-Billings, G. (1995). Toward a theory of culturally relevant pedagogy. *American Educational Research Journal*, *32*(3), 465–491. doi:10.3102/00028312032003465

Lawson, M. A., & Lawson, H. A. (2013). New conceptual frameworks for student engagement research, policy, and practice. *Review of Educational Research*, *83*(3), 432–479. doi:10.1177/1476718X10366727

Lederman, D. (2008). Who produces black PhD's. *Inside Higher Ed*. Retrieved from https://www.insidehighered.com/news/2008/09/02/phds

Leggon, C. B., & Gaines, M. S. (2017). *STEM and social justice: Teaching and learning in diverse settings, a global perspective*. Cham: Springer International Publishing.

Malcom, L. E., & Malcom, S. M. (2011). Reflecting and defying tradition: African-American women in STEM. *Beyond stock stories and folktales: African Americans' paths to STEM fields* (pp. 131–147). Bingley: Emerald Publishing.

Malter, Z. (2016). The big secret of culturally relevant teaching. *American Youth Policy Forum*. Retrieved from http://www.aypf.org/competency-and-performance-based-education/the-big-secret-of-culturally-relevant-teaching/

Martin, D. C., & Arendale, D. R. (1992). *Supplemental instruction: Improving first-year student success in high-risk courses* (Vol. 7). Columbia, SC: National Resource Center for the Freshman Year Experience, University of South Carolina.

Martin, J. (2004). Self-regulated learning, social cognitive theory, and agency. *Educational Psychologist*, *39*(2), 135–145.

Mathabathe, K. C., & Potgieter, M. (2014). Metacognitive monitoring and learning gain in foundation chemistry. *Chemistry Education Research and Practice*, *15*(1), 94–104. doi:10.1039/c3rp00119a

McCormack, K., Togunde, D., Leggon, C. B., de Brito Galvão, C., Clay, K. E., & Burnett, M. (2017). Enhancing global research and education (G-STEM) at Spelman College. *STEM and social justice: Teaching and learning in diverse settings, a global perspective* (pp. 53–74). Cham: Springer International Publishing.

Mercer, S. (2011). Understanding learner agency as a complex dynamic system. *System*, *39*(4), 427–436. doi:10.1016/j.system.2011.08.001

Merisotis, J. P., & Kee, A. M. (2006). A model of success: The model institutions for excellence program's decade of leadership in STEM education. *Journal of Hispanic Higher Education*, *5*(3), 288–308. doi:10.1177/1538192706288422

Metcalfe, J. E., & Shimamura, A. P. (1994). *Metacognition: Knowing about knowing*. Boston, MA: The MIT Press.

Mooring, S. R., Mitchell, C. E., & Burrows, N. L. (2016). Evaluation of a flipped, large-enrollment organic chemistry course on student attitude and achievement. *Journal of Chemical Education*, *93*(12), 1972–1983. doi:10.1021/acs.jchemed.6b00367

National Academy of Sciences. (2011). *Expanding underrepresented minority participation: America's science and technology talent at the crossroads*. Washington, DC: National Academies Press.

Noonan, A., Lindong, I., & Jaitley, V. N. (2013). The role of historically black colleges and universities in training the health care workforce. *American Journal of Public Health*, *103*(3), 412–415. doi:10.2105/AJPH.2012.300726

Office of Communication. (2018). *STEM at Spelman*. Retrieved from https://www.spelman.edu/academics/stem-at-spelman

Office of Institutional Effectiveness. (2017). *Fact book series*. Atlanta, GA: Spelman College.

Office of Institutional Effectiveness. (2018). *Institutional aggregated data on student retention in the major*. Atlanta, GA: Spelman College.

Office of the President. (2010). *Strengthening the core: Strategic plan for 2015, an overview*.

Office of the President. (2017). *Spelman 2022: Imagine, invent, ascend*. Retrieved from https://www.spelman.edu/about-us/office-of-the-president/strategic-planning/strategic-plan-2022

Office of the Provost. (2010). *The Spelman MILE – My integrated learning experience*. Retrieved from https://www.spelman.edu/academics/spelman-mile

Office of the Registrar. (2018). *Institutional aggregated data on course success rates*. Atlanta, GA: Spelman College.

Olson, S., & Riordan, D. G. (2012). *Engage to excel: Producing one million additional college graduates with degrees in science, technology, engineering, and mathematics*. Washington, DC: The President's Council of Advisors on Science and Technology.

Palmer, R. T., Maramba, D. C., & Gasman, M. (2015). *Fostering success of ethnic and racial minorities in STEM: The role of minority serving institutions* (Vol. 56). Abingdon: Routledge.

Potgieter, M., Ackermann, M., & Fletcher, L. (2010). Inaccuracy of self-evaluation as additional variable for prediction of students at risk of failing first-year chemistry. *Chemistry Education Research and Practice*, *11*(1), 17–24. doi:10.1039/C001042C

Rahmawati, Y., Ridwan, A., & Nurbaity. (2017). *Should we learn culture in chemistry classroom? Integration ethnochemistry in culturally responsive teaching*. Paper presented at the AIP Conference, Vladivostok, Russia.

Reeve, J., & Tseng, C.-M. (2011). Agency as a fourth aspect of students' engagement during learning activities. *Contemporary Educational Psychology*, *36*(4), 257–267. doi:10.1016/j.cedpsych.2011.05.002

Rickey, D., & Stacy, A. M. (2000). The role of metacognition in learning chemistry. *Journal of Chemical Education*, *77*(7), 915–920. doi:10.1021/ed077p915

Scherer, R., & Tiemann, R. (2012). Factors of problem-solving competency in a virtual chemistry environment: The role of metacognitive knowledge about strategies. *Computers & Education*, *59*(4), 1199–1214. doi:10.1016/j.compedu.2012.05.020

Schraw, G., Crippen, K. J., & Hartley, K. (2006). Promoting self-regulation in science education: Metacognition as part of a broader perspective on learning. *Research in Science Education*, *36*(1–2), 111–139.

Schraw, G., & Dennison, R. S. (1994). Assessing metacognitive awareness. *Contemporary Educational Psychology*, *19*(4), 460–475. doi:10.1006/ceps.1994.1033

Schultz, P. W., Hernandez, P. R., Woodcock, A., Estrada, M., Chance, R. C., Aguilar, M., & Serpe, R. T. (2011). Patching the pipeline: Reducing educational disparities in the sciences through minority training programs. *Educational Evaluation and Policy Analysis*, *33*(1), 95–114. doi:10.3102/0162373710392371

Schunk, D. H., & Zimmerman, B. J. (2012). *Motivation and self-regulated learning: Theory, research, and applications*. Abingdon: Routledge.

Shin, N., Jonassen, D. H., & McGee, S. (2003). Predictors of well-structured and ill-structured problem solving in an astronomy simulation. *Journal of Research in Science Teaching*, *40*(1), 6–33. doi:10.1002/tea.10058

Smiley, W., & Anderson, R. (2011). Measuring students' cognitive engagement on assessment tests: A confirmatory factor analysis of the short form of the cognitive engagement scale. *Research and Practice in Assessment*, *6*, 17–28.

Smith, J. R., Chungh, M., Sadouq, S., & Kandiah, A. (2015). *Enhancing the chemistry experience of undergraduate pharmacy students*. Paper presented at the International Education Conference.

Stroder, M. (2008). *Effects of culturally responsive teaching practices on the literacy learning of Latino students*. (Masters), Western Kentucky University, Bowling Green, KY. Retrieved from https://digitalcommons.wku.edu/cgi/viewcontent.cgi?referer=https://scholar.google.com/&httpsredir=1&article=1028&context=theses

Student Tracker for Educational Organizations (online database). Retrieved December 19, 2018, from National Student Clearinghouse https://studentclearinghouse.org

Swan, K., Garrison, D., & Richardson, J. (2009). A constructivist approach to online learning: The Community of Inquiry framework. *Information technology and constructivism in higher education: Progressive learning frameworks* (pp. 43–57). Hershey, PA: IGI Global.

Tai, R. H., Sadler, P. M., & Loehr, J. F. (2005). Factors influencing success in introductory college chemistry. *Journal of Research in Science Teaching*, *42*(9), 987–1012. doi:10.1002/tea.20082

Tanner, K. D. (2012). Promoting student metacognition. *CBE Life Science Education*, *11*(2), 113–120. doi:10.1187/cbe.12-03-0033

Tatum, B. D. (2010). Why historically black colleges remain relevant. *Huffington Post Blog, 2016* (May 24, 2010). Retrieved from http://www.huffingtonpost.com/beverly-daniel-tatum/why-historically-black-co_b_471067.html

Thompson, A., & Scriven, O. (2008). African American women in science: Spelman College as a national model for baccalaureate degree production. *Models for Success: Successful Academic Models for Increasing the Pipeline of Black and Hispanic Students in STEM Areas* (3rd ed.). Brooklyn, NY: Thurgood Marshall College Fund, Inc.

Toohey, K., & Norton, B. (2003). Learner autonomy as agency in sociocultural settings. In D. Palfreyman & R. C. Smith (Eds.), *Learner autonomy across cultures: Language education perspectives* (pp. 58–72). New York, NY: Palgrave Macmillan.

Topping, K. J. (1996). The effectiveness of peer tutoring in further and higher education: A typology and review of the literature. *Higher Education, 32*(3), 321–345. doi:10.1007/BF00138870

Tsai, C. C. (2001). A review and discussion of epistemological commitments, metacognition, and critical thinking with suggestions on their enhancement in internet-assisted chemistry classrooms. *Journal of Chemical Education, 78*(7), 970–974. doi:10.1021/ed078p970

Veenman, M. V., Van Hout-Wolters, B. H., & Afflerbach, P. (2006). Metacognition and learning: Conceptual and methodological considerations. *Metacognition and Learning, 1*(1), 3–14. doi:10.1007/s11409-006-6893-0

Vidal-Abarca, E., Mañá, A., & Gil, L. (2010). Individual differences for self-regulating task-oriented reading activities. *Journal of Educational Psychology, 102*(4), 817–826. doi:10.1037/a0020062

Wang, C. Y. (2015). Scaffolding middle school students' construction of scientific explanations: Comparing a cognitive versus a metacognitive evaluation approach. *International Journal of Science Education, 37*(2), 237–271. doi:10.1080/09500693.2014.979378

Wang, J. R., & Chen, S. F. (2014). Exploring mediating effect of metacognitive awareness on comprehension of science texts through structural equation modeling analysis. *Journal of Research in Science Teaching, 51*(2), 175–191. doi:10.1002/tea.21131

Wells, B. (2014). *Chemistry and education: Instructional and assessment strategies to improve knowledge retention.* Paper presented at the Proceedings of The National Conference on Undergraduate Research (NCUR) 2014.

Winfield, L. (2015). *Community-based interactive engagement in an organic chemistry course.* Paper presented at the ICERI 2015: 8th International Conference of Education, Research and Innovation, Seville, Spain.

Winfield, L., & Hibbard, L. (2015). *Lessons from chemists on improving learning behaviors and outcomes.* Paper presented at the SoTL Commons Conference, Savannah, GA. https://digitalcommons.georgiasouthern.edu/sotlcommons/SoTL/2015/63

Winfield, L., & McCormack, K. (2018). *Stimulating visual literacy and comprehension of alkane conformations using iSpartan* (Manuscript in Preparation).

Winfield, L., McCormack, K., & Shaw, T. (2018). Using iSpartan to support a student-centered activity on alkane conformations. *Journal of Chemical Education.* doi:10.1021/acs.jchemed.8b00145

Woods, D. R. (2014). Problem-oriented learning, problem-based learning, problem-based synthesis, process oriented guided inquiry learning, peer-led team learning, model-eliciting activities, and project-based learning: What is best for you. *Industrial & Engineering Chemistry Research, 53*(13), 5337–5354. doi:10.1021/ie401202k

Zhao, N., Wardeska, J. G., McGuire, S. Y., & Cook, E. (2014). Metacognition: An effective tool to promote success in college science learning. *Journal of College Science Teaching, 43*(4), 48–54.

Zimmerman, B. J. (1989). Models of self-regulated learning and academic achievement. *Self-regulated learning and academic achievement.* In B. J. Zimmerman & D. H. Schunk (Eds.), Self-Regulated Learning and Academic Achievement: Theory, Research, and Practice (pp. 1-25). New York, NY: Springer New York.

HIGH-IMPACT EDUCATIONAL PRACTICES THAT PROMOTE STUDENT ACHIEVEMENT IN STEM

Angela W. Peters, Verlie A. Tisdale and Derrick J. Swinton

ABSTRACT

Findings within the last decade reveal a core set of activities that have been correlated to student success metrics such as persistence, retention, and graduation (Kuh, 2008). These research-based activities are called high-impact practices (HIPs). Students who have participated in HIPs have shown gains in retention, in persistence, intellectually and in an overall positive college experience. This chapter provides an overview of 10 HIPs and their importance and benefits to underserved students, that is, first-generation college students, low-income college students, and underrepresented students of color such as African American, Latino/a, and Native American. Findings within the chapter also recognize how HIPs can be extremely beneficial for historically Black colleges and universities to build capacity and to ensure student success, particularly in science, technology, engineering, and mathematics (STEM).

Keywords: High-impact practices; science, technology, engineering, and mathematics (STEM); learning communities; retention; graduation; underrepresented minorities

---

Broadening Participation in STEM: Effective Methods, Practices, and Programs
Diversity in Higher Education, Volume 22, 183–196

ISSN: 1479-3644/doi:10.1108/S1479-364420190000022008

HIGH-IMPACT EDUCATION PRACTICES: OVERVIEW

According to the Association of American Colleges and Universities' 2007 report *College Learning for the New Global Century* (AAC&U, 2007), and the National Survey of Student Engagement 2017 report *High-impact Practices* (NSSE, 2017), there are a core set of "effective educational practices" of valuable importance that are prevalent in higher education. Identified as "high-impact" practices (HIPs), these core activities contribute greatly to student success in college and include the following (Table 1): learning communities, service learning, undergraduate research, first-year seminar and experiences, capstone courses, internships, common intellectual experiences, writing intensive courses/projects, collaborative assignments/projects, diversity, and global learning. These practices are thought to lead to higher levels of student performance

Table 1. High-impact Practices (HIPs).

HIPs	Descriptions
Learning communities	Common courses, common readings, and integration of learning across courses involve students with thought process beyond the classroom
Service learning, community-based learning	Field experiences with community partners give students direct experience with issues studied in class; real-world learning and reflect in classroom. Connected to the curriculum and part of experiential learning
Undergraduate research	Engage students in discipline-specific research that promotes active learning and contributes to discussion, collaboration, presentations, and possibly publications
First-year seminars and experiences	First-year seminars or courses that promote critical thinking, writing, information literacy, collaborative learning and introduce students to campus life and infrastructure. Small groups of students with high faculty interaction
Capstone courses & projects	Culminating experiences at the end of the student's college years, project, presentation, and report that integrates and applies what they learned throughout college
Internships	Provides direct experience in a work setting, related to career interests, supervised by mentor, coach, or professional in the field. Can be project, paper, or for credit. Includes cooperative education and part of experiential learning
Common intellectual experiences	Required common courses. Broad themes within curriculum and could be vertically organized through general education. May also include curricular and cocurricular options
Writing-intensive courses	Courses emphasize writing across the curriculum and provide revisions, feedback, and include writing as a significant part of course
Collaborative assignments and projects	Study groups within a course, team-based project/assignments, collaborative projects/assignments. Combines learning to work, solving problems, and listening to insights of others
Diversity/global learning	Includes study abroad and study away. Explores cultures, life experiences, worldviews, and diversity

than traditional classroom experiences (Brownell & Swaner, 2010; Swaner & Brownell, 2009). This is consistent with the claim made by Kuh (2008), and Sweat, Jones, Han, and Wolfgram (2013), that the outcomes of HIPs include higher grades, higher persistence rates, higher retention rates, intellectual gains, increased interaction with faculty and peers, and an overall positive college experience particularly during the first year. These results equate to deeper learning (moving students beyond memorization), improvements in general education skills (communication, quantitative reasoning), improvements in practical skills (problem-solving, use of technology, work-related skills), and improvements in personal and social development (Kuh, 2008).

Whereas 10 activities have been identified as HIPs, Kuh and O'Donnell (2013) found that these practices share eight important elements:

(1) high-performance expectations that are clearly communicated;
(2) significant investment of time and effort by students over an extended period of time;
(3) interactions with faculty and peers about substantive matters;
(4) experiences with diversity and people they are not familiar with;
(5) frequent, timely, and constructive feedback;
(6) periodic, structured opportunities to reflect and integrate learning;
(7) real-world application of learning;
(8) public demonstration of competence.

IMPORTANCE OF HIPS FOR UNDERSERVED STUDENTS

Activities such as service learning, learning communities, undergraduate research, and bridge programs are not new practices and have been proven as highly successful for the advancement of underrepresented minorities (Urban Institute, 2006, 2011). For example, the National Science Foundation (NSF)-funded Louis Stokes Alliance for Minority Participation and the historically Black colleges and universities undergraduate program (HBCU-UP) has contributed thousands of underrepresented minorities to the science, engineering, technology, and mathematics (STEM) workforce by supporting activities such as bridge programs, service learning, learning communities, and undergraduate research. According to the Urban Institute reports, both programs have a track record for engaging undergraduate students in research that ultimately motivates them to pursue graduate degrees in STEM.

The national focus on college completion rates threatens to overshadow the importance of ensuring that all students have access to high-quality learning experiences (Finley & McNair, 2013). Therefore, the implementation of sound, research-based strategies that recognize the benefits of diversity can be extremely beneficial to HBCUs to build capacity and to ensure student success. Critically important is recognizing that the use of research-based practices can improve academic achievement for all students. Campuses need a diverse repertoire of approaches to teaching and engaging today's students, particularly underserved students (i.e., first-generation college students, low-income college students, and underrepresented students of color such as African American, Latino/a, and Native American).

Even though these practices seem to be widespread in higher education, they are not mainstream in many institutions. Despite the abundance of evidence to support the fact that HIPs promote student achievement and learning outside of the classroom, these practices only serve a fraction of students (Swaner & Brownell, 2009). In fact, there are gaps in research surrounding the relationship between underserved students learning and their engagement in HIPs (Finley & McNair, 2013). For many parents and students, the commitment to achieving educational goals is associated with not only the quality of education but also the cost of education, time to graduate, post-graduation employment, and/or graduate or professional school matriculation. The impetus may not be on acquiring meaningful out-of-classroom activities and life-changing educational experiences that HIPs provide. Moreover, for non-traditional students, the variables for student success are much different and include such dynamics as balancing school and work, balancing family obligations, accelerated degree options, and opportunities for distance learning. This holds true for underserved students from HBCUs whose student population is shifting toward the non-traditional model and whose religious affiliations may be steeped in core denominational values.

Since the General Education programs at HBCUs are foundational and connected to the mission and vision (educational principles) of the institution, it would be beneficial for campuses to vertically integrate the curriculum to create more intentional and integrative scaffolding of student-learning outcomes and goals. Experiential learning, first-year, and senior-year experiences (components of general education programs) are highly beneficial to student success, that is, retention and graduation (Kuh, 2008; Tinto, 1987).

Claflin University, United Methodist-affiliated and private institution, is the oldest HBCU in the state of South Carolina and has played a key role in the awarding of science and mathematics degrees to African Americans. *Forbes* magazine ranked Claflin #1 among the nation's HBCUs for two consecutive years; *US News* and *World Report* ranked Claflin #1 in the South for "Strong Commitment to Teaching" and ranked Claflin #7 nationally among best Black colleges; *Consumers Digest* ranked Claflin #3 for best value top private colleges and universities nationally; *Chronicle* of higher education ranked Claflin #2 for retention rate; and *The Journal of Blacks in Higher Education* ranked Claflin # 2 for graduation rate. The university has a 44% graduation rate (six years) and a retention rate of 79%. Claflin revitalized the General Education Program to incorporate workforce competencies in the curriculum and to allow flexibility, options, and opportunities for minors and certificate offerings. The University adopted a four-year plan, beginning in the academic year 2017–2018, to revitalize the general education curriculum by re-engineering foundational courses, through faculty capacity building such that professors will embed workforce competencies within the curriculum. University faculty voted overwhelmingly to approve the General Education Revitalization Plan partly because of the emphasis placed on the infusion of workforce competencies into the curriculum.

This vote was concluded with more than three years of intense discussion by the entire campus including input and feedback from faculty and students. Specifically, Claflin's student-learning outcomes were aligned with cocurricular activities that included HIPs and twenty-first-century skills that employers highly value such as communication, problem-solving, teamwork/collaboration, social justice, and ethics, innovation, experiential learning, and workforce competency-related skills/professional development (NACE, 2017).

BENEFITS OF HIPS FOR UNDERSERVED STUDENTS

According to researchers such as Gonyea, Kinzie, Kuh, and Laird (2008) and Swaner and Brownell (2009), the outcomes are more prevalent for the general student population and less for underserved student populations. Perhaps the rationale is that underserved students do not have the opportunity to participate in HIPs mainly because their college campus may not offer such practices and/or there are limitations on the HIPs that are being offered. While most of the HIPs described in Table 1 are increasingly found on college campuses, it is only within the last 10–15 years they have become the subject of assessment and evaluation relative to their success with first-year students. Understanding what and why certain practices show greater benefits for students is an ongoing debate for researchers (Kuh, 2008; Swaner & Brownell, 2009; Sweat et al., 2013).

However, the literature and findings are consistent in showing a remarkable increase in student success through HIPs for all students, irrespective of race, gender, religion, age, socioeconomic status, and so on. In addition, notable findings illustrate that HIPs do increase retention and graduation among underserved college students. For example, Zhao and Kuh (2004) showed that participating in learning communities increased academic performance and overall college experience. "All students should have an opportunity to get involved in a structured effort that creates conditions for connected learning and integrates academic and social experiences," says Zhao and Kuh. Moreover, Ishiyama (2001), Pascarella and Terenzini (2005), and He, Scheuch, Schwartz, Gayles, and Li (2008) found positive results when underserved students participated in undergraduate research. There was a direct correlation between faculty-mentored research, retention, and subsequent matriculation in graduate school. These results are especially important for students who major in STEM. Peters (2005) used collaborative learning strategies to increase academic performance for underserved students in biochemistry. In addition, DeLauder and Hollowell (2012) used inquiry-based laboratories, writing, collaboration, and research to increase achievement in general biology and general chemistry for underserved students.

Interestingly, Gonyea et al. (2008) recommend that all students in higher education participate in at least two HIPs: one in their first year and another in their academic major. Benedict College (private HBCU) and University of Wisconsin-Eau Claire both require students to engage in service learning as a graduation requirement. Delaware State University (public HBCU), Fayetteville State University (public HBCU) and Johnson C. Smith University (private

HBCU) have focused their attention on retention and degree completion including such practices as peer mentoring, internships, field experiences, and study abroad. At Claflin University faculty overwhelmingly adopted a resolution in 2016, that the incoming class of freshmen (2017) would have at the least, one mandatory experiential learning opportunity as part of their graduation requirement, specifically internships, study abroad, directed research, service learning, and cooperative learning. This mandatory experience would be placed on the students' cocurricular transcript and would be connected to their major and interests. One prime example of this is the Lilly Endowment-funded United Negro College Fund Career Pathway Initiative (CPI). The CPI utilizes a three-prong approach to strengthening student outcomes, specifically guided career pathways, curricular enhancements, and integrated cocurricular engagement. Among other things, the initiative aims to increase the employment rates of HBCU and PBI (predominantly Black Institution) graduates by at least 15% over seven years (UNCF & Lilly Endowment, Inc., 2015). Because HBCUs serve disproportionately large numbers of underserved students, and they incline to have low graduation and retention rates and high educational debt, initiatives such as the CPI afford a viable option for providing meaningful and intentional pathways for student success. CPI coupled with the new General Education Curriculum affords Claflin students the option of strengthening their liberal arts foundation with in-demand minors and certificate programs making them more competitive for high-paying jobs.

For campuses intending to implement HIPs, there will be a tremendous degree of variability across the board. For example, the capstone course or the service-learning course may fluctuate or be formulated differently from one institution to another. The deliverables for the writing project and the collaborative assignment will also fluctuate from one institution to another. In the case of higher education institutions whose student population is underserved, HIPs may be integrated into two or more practices. For example, at Claflin University, the first-year experience also includes the learning community and the capstone experience also includes undergraduate research. This integration and coupling seem to be the case for many HBCUs particularly because of staffing, resources, and facilities. This explains the importance of beginning the selection and design process of HIPs based on the strategic priorities of the institution. It is safe to speculate that each HIP is unique and distinct in its implementation and formulation. Thus, the outcomes will be institution-specific and determined by the variables within each practice that may or may not have an impact on student success.

KEY CONSIDERATIONS TO ADDRESS

In assessing and addressing the needs of students pursuing any major, the student's academic background should become a part of planning the curriculum for the student. For example, many students attempting to pursue a STEM curriculum have not necessarily been exposed to or achieved the mathematical skills, a major requirement for STEM, required to be successful. Van Soom and Donche (2014) indicate that STEM programs require a strong preparation in mathematics, primarily because mathematics modules of differing complexity are pronounced in most first-year STEM programs. Thus, assessing the student's mathematical skills as well as reading comprehension can play a major role in determining where the student should begin in their course of study. Once an assessment is made the advisor and student should develop a curriculum guide (guided career pathway) so that the student has a clear pathway of study toward a degree and career.

Communication and comprehension are important in promoting student achievement. The student must have a clear understanding of what is expected in the course. Once there is an understanding of expectations, it is the student's responsibility to take advantage of services offered, specifically HIPs that may be available at their institution. Students receiving directed guidance and even proactive advising during the first year will be successful via seminars, peer mentoring, tutoring, supplemental instruction, and experiences related to desired career goals.

In all majors, writing reports, essays, and even journaling drastically improves written communication skills. Often, the first draft is full of errors ranging from changing tenses to incorrect grammar. A Campus Writing Center can be very helpful in assisting students with their writing skills and helping them prepare for oral presentations. Group assignments also play an important role in developing communication skills as well as cooperation, collaborative learning, and teamwork.

DeLauder and Hollowell (2012) have shown that research-based laboratories covering concepts learned in the lecture reinforce the concept, allow students to review the literature, design experiments, and promote project-based learning. Students are engaged, and the laboratories mimic what happens in a research laboratory. Undergraduate research allows the student to work directly in a collaborative learning environment, work one-on-one with the faculty, get a good understanding of the challenges associated with research, and learn how to analyze and present data. As early as 1998, Nagda, Gregerman, Jonides, Von Hippel, and Lerner (1998) found that introducing students to research during the first or second year of study impacts STEM retention. The research project may be part of a course or in a laboratory setting.

Internships/cooperative education, capstone courses, study abroad, and community service are also very important in providing students an opportunity to experience real-world engagement in their area of study, especially for STEM students. The opportunity exists for students to learn project management skills, how to work collaboratively, understand how everyone has a role, and how their responsibility impacts the overall outcome of the project. The capstone courses/projects should capture all aspects of STEM and the student's journey in college.

USING HIPS TO ENHANCE STUDENT LEARNING AND SUCCESS IN STEM

Higher education institutions typically develop new programs to meet the needs of the state's workforce. According to the Georgetown Center on Education and the Workforce, by 2020, 65% of American jobs will require some form of post-secondary degree or credential, but the current rate at which colleges and universities are awarding them will fall short by 5 million (Center on Education and the Workforce, 2013). Interestingly, STEM fields will continue to be in high demand. Strategies and programs that strengthen the connections between the major/minor to experiential learning to social development to leadership development can positively affect college and career success for both students and the communities they live in.

Too numerous to mention, there is a body of work that speaks to the overwhelming success of HIPs and provides examples of campus models and case studies through the LEAP Project (Liberal Education and America's Promise (LEAP)) (www.aacu.org). In addition, there are successful programs such as the Summer Bridge Getting a Leg Up at Emory (GLUE) Program at Emory University that has retained 100% of STEM students over the past years. Grand Valley State University has increased their graduation rates significantly by investing in cooperative education, internships, and mentoring. Lorain County Community College has embedded experiential education in all their programs of study.

Many researchers agree that peer support, academic advising, mentorship, and having a good relationship with faculty are important to retention in STEM for underrepresented minorities (Espinosa, 2011; McPherson, 2014). Moreover, many campuses have adopted initiatives focused on improving STEM retention, especially for underserved minorities (Estrada et al., 2016; Kuh, 2008; Nostrand & Pollenz, 2017; Peters, Mani, Rasathurai, & Greene, 2006). These practices vary based on institutional needs, but the consistent dynamic is the emphasis on student success.

One size does not fit all for institutions hoping to adopt HIPs in one form or another. It is therefore important for institutions to design programs and opportunities based on their needs, goals, and capabilities. Just as accreditation bodies are interested in assessment and key performance metrics, the rapid pace of change in higher education forces institutions to move away from performance measures and toward programs and initiatives that include metrics such as retention, graduation, post-graduation outcomes, and even the production of career-ready graduates (Lee & Keys, 2013).

This chapter ends by providing an example of an institution whose vision is known as "Claflin LEADS" (www.claflin.edu). Claflin LEADS is an acronym for leadership development, experiential learning, academic excellence, diversity and inclusion, and student success. The Long-range Planning Committee proposed "Claflin LEADS" to prioritize student success, experiential learning, and diversity and inclusion. Four of the six goals of the Strategic Plan are central to the CPI: visionary leadership development, experiential learning, academic excellence, and student success. At Claflin University, "Learning Communities and a Collaborative Team-teaching Model" are utilized for all freshmen aimed at preparing them for the demands of college and the demands of the professional world. The results have shown an increase in retention and higher persistence within major.

EXAMPLES OF SUCCESSFUL HIGH-IMPACT PROGRAMS AT CLAFLIN UNIVERSITY

Learning Communities (LC)

At Claflin, learning communities are designed to improve the academic success and retention of students by creating purposive and collaborative student-based learning experiences and problem-based activities centered on common goals and themes. Carrino and Gerace (2016), upon studying a two-year STEM-based learning community, reported that students participating in learning communities attained certain psychosocial learning factors through social interactions taking place within the learning community, which subsequently improved student development. Those psychosocial learning factors include academic self-regulation, STEM identity, metacognition, self-efficacy, interaction with faculty/staff/STEM professionals, and interactions with peers. LCs are effective in that they provide structure for active learning in groups and integrate aspects of social learning theory by encouraging students to learn from one another as well as from faculty mentors who are often selected for their humanism, professionalism, and interest in mentoring students.

With buy-in from university stakeholders, that is, faculty, staff, students, alumni, and community, the university embarked on a major Student Success Initiative (SSI) to improve retention, graduation rates, student-learning outcomes, and stakeholder satisfaction/engagement. Launched in 2011, Claflin's SSI redesigned the approach to managing student support services and success. The initiative targeted six factors as essential to student success: *Quality Enhancement Plan – Panther STEPS* (Students in Transition Engaged and Preparing for Success) enhanced the freshman year experience. The *Financial Challenges* Team developed action items to create a smooth transition through financial challenges for students. The *Service Excellence Initiative* was established to maintain an environment that improves quality and increases consistency in all student transactions. The *Early Alert* Team created strategies and processes to allow Claflin to identify high-risk or dropout-prone students early. The *Advising Team* was established to create a quality advising system that is consistent across campus. The *Learning Communities* approach was adopted as a best practice to increase student learning and achievement through active learning, teamwork, and leadership development. The initiative started in 2011 with six very successful communities. The university decided in Fall 2014 to expand the initiative and place every freshman in a learning community.

Each first-time freshman is placed in a learning community with no more than 25 students, one coordinator, and one peer mentor (upper-class student). Students are segregated by discipline, academic preparedness, and career goals. There is also a learning community for first-generation students (LinCs). For 2016, there were 21 learning communities serving more than 500 students with three common first-year courses. Table 2 provides a sample of the learning communities, followed by a list of actual out-of-classroom activities. Each LC Coordinator had a small budget to plan monthly, low-cost activities. Typically,

Table 2. Sample Learning Communities.

Learning Community (LC)	Common Courses
LinCs (first-generation students)	Freshman year experience (UNIV 101); English I; African American Heritage
Honors College LC	Freshman year experience (UNIV 101); English I; Honors Leadership Development
Careers Explorers LC	Freshman year experience (UNIV 101); English I; African-American Heritage
Mass Communication LC	Freshman year experience (UNIV 101); English I; Intro to Mass Communication
Teacher Education LC	Freshman year experience (UNIV 101); English I; Intro to education
Chemistry LC	Freshman year experience (UNIV 101); English I; General Chemistry I
Biology LC	Freshman year experience (UNIV 101); English I; General Biology I
Computer Science LC	Freshman year experience (UNIV 101); English I; Fundamentals of Computer Programming
Mathematics LC	Freshman year experience (UNIV 101); English I; Calculus I

the culminating activity was a trip to a nearby professional basketball game or something similar.

Specific objectives of LC include the following:

- formation of a robust collaborative partnership between faculty and staff of connected courses;
- student engagement in high-impact educational activities;
- interdisciplinary learning opportunities for students enrolled in connected courses; and
- programmatic assessment of instructor and student engagement.

Below are actual enrichment activities that were implemented for the learning communities. More than just providing food and fellowship, the activities were intentional and purposeful.

Activity: *International Night* – The purpose of this activity was to learn about different cultures. Different learning communities represented the United States along with the Student Government Association. This was a university-wide event.

Activity: *High School Dropout Prevention Outreach* – Students provided volunteer assistance to the local school district to reach out to recent high school dropouts. Teams went to designated homes in the area to talk to students about the dangers of being a dropout and to provide encouragement and information about services that are available to assist them upon their return.

Activity: *Pennies for Posho/Ugandan Thunder* – Pennies for Posho is a 501(c) 3 non-profit organization that raises money to purchase food for 18 orphanages (almost 6,000 children) in Uganda. Students attended a musical concert put on by the group of orphans known as "Ugandan Thunder." Students were able to participate in the concert and talk to the group afterward. Students were exposed to stories of the orphanage life first hand. They also learned a lesson in benevolence and were able to give donations to the group and purchase items for the support of the orphanage. The students were excited about this activity and even returned to class asking for the support of their classmates to sponsor a child!

Activity: *Faculty/Staff vs Students Quiz Bowl* – Students were able to interact with faculty and staff through competition. Quiz bowl questions were submitted by both teams prior to the competition. The categories consisted of general knowledge and academic knowledge to include but not limited to Sports, Science, Music, and Television Shows. The winner had bragging rights over the other. So far, faculty/staff remain the champions.

Activity: *Final Meltdown (Painting "without" a Twist)* – Students participated in a painting meltdown. No skill was needed, the art instructor directed students through a step-by-step process for them to create an art masterpiece of the Claflin "C." Students were able to discover their artistic ability and interact with students from other learning communities. This was a destress activity during final exams week.

Collaborative Team-teaching Model

This approach was utilized for freshman year experience courses (UNIV101-3 credit hours; UNIV102-3 credit hours). Two people team-taught the course and consisted of a professor and a staff member in the classroom (Fig. 1). The professor was also the academic advisor for students within the learning community.

Each teacher is responsible for a portion of the lesson. This model requires good timing and practice between the two teachers. It is also advantageous when both teachers provide different skill sets so that they can complement each

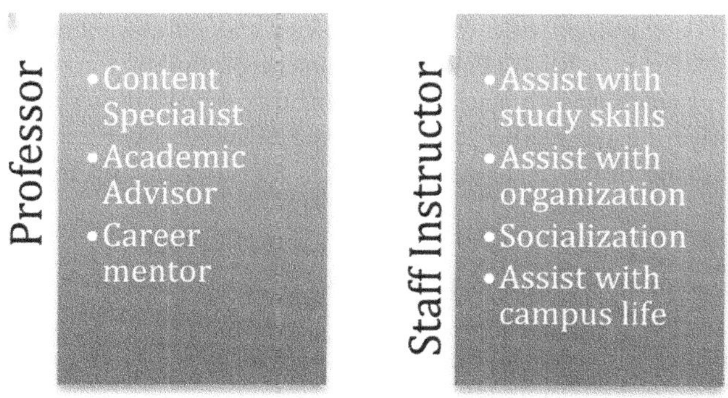

Fig. 1. Structure for Freshman Year Experience Courses.

other in the classroom. Therefore, choosing the right team to co-teach is very important.

Evidence from learning communities combined with the co-teaching model showed the following:

- Students are more engaged within and outside the classroom through the activities of UNIV 101/102 and learning communities.
- Anecdotal evidence suggests that STEM learning communities are more effective than non-STEM learning communities.
- Students expressed greater satisfaction with their overall college experience.
- Between 2014 and 2017, the retention rate increased by 4% (75– 79%).
- Students in extended English and math classes are achieving at a rate comparable to their counterparts in regular sections of the classes.
- In 2017, 95% of all freshmen successfully pre-registered during the pre-registration period for the spring session. These data can ultimately equate to increased retention and persistence in major.

CONCLUSION

The ability to make purposive changes for the betterment of student success depends a lot on the structural flexibility of the system. For example, Claflin University has a history of navigating through organizational decision-making through strategic planning. This involves all stakeholders (students, faculty, staff, community, and alumni). The University uses institutional assessment and models of excellence of peer and aspirant institutions to identify the trajectory for capacity building. One prime example of institutional change involving consensus building and institutional analysis is the intentional decision to make HIPs available to all university students to promote not only deeper learning but also persistence, retention, and graduation. A lesson learned from this decision is the importance of assessment (i.e., mixed-method design using qualitative and quantitative methods) and data collection to further substantiate the effect of HIPs on student outcomes.

Institutions of higher education must develop faculty and staff capacity to implement HIPs and provide relevant professional development opportunities (McNair & Albertine, 2012). The goal should be on building and maintaining a community of teaching and learning. Toward this end, Claflin University is currently investigating the relationship between HIPs and cocurricular educational activities and the impact on post-graduation employment in discipline-specific fields.

REFERENCES

Association of American Colleges and Universities. (2007). *College learning for the new global century*. Retrieved from https://www.aacu.org/publications-research/periodicals/college-learning-new-global-century

Brownell, J. E., & Swaner, L. E. (2010). *Five high-impact practices*. Washington, DC: Association of American Colleges and Universities.

Carrino, S. S., & Gerace, W. J. (2016). Why STEM learning communities work: The development of psychosocial learning factors through social interaction. *Learning Communities Research and Practice*, *4*(1), 3.

Center on Education and the Workforce, Recovery: Job Growth and Education Requirements Through 2020. (2013). Executive Summary. Retrieved from https://cew.georgetown.edu/cew-reports/recovery-job-growth-and-education-requirements-through-2020/

DeLauder, S. F., & Hollowell, G. P. (2012). *Embracing innovation and broadening student engagement for STEM majors*. Washington, DC: Association of American Colleges and Universities.

Espinosa, L. L. (2011). Pipelines and pathways: Women of color in undergraduate STEM majors and the college experiences that contribute to persistence. *Harvard Education Review*, *81*(2), 209.

Estrada, M., Burnett, M., Campbell, A. G., Campbell, P. B., Denetclaw, W. F., Gutiérrez, C. G., et al. (2016). Improving underrepresented minority student persistence in STEM. *CBE-Life Sciences Education*, *15*(5), 1–10.

Finley, A., & McNair, T. (2013). *Assessing underserved students' engagement in high-impact practices*. Washington, DC: Association of American Colleges and Universities.

Gonyea, R. M., Kinzie, J., Kuh, G. D., & Laird, T. N. (2008). *High-impact activities: What they are, why they work, and who benefits*. Presentation at the Annual Meeting of the Association of American Colleges and Universities (AAC&U), Washington, DC. Retrieved from http://cpr.indiana.edu/uploads/AACU_2008_high_impact_practices%20Kuh,%20Gonyea,%20Nelson%20Laird,%20Kinzie%20final.pdf

He, S., Scheuch, K., Schwartz, R., Gayles, J. G., & Li, S. (2008). Reinventing undergraduate education: Engaging college students in research and creative activities. *ASHE Higher Education Report* (Vol. 33, no. 4). San Francisco, CA: Jossey-Bass.

Ishiyama, J. (2001). Undergraduate research and the success of first generation, low income college students. *Council on Undergraduate Research Quarterly*, *22*, 36.

Kuh, G. D. (2008). *High-impact educational practices: What they are, who has access to them, and why they matter*. Washington, DC: Association of American Colleges and Universities.

Kuh, G. D., & O'Donnell, K. (2013). *Ensuring quality & taking high-impact practices to scale*. Washington, DC: Association of American Colleges & Universities.

Lee, J. M., & Keys, S. W. (2013). *Repositioning HBCUs for the future: Access, success, research & innovation (APLU Office of Access and Success Discussion Paper 2013-01)*. Washington, DC: Association of Public and Land-grant Universities.

Liberal Education and America's Promise (LEAP). *Association of Americas colleges and universities*. Retrieved from https://www.aacu.org/leap/index.cfm

McNair, T. B., & Albertine, S. (2012). Seeking high-quality, high-impact learning: The imperative of faculty development and curricular intentionality. *Peer Review*, *14*(3), 4–5.

McPherson, E. (2014, February). *High-impact practices that retain underrepresented STEM Students*. OU Student Success Conference. Troy, MI. Retrieved from https://www.oakland.edu/Assets/upload/docs/UG.../52_HIP_That_Retain.pdf

Nagda, B. A., Gregerman, S. R., Jonides, J., Von Hippel, W., & Lerner, J. S. (1998). Undergraduate student-faculty research partnerships affect student retention. *Review of Higher Education*, *22*(1), 55–72.

National Association of Colleges and Employers (NACE). (2017). *NACE research report: Job outlook*. Retrieved from http://www.naceweb.org

National Survey of Student Engagement. (2017). *High-impact practices*. Bloomington, IN: Indiana University Center for Postsecondary Research.

Nostrand, D. F.-V., & Pollenz, R. S. (2017). Evaluating psychosocial mechanisms underlying STEM persistence in undergraduates: Evidence of impact from a six-day pre–college engagement STEM academy program. *CBE – Life Sciences Education*, *16*(2), ar36.

Pascarella, E. T., & Terenzini, P. T. (2005). *How college affects students: A third decade of research* (Vol. 2). San Francisco, CA: Jossey-Bass.

Peters, A. W. (2005). Teaching biochemistry at a minority-serving institution: An evaluation of the role of collaborative learning as a tool for science mastery. *Journal of Chemical Education*, *82*(4), 571.

Peters, A. W., Mani, D., Rasathurai, S., & Greene, M. (2006). The effectiveness of supplemental instruction and technology in increasing student performance in mathematics. *Journal of Women and Minorities in Science and Engineering*, *12*, 1–9.

Swaner, L. E., & Brownell, J. E. (2009). *Outcomes of high impact practices for underserved students: A review of the literature*. Washington, DC: Association of American Colleges and Universities.

Sweat, J., Jones, G., Han, S., & Wolfgram, S. M. (2013). How does high impact practice predict student engagement? A comparison of White and minority students. *International Journal for the Scholarship of Teaching and Learning*, *7*(2), Article 17.

Tinto, V. (1987). *Leaving college: Rethinking the causes and cures of student attrition*. Chicago, IL: University of Chicago Press.

UNCF and Lilly Endowment, Inc. (2015, October 5). *Announce $50 million initiative to better prepare students for 21st century jobs*. Retrieved from https://www.uncf.org/news/entry/uncf-and-lilly-endowment-inc.-announce-50-million-initiative-to-better-prep

Urban Institute. (2006, April 6). *Final report on the evaluation of the National Science Foundation Louis stokes alliances for minority participation program*. Retrieved from https://www.urban.org/research/publication/final-report-evaluation-national-science-foundation-louis-stokes-alliances-minority-participation-program

Urban Institute. (2011, March 7). *Capacity building to diversify STEM: Realizing potential among HBCUs*. Retrieved from https://www.urban.org/research/publication/capacity-building-diversify-stem-realizing-potential-among-hbcus

Van Soom, C., & Donche, V. (2014). Profiling first-year students in STEM programs based on autonomous motivation and academic self-concept and relationship with academic achievement. *PloS One*, *9*, 1–13.

Zhao, C., & Kuh, G. D. (2004). Adding value: Learning communities and student engagement. *Research in Higher Education*, *45*(2), 115–138.

SUPPLEMENTAL INSTRUCTION LEVELS THE PLAYING FIELD IN STEM AT LOUISIANA STATE UNIVERSITY

Gloria Thomas, Lahna Roche, Melissa Brocato and Saundra McGuire

ABSTRACT

The Center for Academic Success (CAS) at Louisiana State University (LSU), certified as a Center of Excellence by the National College Learning Center Association, has utilized Supplemental Instruction© (SI) for the past 20 years to provide student support for historically difficult courses – those courses with D, F, or withdrawal rates of greater than 30%. In this model, peers called "SI leaders" facilitate study sessions outside of class time to help the enrolled students develop effective learning strategies and better understand and master course concepts. SI relies upon collaboration with faculty and is supported by cognitivism and social constructivism learning theories.

Benefits of the successful model include supporting students to become self-directed independent learners, reducing the stigma associated with using academic support and reducing the demands for tutoring. Outcomes observed at LSU include positive correlations between the course-passing rates and six-year graduation rates of women, underrepresented minorities and first-generation college students who participated in SI compared to the peers who participate less frequently and those who do not participate.

Keywords: Supplemental Instruction; learning strategies; metacognition; retention; academic support; peer-led learning

Broadening Participation in STEM: Effective Methods, Practices, and Programs
Diversity in Higher Education, Volume 22, 197–208

ISSN: 1479-3644/doi:10.1108/S1479-364420190000022009

INTRODUCTION

Supplemental Instruction (SI)

Supplemental Instruction© (SI) was started in 1973 by Dr. Deanna Martin to decrease high attrition rates of minority students in the health professions schools at the University of Missouri-Kansas City (UMKC) (Arendale, 2002; Martin & Arendale, 1990). In this model, peers called "SI leaders" facilitate study sessions outside of class time to help enrolled students develop effective learning strategies and better understand and master course concepts (Hurley, Jacobs, & Gilbert, 2006). SI leaders are ideally students who have previously excelled in a course and are recommended by the instructor to conduct study sessions for currently enrolled students. Potential SI leaders participate in a rigorous selection process to determine their interpersonal skills, problem-solving abilities, and presentation skills. Those who are selected participate in training that focuses on group dynamics, active learning strategies, learning strategies, record–keeping, and other topics. Additionally, the leaders conduct mock first-day presentations. Each SI leader holds a total of three hours of group study sessions weekly.

The SI model has been used successfully across a wide variety of undergraduate institutions and contexts. Today, the International Center for Supplemental Instruction oversees programs at thousands of institutions on five continents. The efficacy of the model has proven to be robust, and it is certified by the National Diffusion Network (NDN) of the US Department of Education as an Exemplary Educational Program. The data collected by the program demonstrate that SI improved students' academic achievement (indicated by higher grades) as well as students' attitudes and behaviors (indicated by fewer withdrawals from courses).

Strategy, Context, and Motivation for Pursuing This Stratagem

As noted, SI was first introduced in the health sciences, but its implementation soon spread to other areas, including science, technology, engineering and mathematics (STEM) discipline students and particularly among those from underrepresented groups in STEM (Peterfreund, Rath, Xenos, & Bayliss, 2008; Renzulli, Grant, & Kathuria, 2006). The Center for Academic Success (CAS) at Louisiana State University (LSU), certified as a Center of Excellence by the National College Learning Center Association, adopted the SI model in 1998 for that purpose. Dr Rhonda Atkinson, CAS Director (1989–1999), found the SI approach particularly attractive because SI does not target high-risk students, but instead targets historically difficult courses (those courses with D, F, or withdrawal rates of greater than 30%). By inviting all students enrolled in these courses to participate in SI, stigmas associated with participation can potentially be reduced (Atkinson, 2017). Additionally, because SI focuses on teaching students effective learning strategies during the course of a peer-led cooperative learning session, SI sessions enable students to develop into self-directed independent learners.

THEORETICAL UNDERPINNINGS

Two well-established learning theories – cognitivism, and constructivism, more specifically social constructivism – underpin Supplemental Instruction (McGuire, 2006). These theories emphasize information-processing and student-centered learning activities in contrast to B. F. Skinner's earlier theory of behaviorism that describes learning as a change in the learner's behavior that could be brought about by training the learner to respond appropriately to a set of stimuli (Ertmer & Newby, 1993).

Cognitivism

Cognitivists proposed that the amount of learning could not be judged by simply observing changes in the learner's behavior, but that what was happening in the mind of the learner was crucial to learning (Bodner, 1986; Flavell, 1976). Learners are thus viewed as active processors of information who are receiving, processing, storing, and retrieving information for use in learning tasks such as problem-solving and critical thinking. Cognitivism asserts that learners must actively engage in learning activities in order for them to process the information that they are receiving. Activities such as brainstorming, games, reciprocal teaching, and active discussion, which are employed during SI sessions, are necessary for learners to fully process and retain information.

Constructivism

Constructivists, such as Vygotsky (Vygotsky, 1992), view learning as the process during which learners construct their own understanding of information. In order for learning to occur, learners have to construct their own meaning from information they are receiving. They construct this understanding by integrating new information with information that they already know. The learner's prior knowledge is the scaffold onto which new knowledge is anchored. The goal is for the learner to construct his or her own conceptual framework to represent the information that is being learned. Social constructivists add that individuals learn best when they are in a setting where they can interact with others to discuss ideas, question each other, and construct new knowledge. The cooperative learning environment in SI is an example of social constructivism at its best; the peer-led, cooperative learning activities allow learners to examine what they already know, and build new understandings in discussion with their peers (Table 1).

STRATAGEM STRUCTURE, IMPLEMENTATION, AND EVOLUTION DUE TO INSTITUTIONAL CONTEXT

SI Leaders, Sessions, and Structure

SI sessions are free, voluntary, peer-led study sessions that are facilitated by students who have previously excelled in the course. These SI leaders receive training on guiding collaborative group study sessions and develop working

Table 1. Summary of Learning Theories.

Learning Theory	Learning Process	Learning Activities
Cognitivism	Learner receives, processes, stores, and retrieves information	Active learning activities, like puzzles
Constructivism	Learner constructs understanding by integrating new information with what s/he already knows	Predicting solutions to new problems based on information already known
Social Constructivism	Learner constructs understanding best when interacting with others	Think–pair–share, reciprocal teaching

relationships with the faculty teaching the assigned course. During the semester, the SI leader attends each instructor-led class period and then holds regularly scheduled SI sessions that promote active learning and effective study strategies along with "office hours." Leaders do not provide lectures or introduce new content, and although the SI setting is intended to be informal, the sessions are well-planned in advance.

The SI model is built upon engaging the students in not only "what to learn" but also "how to learn" (Arendale, 2002). As such, the SI session generally is structured in three parts: opening the session, conducting the session, and closing the session. Opening a session consists of establishing priorities and expectations for the work to be accomplished in that session. It may also include an activity that eases the group into an academic mindset. Conducting the session consists of engaging students meaningfully in the learning process with the bulk of interactions occurring student-to-student, as opposed to leader-to-student. Effective strategies include think-pair-share discussions, predicting test items, comparing class notes, and developing concept maps. During these sessions, SI leaders don't provide direct answers to questions, but instead redirect the discussion to the group or provide prompts to encourage students to critically think through challenges. At the end of the session, a closing activity helps the group summarize what they have learned, provides next steps and connects learning back to the "big picture."

In the SI model, the relationship between class instructor and SI leader is critically important. The leader will meet with the instructor on a weekly basis to gain insight into topics that students generally have difficulty mastering. The leader will focus on these topics in sessions and provide feedback to the instructor about the general student progress. In order to alleviate conflicts, SI leaders are not involved in the grading aspects of the course, do not help construct test questions or have access to the exams before they are administered.

Just as SI is intended to cultivate metacognition within students, the SI leaders are guided toward thinking about their thinking in the model as well. Critical to the process is the planning phase that SI leaders engage in to prepare for their sessions and also a critical reflection phase after the session. Leaders are also formally observed during SI sessions by their peers and supervisors to ensure best practices are being maintained. The debrief that occurs as a part of the observation process includes feedback from the observer and also feedback from the SI leader.

Implementation and Evolution

The original goals pursued at LSU, and the goals that persist, are to decrease DFW rates, to decrease the achievement gap between underrepresented students and women as compared to majority students, and to ultimately increase retention and graduation rates at the university. Given the success and national recognition of the SI program and its effectiveness with minority students, LSU invited UMKC Professionals to conduct a training session at LSU for faculty from several STEM departments, as well as staff from the CAS. In addition to the onsite training, LSU professional staff also attended training at UMKC for SI coordinators.

Implementation began in the Fall of 1998 with a coordinator overseeing seven leaders in mathematics, biology, and chemistry, courses with historically high rates of D and F grades and course withdrawals. With a three-year funding commitment from the university, the program expanded to 35 leaders supporting eight departments between Fall 2000 and Spring 2002. In general, courses with enrollments greater than 80 students and a DFW rate of at least 15% were targeted as critical courses that may get benefit from SI.

After funding changes prompted by turnover in the upper administration, the CAS restructured its efforts by combing SI and tutoring programs under one coordinator. This new Academic Support Coordinator was responsible for developing training programs as well as observations of the SI leaders. In the intervening years between 2002 and 2011, most of the program support came directly from the departments whose courses were being supported, with most of the program occurring in chemistry due to the department's desire for the resource and the joint appointment of Dr Saundra McGuire who was CAS Director (1999–2009) and Professor of Chemistry.

Beginning in 2011, then CAS Director Melissa Brocato (2009–2017) was able to obtain funding from various humanities departments to pilot SI support in geography and history courses. From that modest beginning, the number of humanities courses steadily increased. In the same way, support from the College of Business provided SI for economics sections, as well as for an accounting course. With this growth of the program, the academic support staff expanded to include graduate assistants to help with observations of SI leaders and logistics. In all, using research, data, and results have been key to making the case for the program's effectiveness in meeting the student's needs.

Funding somewhat stabilized after the university levied a student fee aimed at providing improved academic experiences for students during the 2015–2016 academic year. As a result, two of LSU's largest academic colleges requested additional support for their courses. From 35 leaders in the preceding fall semester, the program expanded to 63 leaders supporting courses in 11 departments. To date, five (5) colleges have contributed on a continuous basis over the years and as many humanities sections are now being supported as STEM sections.

Initially, funding proposals were submitted each semester to the colleges; however, many colleges have now opted to fund the program on an annual basis, allowing for more continuity and efficiency. Additional internal and external proposals have been submitted over the years in order to sustain the program.

STRATAGEM OUTCOMES

Evaluation

Although it is generally accepted that SI is effective, many studies have sought to empirically prove that SI is effective when motivation, as distinct from prior academic achievement, is controlled for (Dawson, van der Meer, Skalicky, & Cowley, 2014; Wilson IV & Steele; Wilson, Waggenspack, Steele, & Gegenheimer, 2016). One positive correlation demonstrating the program's success is the comparison of the graduation rates of students who use the service with those who do not. When evaluating the 2009 cohort at LSU, the likelihood of graduating from college for students using SI was 4.6 times higher than for those who did not use the service (Huang, Roche, Kennedy, & Brocato, 2017). This cohort also had an 86% six year graduation rate for those who use the service as compared to a 57% graduation rate for the students who did not use the services and an overall graduation rate of 65% for the cohort. Among this cohort, a strong correlation was also found among women, underrepresented minorities and first-generation college students who participated in SI (Fig. 1).

Another positive correlation can be seen when comparing the course-passing rate of students who regularly participate in SI with that of students who participate less frequently and those who do not participate. To examine this

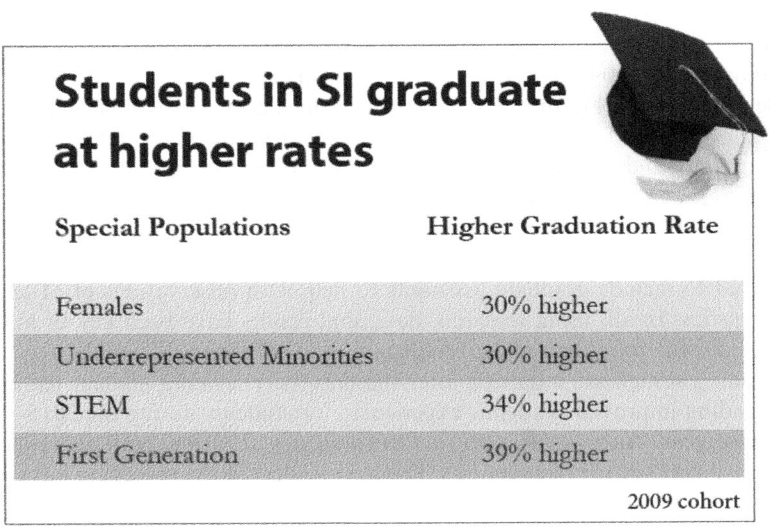

Fig. 1. Comparison of Graduation Rates of Special Populations, 2009 Cohort.

correlation, for the Fall 2017 and Spring 2018 semesters, courses which met two conditions were examined: (1) lower passing rates (grades of A, B or C) were observed among non-SI participants from underrepresented racial and ethnic groups, and (2) adequate numbers of students from various racial and ethnic groups participated in SI at various rates. SI participation was characterized as no participation, participation in one to three sessions, and participation in four or more sessions. Attendance in four sessions was selected as a threshold for the levels of participation to represent students who participated more frequently than those who attended sporadically or only for exam reviews (Table 2).

In general, students who participated in four or more SI sessions performed, on average, better in the course than their less frequently and non-SI-participating counterparts. Notably, in BIOL 1201 and CHEM 2261, the achievement gaps between students from historically underrepresented groups and their majority counterparts were considerably decreased among students participating in four or more sessions (Fig. 2).

Reduction in Tutoring Demand

As the SI program has grown in popularity and with proven, effective results, students and faculty have begun to demand an increase for SI in courses across disciplines. And, administrators are investing in the program. As more courses offered the SI model for academic support, students have begun using tutoring and the tutorial center less. The SI model in our experience and through data is proven to be a far superior model to tutoring. We believe this is due to the perceived stigma associated with individuals seeking tutoring services compared to those who seek SI offered to all students enrolled in a course acknowledged to be historically difficult. Additionally, SI utilizes methods that increase the overall learning rate that is not associated with the content of that particular course.

LESSONS LEARNED AND RECOMMENDATIONS

Structure and Climate

Prior to implementing new programs, understanding the current campus structure and climate is necessary to determine the best approach for gaining university support. When implementing any academic support program, the following questions arise: Is there an established centralized learning center on campus to provide academic support services? Are there colleges or departments who are currently running their own academic support functions in a decentralized fashion? Is there funding available to provide new services or enhance the services that already exist? The answers to these questions will guide the methods and processes for implementing new initiatives.

Centralized learning centers are the ideal way to implement academic support programs that address gateway courses. Due to their nature, gateway courses pose challenges for many students, are usually required for students regardless of major, and can pose obstacles to timely graduation. Centralized learning centers should have the ability to collect longitudinal data across disciplines to

Table 2. Numbers and SI Participation of Students from Various Racial and Ethnic Groups in SI-related Courses.

Course	Total				No SI (%)				1–3 SI (%)				4 or more SI (%)			
	BNH	HIS	ASIAN	WNH	BNH	HIS	ASIAN	WNH	BNH	HIS	ASIAN	WNH	BNH	HIS	ASIAN	WNH
BIOL 1001	296	118	57	1,151	45	46	52	57	21	22	18	19	6	5	7	7
BIOL 1201	134	62	57	779	63	45	53	52	28	40	35	35	8	15	12	13
CHEM 1201	422	171	141	2,095	27	30	41	31	26	26	22	25	15	19	18	18
CHEM 1202	234	123	116	1,331	22	18	10	16	14	17	10	12	10	4	3	4
CHEM 2261	157	69	95	717	15	15	19	14	22	21	15	22	26	34	41	36
CHEM 2262	107	50	75	535	7	9	4	6	15	9	7	8	24	9	18	11

Notes: BNH = Black Non-Hispanic, HIS = Hispanic, WNH = White Non-Hispanic.
As is illustrated in Fig. 2, achievement gaps are observed among students from underrepresented groups who did not participate in SI when compared to their majority and Asian counterparts within the selected courses. On average, SI-participating students from all groups passed the course at higher rates than students from their respective racial and ethnic groups who participated less frequently or did not participate at all. In addition, the achievement gap in BIOL 1201 and CHEM 2261 was significantly reduced among students who attended 4 or more SI sessions.

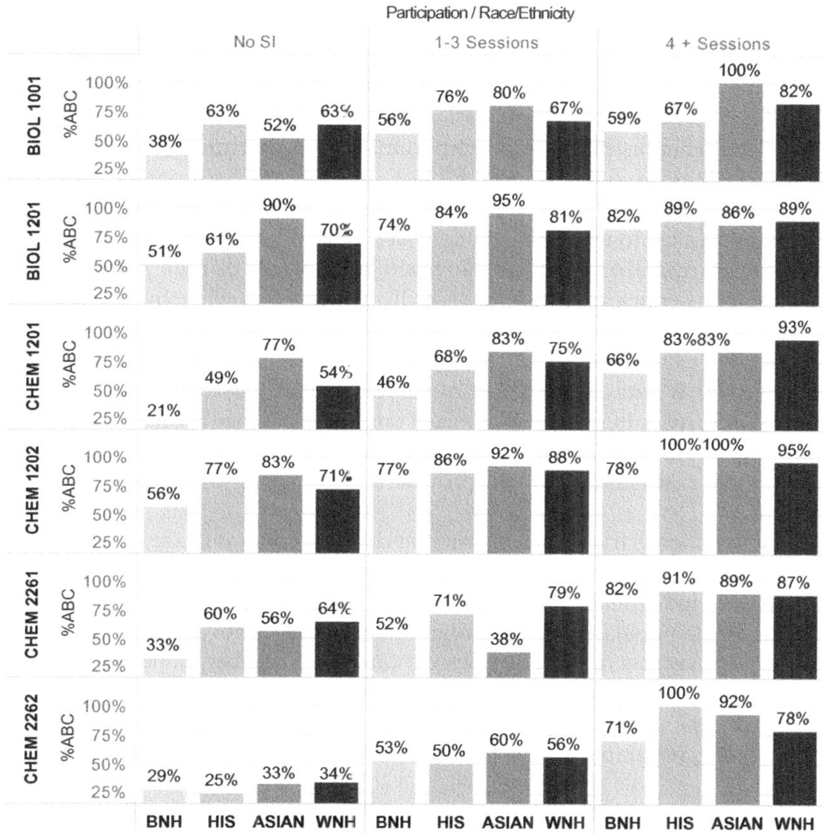

Fig. 2. Comparison of ABC Rates (Grades of A, B or C) of Students in SI-related Courses. *Note*: The average pass rating within each group of students is shown as a dashed line in each pane of data.

determine which courses are critical for students, which have the highest DFW rates, and which have the largest number of students enrolled. However, if academic support is implemented utilizing a decentralized structure, then academic affairs, the institutional research office, or each department or college would need to analyze the data to determine which courses to target. If no academic support structure exists, it is highly recommended to establish a centralized learning center within academic affairs, based on the best practices in the field.

Once the criteria are determined for course selection, and funding identified to pay the SI leader wages, a pilot scope will need to be determined. Determining where the program will be housed; meeting with deans, department heads; and working with the willing faculty in these departments is key to initial success, as all, particularly faculty, play a key role in the success of SI.

Assessment and Evaluation

Assessment and evaluation play critical roles in determining the size and scale for offering the SI program to the campus community. Identifying needs, primarily for students, while also considering the needs of faculty, departments, colleges, and administrators, is an important step in determining priorities, particularly if funds are limited. In addition to offering the SI program for critical path courses, or those necessary for students to persist toward graduation in their major, using longitudinal data to establish criteria for selecting courses allows for an equitable process while meeting the needs of each group. Typical metrics such as courses that are historically difficult and whereby students earn high levels of DFW grades indicate needs for all parties, specifically students. Additionally, examining large classes is important, because while some may have lower DFW rates, the number of students affected by not having the intervention can surpass higher DFW rates of smaller enrollment courses. Therefore, when conducting a thorough needs assessment, in addition to gaining stakeholder input, examining data can provide a fair starting point for offering the program.

The low-hanging fruit when examining data is program usage – the number of students utilizing the services and the number of times they do so. Capturing both unique (individual) student visits and the overall visits can give a quick snapshot of the popularity of the program and the need for students to gain the extra help for success in these courses. At times, the number alone can make the case for continuation and scaling of the program. However, building SI programs take time; therefore, lower participation numbers can be expected at the outset of the program or when offering the program in a new course or discipline.

The International Center for SI provides many resources to help create an effective SI program. Once established, it is imperative to assess the efficacy of the program to ensure the model is being implemented effectively. Developing clear learning outcomes for both student participants and the SI leaders facilitating the program is important to determine whether students are gaining the skills and services expected at the program. Measuring these outcomes using both direct and indirect measures is helpful. For instance, through surveys of student participants and SI leaders, questions related to the learning outcomes can help centers determine whether targeted learning outcomes are being met. Similarly, DFW rates, and conversely, ABC rates of the courses can detect program effectiveness, particularly if examined in the first year or two following implementation. In addition to determining course grade performance, data examining student performance can be used to determine SI program effectiveness. Comparing the grades of students who attended the SI program with those who did not is a good starting point. Further, examining the number of times students participate in relation to their academic performance to identify trends also provides more information about the correlation.

Similarly, institutional data metrics, such as retention and graduation rates, can make the case for program efficacy and scaling. Using data to track student attendance is vital in the ability to examine participants versus non-participants and compare their grade-point averages, persistence rates, and completion rates. Comparing students who participate in the SI program with those who do not is important; therefore, having access to all students enrolled in SI courses, for data analysis, is a necessity. Data collection for an SI program can be complex, and while our Center has a database specifically designed for this purpose, many programs utilize paper sign in sheets and Excel spreadsheets to track their attendance and results. To gain the results, programs must have the ability to either request that data are analyzed from the institution or complete analyses in-house.

Finally, satisfaction data are important in telling the story regarding program effectiveness. Student participant, SI leader, and faculty inputs are vital to determine how the primary players are interacting to create a solid program. Administrator input is also important when considering whether or not to provide more or less fiscal support for the future of the program. And, typically college administrators rely on the opinions of their respective faculty to make the decision. The LSU CAS does a mid-semester check-in with faculty to make sure that they are satisfied with the performance of the SI leader in order to provide ample time to make changes if needed during the semester. If both the faculty and students are having positive experiences with the program, then administrators are more likely to continue funding and/or scale the program. Therefore, gathering satisfaction data from both faculty and students is important in making the case for program continuations.

University Buy-in

In times of limited resources in higher education, proving the value of a program is very important. The national emphasis on college completion provides a good platform to educate the university community on high-impact educational practices, such as SI. There is an abundance of research that supports the efficacy of SI, including the SI return on investment for institutions (Congos, 2001), that can be used to demonstrate the effectiveness of the program at other institutions. In addition to educating the community of the value of the program as a whole, gaining institution-specific data findings will enhance the case for offering the program. Therefore, starting the SI program as a pilot at the institution to gain this information is recommended. With the proper information from research and the institution, the case can be made that SI increases student learning, performance, satisfaction, persistence, and degree completion and decreases DFW rates, student time to graduation, and debt accrued. Typical resistance to implementing the program includes faculty doubt regarding program effectiveness, limited funding, and/or arguments for selection bias. Therefore, anticipating potential resistance and understanding how SI works and the research that supports it will help when making the case to gain university buy-in.

Funding Sources

When institutional funds are limited, finding internal and external ways to raise funds for this effective intervention have been imperative for keeping the services free for students. Seeking investments from unique sources such as colleges and departments of the students and faculty who get the most benefit from the services has been the primary focus of the SI program at LSU. While this has been one strategy, stable funding from a permanent source such as general university budget funds or an endowment of the program would be ideal.

REFERENCES

Arendale, D. (2002). History of supplemental instruction (SI): Mainstreaming of developmental education. *Histories of developmental education*, 15–28.

Atkinson, R. (2017, December 1). Email.

Bodner, G. M. (1986). Constructivism: A theory of knowledge. *Journal of Chemical Education*, *63*(10), 873.

Congos, D. (2001). How supplemental instruction (SI) generates revenue for colleges and universities. *Journal of College Student Retention: Research, Theory & Practice*, *3*(3), 301–309.

Dawson, P., van der Meer, J., Skalicky, J., & Cowley, K. (2014). On the effectiveness of supplemental instruction: A systematic review of supplemental instruction and peer-assisted study sessions literature between 2001 and 2010. *Review of Educational Research*, *84*(4), 609–639.

Ertmer, P. A., & Newby, T. J. (1993). Behaviorism, cognitivism, constructivism: Comparing critical features from an instructional design perspective. *Performance improvement quarterly*, *6*(4), 50–72.

Flavell, J. H. (1976). Metacognitive aspects of problem solving. *The nature of intelligence*, 231–235.

Huang, L., Roche, L. R., Kennedy, E., & Brocato, M. B. (2017). Using an integrated persistence model to predict college graduation. *International Journal of Higher Education*, *6*(3), 40.

Hurley, M., Jacobs, G., & Gilbert, M. (2006). The basic SI model. *New Directions for Teaching and Learning*, *2006*(106), 11–22.

Martin, D. C., & Arendale, D. R. (1990). Supplemental Instruction: Improving Student Performance, Increasing Student Persistence.

McGuire, S. Y. (2006). The impact of supplemental instruction on teaching students how to learn. *New Directions for Teaching and Learning*, *2006*(106), 3–10.

Peterfreund, A. R., Rath, K. A., Xenos, S. P., & Bayliss, F. (2008). The impact of supplemental instruction on students in STEM courses: Results from San Francisco State University. *Journal of College Student Retention: Research, Theory & Practice*, *9*(4), 487–503.

Renzulli, L. A., Grant, L., & Kathuria, S. (2006). Race, gender, and the wage gap comparing faculty salaries in predominately white and historically black colleges and universities. *Gender & Society*, *20*(4), 491–510.

Vygotsky, L. (1992). *Educational Psychology* (R. Silverman, Trans.). Boca Raton, FL: St Lucie Press.

Wilson, C., Waggenspack, W., Steele, A., & Gegenheimer, J. (2016). *The unsubstantiated cutoff: Deeper analysis of supplemental instruction sessions on engineering courses.* Paper presented at the Proceedings of the 2016 ASEE Annual Conference, New Orleans, LA.

Wilson IV, M. C. A., & Steele, A. Curiosities Regarding Exam Review Sessions at LSU's College of Engineering.

HISTORY AND EVOLUTION OF STEM SUPPLEMENTAL INSTRUCTION AT SAN FRANCISCO STATE UNIVERSITY: A LARGE, URBAN, MINORITY-SERVING INSTITUTION

A. Alegra Eroy-Reveles, Eric Hsu, Kenneth A. Rath, Alan R. Peterfreund, and Frank Bayliss

ABSTRACT

Supplemental Instructions (SIs) were introduced into the San Francisco State University College of Science & Engineering curriculum in 1999. The goal was to improve student performance and retention and to decrease the time to degree in STEM majors. While for the most part we followed the structure and activities as developed by the International Center for Supplemental Instruction at the University of Missouri, Kansas City, we discovered several variations that significantly improved our outcomes. First and foremost, we created SI courses that require attendance, which results in higher students' performance outcomes compared to drop-in options. Second, at SFSU the SI courses are led by pairs of undergraduate student facilitators (who are all STEM majors) trained in active learning strategies. Each year, more than half of our facilitators return to teach for another year. Thus, each section has a returning "experienced" facilitator who works with a new "novice" facilitator. Third, the SI courses were created with a distinct course prefix and listed as courses that generate revenue and make data access

Broadening Participation in STEM: Effective Methods, Practices, and Programs
Diversity in Higher Education, Volume 22, 209–235

ISSN: 1479-3644/doi:10.1108/S1479-364420190000022010

available for comparison studies. Results are presented that compare SI impact by gender and with groups underrepresented in STEM disciplines.

Keywords: Supplemental instruction (SI); facilitated study groups (FSG); near-peer instruction; STEM retention; STEM workshops; implementation of SI

INTRODUCTION

Retention of STEM majors is a national concern. Students leaving STEM majors are not leaving because STEM majors are too hard. They are leaving because they are not engaged in the classroom and what they are learning in lower-division STEM courses has little apparent value in the context of their life. To increase engagement and build community in STEM classes, the Supplemental Instruction (SI) program at SFSU was established to support learning in large-lecture science and math classes. Since establishment in 1999, this program has evolved to reflect the changing student needs and demographics, along with increased knowledge of factors affecting student success, particularly success of historically underrepresented (HU) students as defined by the National Science Foundation (Hispanic/Latino, African American, Native American, Pacific Islander).

The suggestion that minority populations leave STEM because they are lacking in motivation and academic preparation has been greatly exaggerated (Schuman, Steeh, Bobo, & Krysan, 1997) and the focus on student deficits has been called into question (Bensimon, 2007; Ford & Grantham, 2003; Trujillo & Tanner, 2014). In fact, some studies find that HU students are both motivated to pursue science careers and academically well prepared (Hurtado et al., 2006; Seymour & Hewitt, 2000). The problems appear to be related to engagement and value, both of which we have tried to address through SI.

SI is an international model for student support that began at the University of Missouri–Kansas City in 1974. As explained in several sources (Arendale, 1994; Martin & Arendale, 1992), SI was conceived as a means of increasing student performance by targeting difficult courses rather than high-risk students. The idea of using SI to support students from HU groups more properly specifically stems from the work done by Uri Treisman while teaching at the University of California, Berkeley (Hsu, Murphy, & Treisman, 2008; Treisman, 1992). Treisman found that HU students were underperforming in their classes compared with their peers despite being a highly motivated and select group. Through his lecturer position at the university, his particular focus was African American students in calculus. While introducing an SI class that was open to everyone, he specifically recruited HU students, and he was able to raise their performance in calculus to a level on par with or better than the average performance among all other groups at the university.

SI classes are cooperative learning environments where students participate in learning activities that complement the course material, focusing on student misconceptions and difficulties, construction of a scaffolded knowledge base, applications involving problem-solving, and articulation of constructs with peers (Gasiewski, Eagan, Garcia, Hurtado, & Chang, 2012; Rath, Peterfreund, Xenos, Bayliss, & Carnal, 2007). The students enrolled in the SI classes benefit by interacting with role models, often senior undergraduate STEM majors, who help them in many ways. This peer support has been used to improve study and problem-solving skills by demonstrating strategies that were effective for them, develop teamwork and oral communications skills, and, for HU students, provide validation that students "like them" can succeed in science majors. All of these benefits contribute to the overall increase in first-year students' success and persistence in their STEM majors (Arendale, 1997; Arendale, 2004; Martin & Arendale, 1992).

San Francisco State University (SFSU) is a large urban Minority-serving Institution. Since 1999, SFSU has offered Science SI, a series of voluntary one-unit courses supplementing the STEM major gatekeeper courses. Facilitators create learning communities that focus on conceptual understanding and mutual support. Instruction is mainly through hands-on, interactive, small-group work, and course size is limited to 20.

HISTORY OF SI DEVELOPMENT AND IMPLEMENTATION PLAN AT SFSU

The initial efforts to address the high failure rates (>40%) of SFSU STEM students in first- and second-year major's courses was to develop "Facilitated Study Groups" (FSG). These FSGs were modeled after those described by Treisman, "Our idea was to construct a hybrid of the regular discussion sections and the 'Math Workshops'" (Treisman, 1992). While these were active efforts to create a mechanism to resolve the same challenges as those addressed by SI programs like the one at U Missouri at Kansas City (UMKC), we were not aware of those efforts or the name used until several years after we started. We developed many of the same approaches as the UMKC SI; however, it turned out to be fortuitous in that in our design we included a very important additional component that has had a significant impact, namely mandatory attendance. Mandatory attendance would otherwise have been overlooked if we had been aware of and adopted the UMKC SI model as they do not have this requirement. That is, we created "FSG workshops" that, unlike the UMKC model, required regular attendance like a formal class. A review article reporting the outcome of 27 peer-reviewed SI research publications (Dawson, van der Meer, Skalicky, & Cowley, 2014) reports that the effectiveness of the SI programs is directly correlated with the number of class meetings attended, with better outcomes from attending more meetings. Over the first six years of offering the SFSU FSG Workshops, we discovered the SI literature and eventually changed the name to SI (Peterfreund, Rath, Xenos, & Bayliss, 2008; Rath, Peterfreund, Xenos, Bayliss, & Carnal, 2007; Rath, Peterfreund, Bayliss, Runquist, & Simonis, 2012).

Interestingly, we found that the SFSU SI participants performed better overall than the participants described in the UMKC SI data provided on their webpage. This observation has been reinforced by Dawson et al. (2014) in their comparison of a number of SI programs that increased attendance leads to increased success (pp. 625–626). We assigned the SFSU SI sections with a new college-wide unique course prefix (SCI) that did not conflict with department designations and had the same course number as the corresponding Biology, Chemistry, Physics, and Mathematics courses they supported. The creation of a course title for each of the one-unit SI sections provided the opportunity to track each student who participated or did not participate in SI, allowing us to compare performance in the respective courses. In one study, we were able to compare the performance of 2,164 SI participants out of 10,259 students enrolled in 137 SI sections from 1999 to 2005 (Peterfreund et al., 2008). The SI groups performed better than the non-SI groups in all but one case (pre-calculus). Also, in most cases, the proportion of SI participants passing the course was substantially higher than that of the non-SI participants. We have included, in the sections "Program Outcomes" and "Stabilization of SI Budget Support", the results obtained for another SI group of 7,451 that includes data on the performance of HU students as well as the differences in performance by gender.

Implementation Plan

Despite the favorable outcomes for students who participate in SI programs, these programs are often considered expensive, and it has been a challenge to incorporate the program into the institutional structure permanently. When external funding runs out, and SI offerings drop, participation and success rates decrease dramatically (Fullilove & Treisman, 1990).

The SI program at SFSU was very successful, and students in the SI courses showed increased students' performance and progression through subsequent courses in a sequence despite having lower academic indicators compared to non-SI students (Rath et al., 2007). The SI benefits were manifold, especially for HU students who benefitted the most (Rath et al., 2007). From 1999 to 2009, the SI program at SFSU was serving ~500 students/semester with funding from a National Institutes of Health (NIH NIGMS MBRS RISE grant GM059298) grant. Then, due to loss of funding and the fiscal crisis, the program was not offered for one academic year. Unfortunately, during the hiatus, there was a significant loss of workshop materials and knowledge as continuity was broken. In 2010, the program was restarted but served many fewer students. Then in 2014, the program was funded by a new grant, (NIH BUILD grant UL1GM118985-03, 5TL4GM118986-03, 5RL5GM118984-03), to create undergraduate student teaching and mentoring opportunities of peers (as well as professional development around social contextual factors that are particularly relevant for HU students. These changes were taking place while the undergraduate student population was shifting, with significant growth in the number of first-year students, STEM majors, and Hispanic and Asian students.

Program Evolution

In 2009–2010, when the grants that supported SI came to an end, the Student Enrichment Opportunities (SEO) office transferred responsibility for the management and funding of SI to the College of Science and Engineering. Unfortunately, the California State University system suffered very large budget cuts that year, to the extent that the system took the unprecedented step of furloughing all staff and faculty for 10% of their time for the academic year. Due to this crisis, the College put the SI program on hiatus until funding could be found to run the program. The following year, Eric Hsu, Professor of Mathematics and the new director of the Center for Science and Mathematics Education (CSME), took over management of SI and re-launched it. Despite the lack of dedicated funding, each year CSME and the Dean of the College found one-time sources of funding to keep the program running in what was called "caretaker mode."

The SI program was on hiatus for only a year, but it was a surprisingly disruptive break, which damaged institutional processes, curriculum, and culture. Part of this was due to the break coming as the program passed in leadership to the College, which caused disruption and a re-negotiation of basic processes. The disciplinary departments demanded that the college (and thus CSME) take over the responsibility for deciding which classes to offer, acquiring space to house classes, scheduling classes, selecting instructors, hiring instructors, paying instructors, training instructors, supervising instructors, publicizing classes, and coordinating instructors of the main course and SI.

In addition to this disruption, the departments asked that the workshops be taught by graduate instructors, which is the standard model for other labs, and not by successful undergraduates, which was our preferred model. Furthermore, departments dropped the expectation that faculty would work closely and regularly with the SI instructors.

Another effect of the hiatus included literal loss of curriculum materials. Without a unit having responsibility for SI, workshop curricula items (such as worksheets and assignments) were lost due to retirements of SI supervisors and turnover among instructor without orderly handoffs. Thus, SI instructors, mostly graduate students, had to come up with their own materials without the support of the main course instructor.

In the re-start year, the program shrunk dramatically. Instead of offering seats for 20–40% of students in introductory science and math classes, there were only seats for about 10% of students. As a result, our data analysis revealed that SI participants and non-participants had similar performance. This indicated a mild positive effect, as the SI participants had lower SAT I test scores and high school GPAs. However, CSME always intended for this caretaker mode to be temporary until further funds could be found to support something closer to the original model.

In 2014, a team of science education and STEM faculty were granted federal NIH funds (SF BUILD) that were used to revamp the SI program. The following two significant changes were made: (1) undergraduates worked together in dyads to facilitate the SI workshops and (2) the SI facilitator training piloted workshops about the psychosocial factors important to success in STEM courses, including stereotype threat, implicit bias, and growth mindset.

THEORETICAL UNDERPINNINGS OF SI

The current research on why HU students do not achieve academic equity or persist in STEM higher education settings is grounded in three theoretical constructs. The work on stereotype threat and critical-race theory contends that the academic setting is fraught with cues that convey threat, non-inclusion, and even inferiority to HU students (Estrada et al., 2018). While much work has described the threatening environment for HU students, this chapter will focus on theories and strategies that decrease the threat experienced by HU students by affirming their goals and values, validating their experiences, and using near-peer instructors/mentors to personalize the learning environment through social and cognitive congruence.

Communal Goal Affirmation Theory

Communal goal affirmation theory (also called *goal congruency theory*) suggests that HU students and women may not pursue STEM fields because the goals of these fields, along with the environments where the training occurs, are not congruent with their personal goals and values (Diekman, Brown, Johnston, & Clark, 2010; Diekman, Clark, Johnston, Brown, & Steinberg, 2011). Research, examining why HU students persist, has called into question the educational environment's emphasis on individualistic goals that focus on personal gain (Abele & Wojciszke, 2007; Diekman et al., 2010; Stephens, Fryberg, Markus, Johnson, & Covarrubias, 2012). The (Western) culture of science in the United States is also predominantly individualistic and lacks emphasis on communal goals (Diekman et al., 2011). Communal goals, in contrast, are based on interdependence and the benefits of working with and helping others (Diekman et al., 2011). Goal congruency theory suggests that people pursue and persist toward goals that match their values (Sansone, Sachau, & Weir, 1989). When there is congruence, there is increased motivation (Isaac, Sansone, & Smith, 1999) and, interestingly, increased feelings of belonging in academic contexts. When applied to STEM field progression, research indicates that some students perceive and then avoid STEM fields that emphasize agency and personal success as opposed to helping others and having community impact (Abele & Wojciszke, 2007; Diekman et al., 2010; Smith, Cech, Metz, Huntoon, & Moyer, 2014). Smith et al. (2014) conclude their work by stating that it is critical for STEM programs to "foster an environment where communal, as well as individualistic, work goals can be afforded" (p. 425).

Validation Theory

Rendón, Linares, and Munoz (2011) define validation as the "intentional, proactive affirmation of students by in- and out-of-class agents (i.e., faculty, students, and academic affairs staff, family members, and peers) in order to (1) validate students as generators of knowledge and as valued members of the college learning community; and (2) foster personal development and social adjustment" (p. 12). Rendón (Rendón, 1994) found that students who experienced validating educational situations were more likely to persist despite academic and social challenges. Rendón's theory for validation proposes that students are most likely to succeed in college if they are empowered to view themselves as capable learners through academic and interpersonal validation by agents inside and outside the classroom. Thus, validation can be seen as a precondition for academic integration (Tinto, 1997), which predicts higher rates of student retention and persistence (Braxton, 2002).

Affirming students as knowledgeable and valued provides students with a sense of self-worth that helps them succeed (Pérez & Ceja, 2010). Numerous studies have emphasized the importance of support and encouragement (Crisp, Taggart, & Nora, 2014; Nora &Cabrera, 1996) and demonstrated the critical role of affirmation for Latino students. In a longitudinal investigation of 146 Latino STEM majors, strong relationships with the faculty led to better academic performance (Cole & Espinoza, 2008). The study suggests that faculty play an important role in creating an engaging environment for Latino students and that departure from STEM majors is because of a "chilly academic environment." It found that faculty accessibility and efforts to engage students outside of the classroom are positive indicators of the environment of the student's chosen field.

While many qualitative studies use Rendón's (1994) theory of validation as a foundational framework to examine the experiences of non-traditional students, there is only one report of validation theory being used in an empirical study. In a study of community college students, Barnett (2011) tested the validation construct as a type of faculty/staff interaction that predicts students' academic integration and intent to persist in college. To do this, Barnett (2011) operationalized students' perception of faculty validation into four constructs: students being known and valued, caring instruction, appreciation for diversity, and mentoring (examples of items are shown in Table 1). Within a sample of 333 community college students, faculty validation's effect on intent to persist

Table 1. Examples of the Four Constructs of Faculty Validation (Barnett, 2011).

Known and Valued	Caring Instruction
• I feel accepted as a capable student by my instructors • My instructors show that they believe in my ability to do the class work	• It seems my instructors really care about whether I am learning • My instructors seem to genuinely care how I am doing
Appreciation for Diversity	**Mentoring**
• People of color are encouraged to contribute to the class discussion • Women are encouraged to contribute to the class discussion	• I have had at least one instructor in this college who helped me to believe in myself • At least one instructor has talked with me about my personal goals at this college

was indirect, mediated through students' sense of academic integration. The strongest predictor of a sense of academic integration was caring instruction, though the other three constructs were also significant predictors.

Social Development Theory

Vygotsky recognized that learning takes place through interactions and communications with the social environment including peers, teachers, and other experts (1962). According to Vygotsky (1980), learning is optimized if a distance between what is already known and understood and what must still be learned is just enough to stimulate active inquiry by the student, a distance called the "zone of proximal development." Near-peers that function as "more knowledgeable others" (Vygotsky, 1980) may sense this zone of proximal development much more easily than content experts, who may not always understand the cognitive problems student experience when processing new information (Topping, 2005). The cognitive congruence between peer and near-peer can help students construct knowledge.

Near-peers also differ from formal teachers because they cannot use the same rewards/punishments to motivate students. Instead, they offer interpersonal rewards such as friendship and role modeling. The social congruence between a student (peer) and more knowledgeable other (near-peer) can result in a trusting relationship that might facilitate self-disclosure of ignorance and cognitive errors, enabling subsequent diagnosis and correction (Topping, 2005). Near-peers are more interested in the daily lives, study experiences, and personalities of those being taught than regular teachers (Moust & Schmidt, 1995). Role models build confidence (Lockspeiser, O'Sullivan, Teherani, & Muller, 2008) and also can help reveal the unwritten rules to survive/persist ("hidden curriculum") in a STEM major.

Summary

The research on communal goal affirmation, validation theory, and social development theory describe how the social context through peers, near-peers, and instructors can better affirm students, connect to their values, and develop a more validating, supportive, and personal learning environment. Further research is needed to better understand the nuances of how HU and majority students are alike and differ with regard to these relationships. These theories provide strong evidence that interventions that provide affirmation and validation cues along with social, cognitive, and goal congruence can connect students to their academic community, which relates to greater persistence and more equitable outcomes.

INSTITUTIONAL CONTEXT

San Francisco State University Context

Founded in 1899 as a teacher's college, SFSU is a comprehensive public university that awards baccalaureate and master's degrees. With a total enrollment of

29,465 in Fall 2014, SFSU is the sixth largest of the 23 campuses in the California State University (CSU) system and the seventh largest of all public master's granting colleges and universities in the nation. SFSU and the other CSU campuses continue to provide the most affordable university education in California and frequently represent the *only* affordable option for economically disadvantaged students. A total of 47% of SFSU undergraduates receive Federal Pell Grants, and 78% of all SFSU students worked part- or full-time to help meet the cost of their education.

SFSU is one of the nation's most ethnically and culturally diverse campuses. Of those declaring their ethnicity in Fall 2014, students of color comprised 68.2% of the undergraduates and 42.7% of the graduate students. The combined undergraduate and graduate student population are as follows: 5.4% African American, 0.3% Native American, 27.3% Hispanic, 32.3% Asian/Pacific Islander, 28.3% White, and 6.5% "two or more races." In total, 10,235 of these are from the four federally designated underrepresented ethnic minority (URM) groups. Approximately 57% of SFSU undergraduate students are female.

SFSU, like the state of California and the nation at large, has witnessed significant growth in the Latino population. In Fall 2014, SFSU reached 25% enrollment of Latino undergraduates to become designated a Hispanic-serving Institution (HSI). In Fall 2016, Latino students accounted for 34.5% of the undergraduate population. This university and HSIs like it are challenged to support and graduate a population that is largely low-income and first-generation college-going (Table 2).

SFSU has also seen incredible growth in the number of science, technology engineering, and math (STEM) majors. While the campus has not grown significantly in the number of total undergraduates, there has been growth in the number of first-time first-year students. Many of these students matriculate with the intention of pursuing a STEM major, with significant growth coming from the Hispanic/Latino students (Fig. 1). With this changing STEM student profile, the SI program at SFSU plays an even more important role in providing support for students, particularly HU students, to successfully excel in and complete a STEM major.

Table 2. SFSU Undergraduate (UG) Enrollments in STEM.

SFSU Undergraduate Enrollments in STEM	F 2013	F 2014	F 2015
Total UG enrollment SFSU	25,176	25,055	25,776
Total UG enrollment SFSU, Hispanic/Latino*	6,144	6,244	7,563
Percentage Hispanic/Latino UG enrollment, Pell eligible*	57.4%	58.1%	60.3%
Total UG enrollment, SFSU Hispanic/Latino STEM*	1,166	1,204	1,457
Percentage Hispanic/Latino in STEM UG enrollment – Pell eligible*	58.5%	60.2%	62.9%

Notes: All figures exclude non-matriculated (transitory) students and 2nd BAs; *Hispanic headcounts are US Citizens or Permanent Residents only.

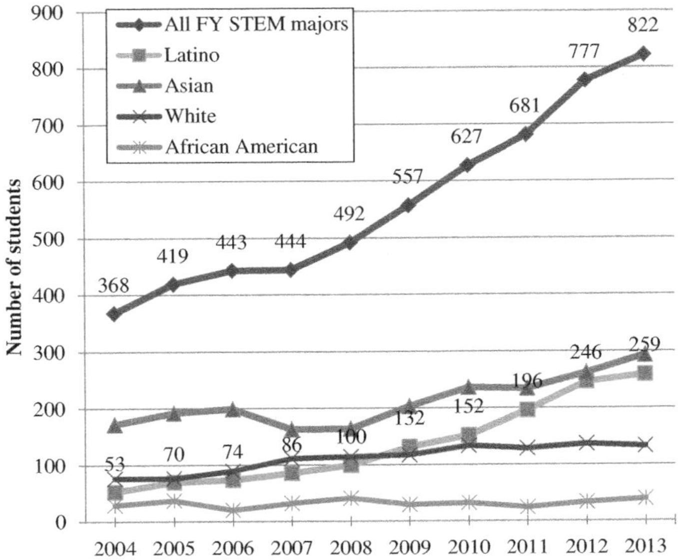

Fig. 1. Growth in the Number of First-year STEM Majors at SFSU from 2004 to 2013. The Numbers for Each Year are the Number of Students, all FY STEM Majors (Top Numbers) and Latino FY STEM (Lower Numbers).

PROGRAM GOALS, STRUCTURE AND COMPONENTS, AND IMPLEMENTATION, ADAPTATION DUE TO INSTITUTIONAL CONTEXT

Program Goals

The goal of the SI program is to support student success, particularly for HU students, in STEM courses that may be thought of as "high risk" (Martin & Arendale, 1993). At SFSU, these high-risk courses are characterized by large classes in which each student has little opportunity to interact with the professor, voluntary or unrecorded class attendance, and infrequent examinations that focus on higher-order cognitive skills that are practiced with limited feedback. The changing undergraduate STEM student profile at SFSU, along with the development of knowledge about social–contextual factors important to learning, brought about a rethinking of how we should teach and serve our SI students. Beginning in 2014, we changed the SI instruction model from lecturers and Graduate Teaching Assistants focusing on content and cooperative group work (1999–2009), to undergraduate leaders who work in pairs to facilitate the SI section. These undergraduates go through extensive SI leader training (30 plus hours) that now focuses on social contextual factors that affect performance that lead to inequities (such as stereotype threat and implicit bias) and provide teaching and learning strategies to overcome them (affirming communal goals, growth mindset, "giving back").

Structure and Components

At SFSU, the SI program exists in the form of one-unit supplemental courses that primarily support large-lecture science courses. These courses meet for 75 minutes once per week to go over concepts and do problems in groups based on material covered in the lecture course. Students who enroll in the supplemental course commit to attending workshops each week for the entire term, and their grade in the supplemental course is based solely on attendance and participation.

At SFSU, the SI courses are led by pairs of undergraduate student facilitators (who are all STEM majors) trained in active learning strategies. When selected, the facilitators make a one-year commitment to the program. This year-long commitment gives facilitators time to develop into strong leaders and also builds a tight-knit, supportive community of facilitators. Each year, more than half of our facilitators return to teach for another year. Thus, each section has a returning "experienced" facilitator who works with a new "novice" facilitator. The experienced facilitator mentors the novice facilitator in how to structure and run the SI classroom and after the first term, the two facilitators have become equals. Notably, these partnerships have decreased the amount of time that facilitators spend with program staff discussing classroom issues and barriers, thus allowing more time for professional development during staff meetings.

The partnerships are also very valuable for the SI students. They get to see two different perspectives on how to view the same question, which shows students that diversity in experience is valuable to the scientific enterprise. Frequently, the two facilitators are not from the same major, and they can tell their students how a concept will be applied in different contexts. For most students, this is the only time they will see two different instructors approaching the same content and sharing their own understanding. Students also benefit from having two near-peer role models who share their own experiences for how to successfully navigate the institution. These two role models also share more similar career goals to the students, compared to instructors who are disciplinary experts and add even greater value to the SI student.

Like other SI programs, the SI program at SFSU relies on a program coordinator, faculty to lead each disciplinary group (Biology, Chemistry, Computer Science, Math, Physics), and the cooperation of instructors to ensure that what is taught in the SI section matches the lecture. The apprenticeship model where experienced facilitators work to train novice facilitators in the classroom, along with more than half of our facilitators returning to facilitate for a second or third year, has resulted in an efficient transfer of materials and knowledge-base for how to successfully run an SI section.

Implementation and Adaptation to Institutional Context

At SFSU, the SI program has been exclusively administered through the College of Science and Engineering (CoSE). The SI program is not drop-in. Students self-select to enroll in a one-unit course that is supplemental to the lecture course. Each course is listed under the SCI prefix, with the course number frequently matching the lecture course number. For example, the SCI 115 course supports the CHEM 115: General Chemistry 1 course. Enrollment of students into the SCI courses allows the program and institution to follow students as they progress through their major.

PROGRAM OUTCOMES

In addition to the extensive data analysis provided for the SFSU SI program in three peer-reviewed publications (Peterfreund et al.; 2008; Rath et al., 2007; Rath et al., 2012) and a review by Dawson et al. 2014, we present additional unpublished data in this chapter that addresses issues of the impact of SI by gender and comparison of HU versus majority students' performance. The data presented and discussed were derived from the participants in the courses listed in Table 3. All of the SI sections have a course number, attendance is required, and the SI sections are graded, allowing the use of institutional data for analysis (which involves grades by semester and SI status for the supported STEM courses between Fall 1999 and Spring 2005). A participant was excluded if they did not complete (receive a grade) in the SI section or course. A total of 7,451 students were enrolled in the courses, and 18% (1,345) participated in SI. The courses and SI analyzed include Biology I, Chemistry I, Genetics, Organic Chemistry I, and Calculus I. Physics I was not included in this report.

In an attempt to understand the relative preparation of each group (SI and non-SI), we compared the SAT scores and high school GPA for each group to control for any obvious bias that might occur when students self-select to participate in the SI sections. As is apparent in Table 4, there was a significant difference in the SAT I Math and Verbal scores ($p < 0.005$ % value), indicating a statistically significant difference favoring the non-SI group enrolled in Biology I. There is no significant difference between the groups relative to high school GPA. A comparison of SAT I Math scores correlated with the course grade is presented in Table 4. As you can see for the students receiving an "A" or "B" grade, the SI students had numerically lower but not significantly different SAT I Math scores; however, a comparison of SI and non-SI students receiving a "C" grade reveals a significantly ($p < 0.005$) lower SAT Math score for the SI participants. The pattern of SI participants having significantly lower SAT

Table 3. SI Supported Courses.

Course	Semesters SI	# SI	# Non-SI
Introduction to Biology I	13	394	990
Genetics	11	195	568
General Chemistry I	10	278	1,511
Organic Chemistry I	11	209	626
General Physics I	12	113	1,216
Calculus I	6	156	1,195
		1,345	6,106

Table 4. Biology I: Student Background.

		SI	Non-SI
SAT I Math		**492**	**518**
SAT I Verbal		**471**	**498**
High school GPA		3.20	3.17
Grade in Class	SI SAT Math	Non-SI SAT Math	Difference
A	570 ($n = 22$)	581 ($n = 46$)	−11
B	527 ($n = 39$)	536 ($n = 72$)	−9
C	447 ($n = 67$)	510 ($n = 123$)	**−63**

Note: Bold = $p < 0.005$.

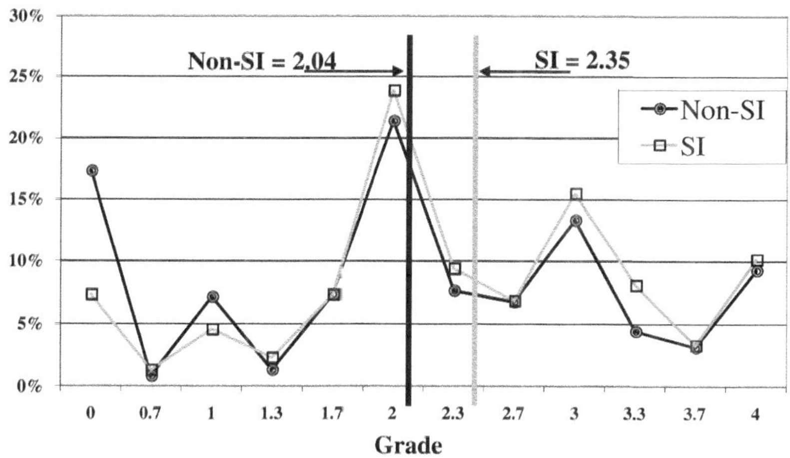

Fig. 2. Intro Biology – Course Grades. *Note*: SI: "C−" or better – 85%; non-SI: "C−" or better – 73%.

scores in comparison with non-SI participants occurs repeatedly and, in our opinion, reflects the benefit of using SI rather than a remedial approach to the successful launching of entry-level STEM majors. That is, it appears the SI experience is compensating for deficiencies that would otherwise limit a passing performance, which is critical to avoiding repeating the course and/or dropping out or changing the major to a non-STEM discipline.

The distribution of Biology I course grades is presented in Fig. 2. While at first glance it may not seem like much of an effect, the decrease in the number of "F" grades from 18% for non-SI to 8% for SI is apparent. In comparing the performance by gender (Fig. 3), we see a decrease for males in "F" grades from 20% to 4% while also seeing an increase in "B+" grades from 4% to 12% and "A" grades from 7% to 11%. This sort of granularity is not as apparent in

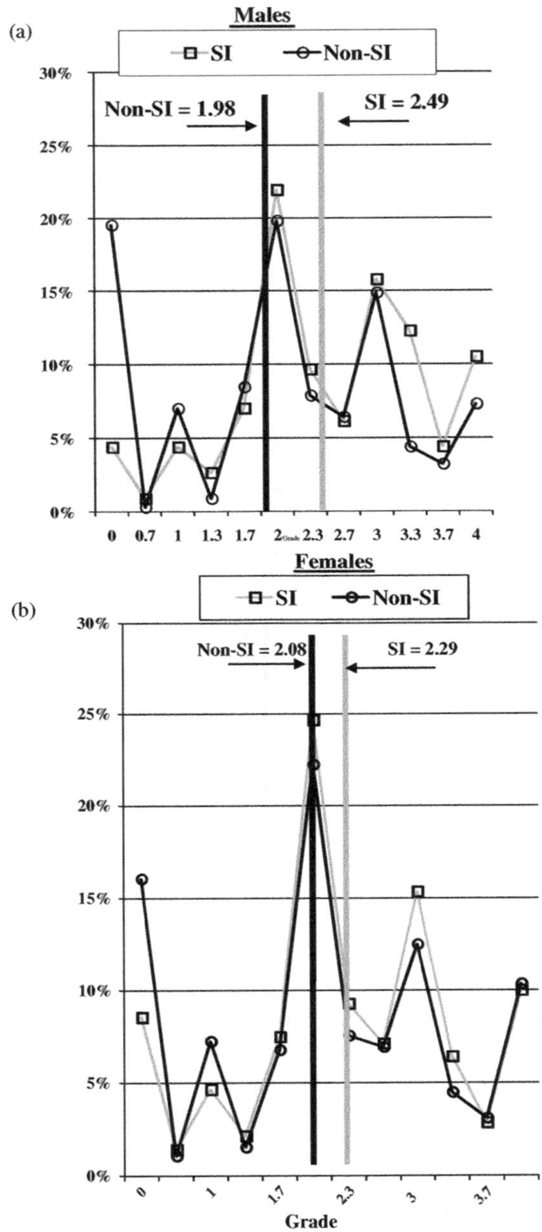

Fig. 3. Biology I: Gender Differences: (a) Males and (b) Females. *Note*: SI – 29% Males, non-SI – 35% Males; SI – 71% Females, non-SI – 65% Females.

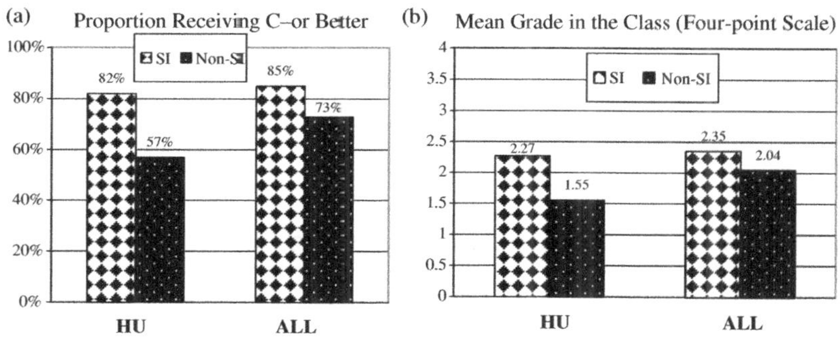

Fig. 4. Biology I: Historically Underrepresented (a) Proportion Receiving "C−" or Better and (b) Mean Grade in the Class (Four-point Scale).

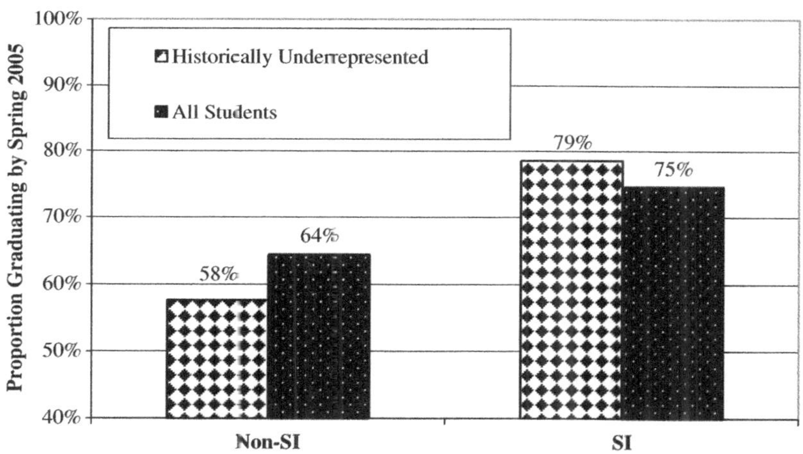

Fig. 5. Graduation Pattern Intro Bio SI Status: HU and All Students based on Students Enrolled Fall 99–Spring 01.

Fig. 1, perhaps because 71% of the SI participants were female. For females, while not as dramatic, the "F" grades drop from 16% to 8%.

As is summarized in Fig. 4, the proportion of students receiving a "C−" grade (minimal grade required to advance) is significantly higher ($p < 0.005$) for both the HU and majority students participating in SI. This is also true for the mean grade in the class. In addition to SI impacting the immediate performance in a class, it also has a cumulative impact on time to and rate of graduation for the participants as seen in Fig. 5. As you can see, 58% ($N = 52$) of the non-SI HU students graduated as compared to 79% ($N = 28$) of the SI HU participants in the same timeframe. In comparison of all non-SI with all SI participants (included the HU students), we see an increase in graduation from 64% ($N =$

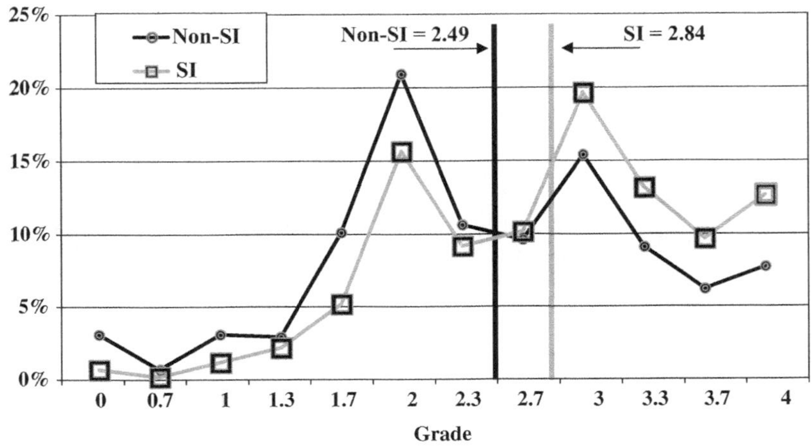

Fig. 6. Genetics: Course Grades. *Note*: SI: "C−" or better – 96%; non-SI: "C−" or better – 91%.

312) to 79% ($N = 126$). Enrolling in SI for Biology I has a dramatic impact on passing the course and graduating sooner (Figs. 4 and 5).

We have also included two upper-division courses (Genetics and Organic Chemistry) to determine how SI might impact student performance. As one might expect after successfully navigating two years of undergraduate science and math courses, we do not see much difference among the participant's backgrounds; however, we do see a positive impact of SI on grades. As you can see for Genetics in Fig. 6, there is a clear shift to higher grades by the SI participants. For the SI participants, the "B" grades are 20% compared to 15% for the non-SI, "B+" grades at 14% compared to 9%, "A−" grades 10% compared to 6%, and "A" grades 13% compared to 8%.

A review of the grades for Genetics broken out by gender again reveals a different picture for males in comparison with females. As was the case for Biology I, the enrollment of females in SI was higher than males (69% in SI and 65% in the non-SI). In a review of Fig. 7a for males, we see an overall shift to higher grades for males that is quite dramatic for the "B+" from 6% for non-SI to 13% SI, "A−" from 8% to 12% and "A" from 4% to 18%. For the female SI participants (Fig. 7b), there is also a positive shift to higher grades but less dramatic with an increase from 16% "B" for non-SI to 24% for SI. Both the mean grade in the class ($p < 0.005$) and the proportion of SI recipients receiving a "C−" grade ($p < 0.005$) were significant (Fig. 8).

The two most challenging first-year courses for entering STEM majors are chemistry and calculus. As you can see in Table 5, the students enrolling in Chemistry I SI have significantly lower SAT I Math and Verbal scores and similar high school GPAs. That said, the Chemistry I students who enrolled in SI did significantly better ($p < 0.005$) than the non-SI participants. As you can see in Fig. 9, there is a clear shift to higher grade performance by SI participants. SI

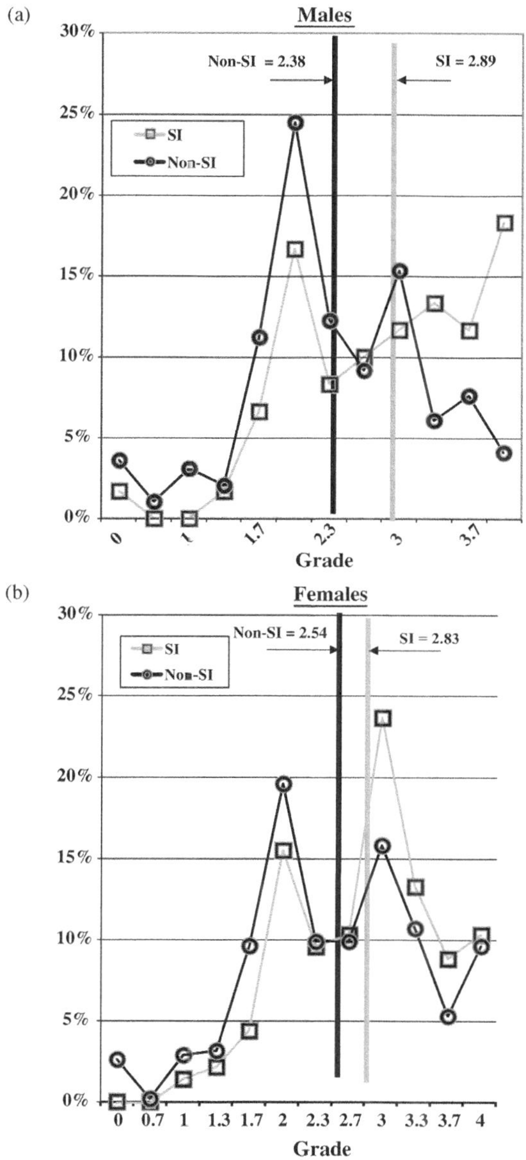

Fig. 7. Genetics: Gender Differences: (a) Males and (b) Females. *Note*: SI – 31% Males, non-SI – 35% Males; SI – 69% Females, non-SI – 65% Females.

had 14% "B" grades as opposed to 10% non-SI, 11% "B+" compared to 9% and 11% "A−" compared to 6% for non-SI. The comparison of male and female participants demonstrates the pattern we have seen with Biology I and Genetics with a more dramatic change for males at the higher grade levels (13% "B"

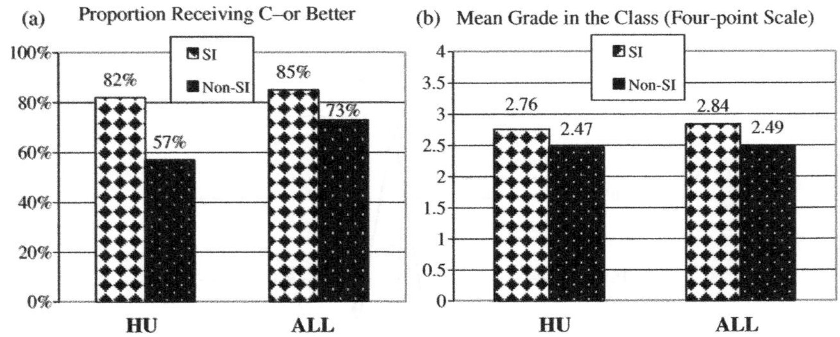

Fig. 8. Genetics: Historically Underrepresented (a) Proportion Receiving C- or Better and (b) Mean Grade in the Class (Four-point Scale).

Table 5. Chemistry I, SAT I and High School Grades.

	SI	Non-SI
SAT I Math	**488**	**523**
SAT I Verbal	**472**	**490**
High school GPA	3.14	3.19

Note: Bold = $p < 0.005$.

grade for SI and 9% for non-SI, 14% "A−" for SI and 7% for non-SI, 5% "A" for SI and 5% for non-SI); but, the female participants show a strong, consistent overall shift of the entire curve to higher grades (Fig. 10). Both the mean grade in the class and the proportion of SI recipients receiving a "C−" grade were significant ($p < 0.05$; Fig. 11).

Organic Chemistry I, like Genetics, is a challenging upper-division course we included in the SI program to determine whether SI would have a differential impact. As you can see in Figs. 12 and 13, SI does have a significant ($p < 0.005$) effect. The threefold drop in "F" grades from 9% for non-SI to 3% for SI participants with a concomitant increase in "A−" grades for participants from non-SI 7% to SI 14% and "A" grades for non-SI of 8% to 13% for SI is dramatic. Both the mean grade in the class ($p < 0.005$) and the proportion of SI recipients receiving a "C−" grade were significant ($p < 0.05$) (Fig. 13).

Calculus has been one of the more difficult courses to support SI because of the number of different instructors (and approaches) involved. While the following results do demonstrate a moderate overall impact on grades, what has not been readily apparent before is the impact SI has had on many individuals with low SAT I Math scores. As you can see in Table 6, the SI participants have significantly ($p < 0.005$) lower SAT I Math scores and according to our Math faculty would not likely pass Calculus I with a 482 SAT I Math score. This score alone would not qualify a student for calculus; rather, it would call for remedial

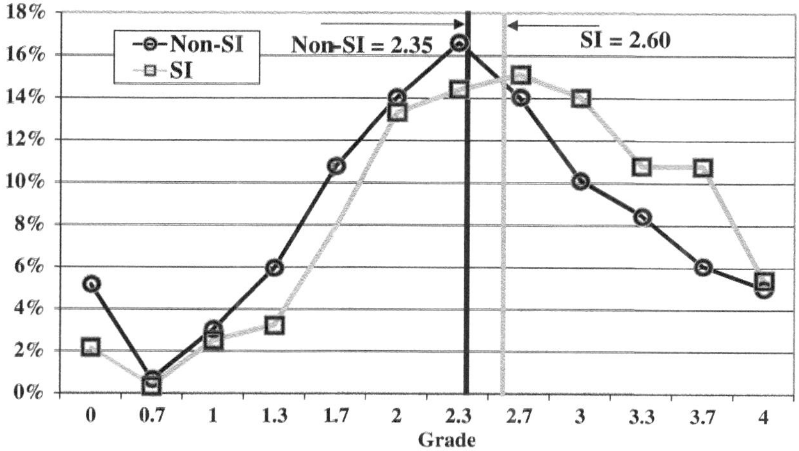

Fig. 9. Chemistry I: Course Grades. *Note*: SI: "C−" or better – 92%; non-SI: "C−" or better – 85%.

placement in Algebra. One way to consider the impact of SI is to consider how much ground was made up *during* the SI and course participation. A review of Table 7 provides some insights into this point. When we showed Table 7 to our Math department chair (on the faculty for 40 years), he told us it was hard to believe that 12 students with an average SAT I Math score of 470 could even pass let alone achieve a "B" grade. At this point, after being uninterested in SI, he became a believer and took active measures to support the SI program. He was also specifically looking at the 62-point difference from the non-SI students receiving a "B" grade, and he was amazed that the SI students could make up for the difference during the course.

Calculus I: Student Background

The overall performance in Calculus I shows a slight improvement in grades for SI participants with a higher pass rate (Fig. 14). The performance of HU SI participants was significantly higher ($p < 0.05$). There was no significant difference in the mean grade between SI and non-SI (Fig. 15).

In summary, we have found a number of significant outcomes and several unexpected insights relative to the participation and performance of HU students as well as gender differences. Relative to participation in SI, we see:

- Higher course performance (passing and progressing).
- SI takers tend to come to SFSU with weaker academic indicators.
- Takers show higher rates of taking subsequent courses in the major.
- Women are more represented, however, men demonstrate more dramatic improvement.

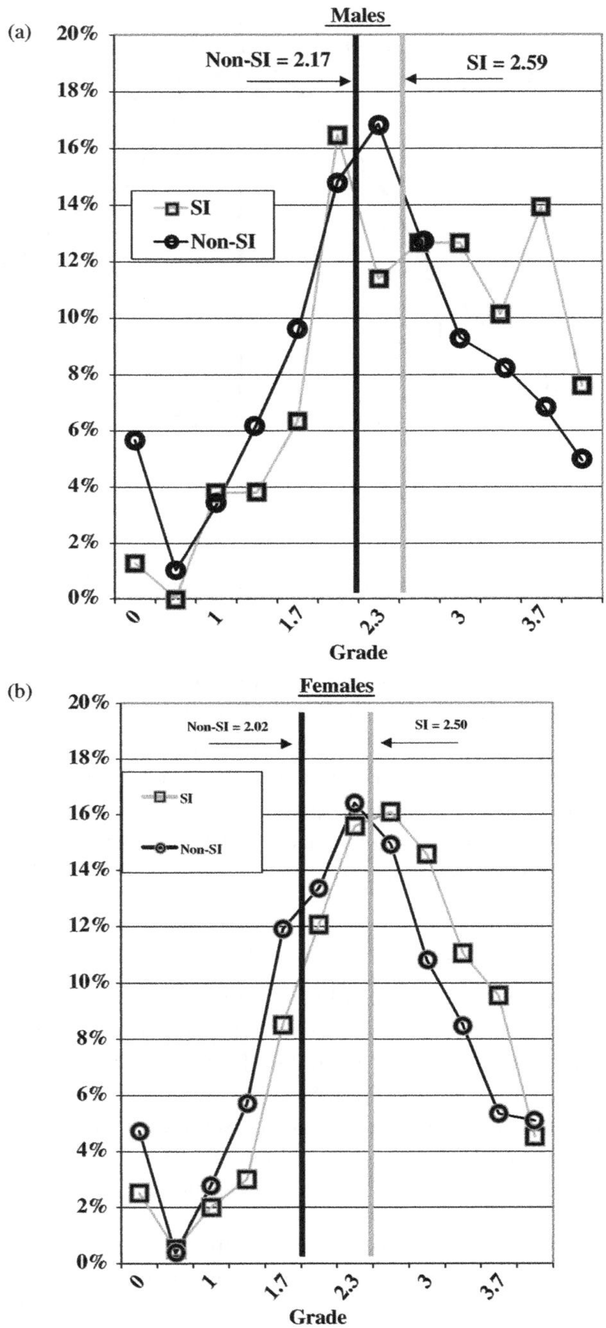

Fig. 10. Chemistry Gender Differences: (a) Males and (b) Females. *Note*: SI – 28% Males, non-SI – 46% Males; SI – 71% Females, non-SI – 54% Females.

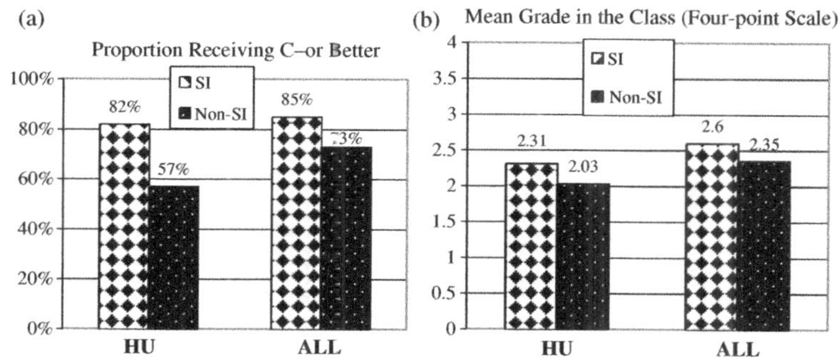

Fig. 11. Chemistry I: Historically Underrepresented (a) Proportion Receiving "C–" or Better and (b) Mean Grade in the Class (Four-point Scale).

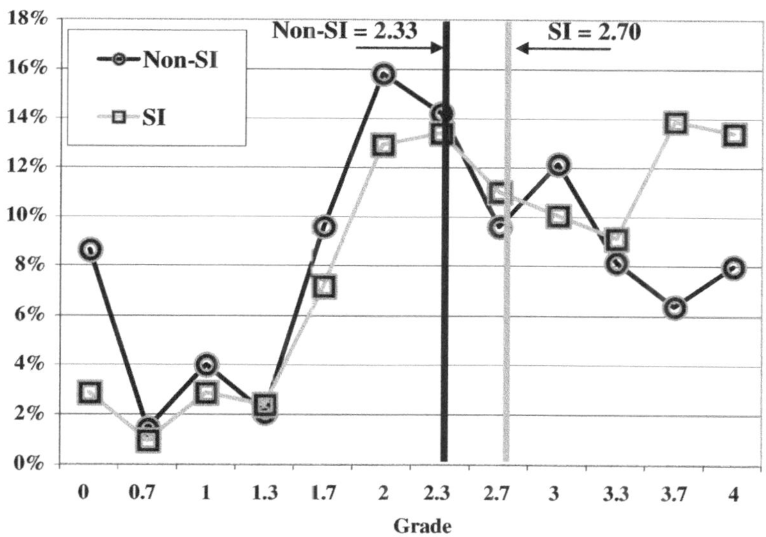

Fig. 12. Organic Chemistry I: Course Grades.

- Performance levels for HU students taking SI to reach and often surpass non-SI takers.
- HU students participate in SI more frequently than majority students.
- SI has a significant impact on upper-division courses resulting in higher grades.

The SI efforts described for these 7,451 participants cost an average of US $115 per SI participant. We calculated that the SI intervention prevented at least 170 students from D/F grades (assuming the failure rate of non-SI students) and also played a role in these students staying in the major as well as college. The

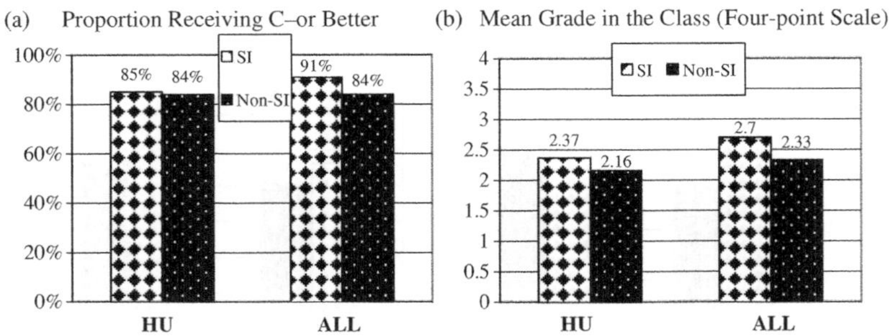

Fig. 13. Organic Chemistry I: Historically Underrepresented (a) Proportion Receiving "C−" or Better and (b) Mean Grade in the Class (Four-point Scale).

Table 6. SAT I and High School GPA.

	SI	Non-SI
SAT I Math	**482**	**526**
SAT I Verbal	466	480
High school GPA	3.14	3.20

Note: Bold = $p < 0.005$.

Table 7. SAT I Math Score Versus Course Grade.

Grade in Class	SI SAT Math	Non-SI SAT Math	Difference
A	519 ($n = 12$)	564 ($n = 118$)	−45
B	470 ($n = 12$)	532 ($n = 84$)	**−62**
C	459 ($n = 11$)	508 ($n = 125$)	−49

Note: Bold = $p < 0.005$.

savings in lost tuition and the cost of more recruitment to replace these students more than justify the costs.

STABILIZATION OF SI BUDGET SUPPORT

Since 2010, SI had been funded by finding one-time money sources each year. Finally, in 2018, an important change in university accounting allowed the institutionalization of the program into the regular college budget. Up until 2018, SFSU colleges had been (under)funded with a fixed sum and then given augmentation money to cover the late addition of the most needed classes. Notably, the budget did not scale with students served. The new policy was to fund colleges in proportion to the number of units taken by students.

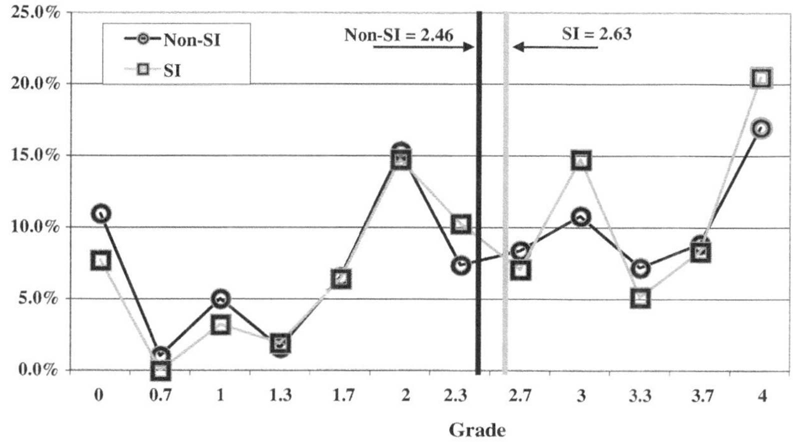

Fig. 14. Calculus I: Course Grades. *Note*: SI: "C−" or better – 87%; non-SI: "C−" or better – 81%.

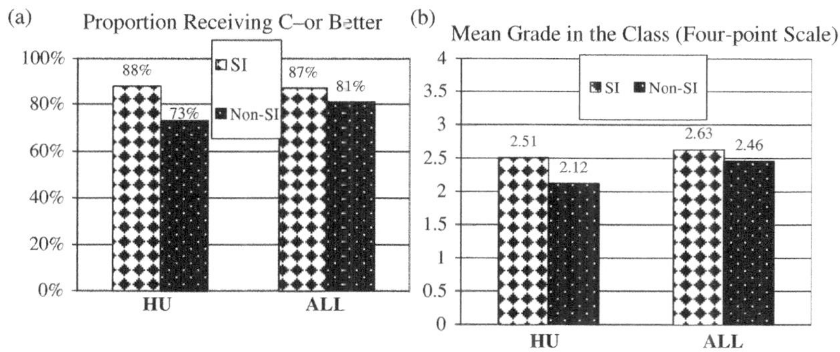

Fig. 15. Calculus I: Historically Underrepresented (a) Proportion Receiving "C−" or Better and (b) Mean Grade in the Class (Four-point Scale).

This was a sensible change and very beneficial for SI. SI had been set up as a proper course, with formal units and instructors of record. It had been run at high financial efficiency during the persistent underfunding. Thus, it turned out that the new formula, where funding was proportional to units taken, gave the College of Science and Engineering more than enough money to pay the student facilitators and instructors of record.

In fact, the formula gave sufficient funds to give the facilitators a small raise, to pay for the program director, and also give support for each department to provide release time for a department-level SI coordinator, who manages the relationships between the SI staff and the instructors of record for the parent courses. Beginning in Fall 2018, SI became a regular part of the college budget.

LESSONS LEARNED AND RECOMMENDATIONS

Undergraduates Are Fantastic SI Facilitators, Particularly When They Work in Pairs

Undergraduate SI facilitators bring a much-needed perspective to the lower-division science courses. As near-peer facilitators in the SI program, they facilitate the workshop and foster a collaborative learning environment where students engage in group work, peer teaching, and problem-solving (Arendale, 2004). They are also able to help students master content while developing and integrating learning and study strategies. While they are not content experts, they can provide personal relevance of content (for instance, they can highlight content that may be important in future classes) and personal connection that lecturers and graduate students are unable to provide.

Recommendations

- *Broadcast success widely across the institution.* A program like SI is seen by many administrators as supplemental and not a necessity and therefore not a funding priority. SI has gained traction at SFSU because of the recent CSU system-wide emphasis on improving graduation rates, which has put the spotlight on campus programs that improve student success. The current SI model is very effective and has remarkably consistent and positive data behind it as provided by the peer-reviewed publication of the SFSU SI program and results (Peterfreund et al., 2008; Rath et al., 2007; and Rath et al., 2012). Sharing these data has led the administration to view SI as a higher funding priority and a model to be spread to other subject areas.
- *Students make great ambassadors for the program.* Just having students and the facilitators sharing their inspiring personal experiences can be very effective, if you can get them in front of key decision-makers. Administrators appreciate having students telling their stories and putting a face and narrative to the numbers.
- *Work together* with institutional research, science education faculty, and evaluation specialists to evaluate the impact of the program. Every program benefits from formative assessment, and campus institutional research is usually the only unit that can get fine-grained student performance data. Science education faculty can help shape the data analysis queries and interpret the results. Furthermore, they can provide perspectives and instruments to be applied at the student level to get qualitative information about the workshops.
- *Program continuity is very important.* The SFSU program was only officially on hiatus for a year, but that gap and change in leadership caused a huge disruption in resources, culture, and practice as discussed above.
- *Formalize SI as much as possible.* In our case, we made the workshops proper courses in the catalog with proper instructors of record and proper units for students enrolled. This helped put the work into a format that the university bureaucracy could understand, which helped with salary, credit toward

tenure, and student incentives. It also allowed SI to quickly take advantage of improvements in the budget process to institutionalize it as part of the college budget.

ACKNOWLEDGMENTS

We thank Tatiane Russo-Tait and Jessica Fielder, former and current SI Directors, who were instrumental in changing the model to two student facilitators/section as well as in gaining budgetary support. These changes would not have been possible without them.

REFERENCES

Abele, A. E., & Wojciszke, B. (2007). Agency and communion from the perspective of self versus others. *Journal of Personality and Social Psychology*, *93*(5), 751–763. doi:10.1037/0022-3514.93.5.751

Arendale, D. (1994). Understanding the supplemental instruction model. In D. Martin & D. Arendale (Eds.), *Supplemental instruction: Increasing achievement and retention* (Vol. 60, pp. 11–21). San Francisco, CA: Jossey-Bass.

Arendale, D. (1997). Supplemental instruction (SI): Review of research concerning the effectiveness of SI from the University of Missouri-Kansas City and other institutions from across the United States. In S. Mioduski & G. Enright (Eds.), *Proceedings of the 17th and 18th annual institutes for learning assistance professionals: 1996 and 1997* (pp. 1–25). Tucson, AZ: University Learning Center, University of Arizona. Retrieved from http://www.lsche.net/?page_id=1044

Arendale, D. R. (2004). Pathways of persistence: A review of postsecondary peer cooperative learning programs. In I. M. Duranczyk, J. L. Higbee, & D. B. Lundell (Eds.), *Best practices for access and retention in higher education* (pp. 27–40). Minneapolis, MN: Center for Research on Developmental Education and Urban Literacy, General College, University of Minnesota. Retrieved from http://www.cehd.umn.edu/CRDEUL/pdf/monograph/5-a.pdf#page=37

Barnett, E. A. (2011). Validation experiences and persistence among community college students. *The Review of Higher Education*, *34*(2), 193–230. doi:10.1353/rhe.2010.0019

Bensimon, E. M. (2007). The underestimated significance of practitioner knowledge in the scholarship on student success. *The Review of Higher Education*, *30*(4), 441–469. doi:10.1353/rhe.2007.0032

Braxton, J. P. (2002). Introduction. In J. M. Braxton (Ed.) *Reworking the student departure puzzle*. Nashville, TN: Vanderbilt University Press.

Cole, D., & Espinoza, A. (2008). Examining the academic success of Latino students in science technology engineering and mathematics (STEM) majors. *Journal of College Student Development*, *49*(4), 285–300. doi:10.1353/csd.0.0018

Crisp, G., Taggart, A., & Nora, A. (2014). Undergraduate Latina/o students: A systematic review of research identifying factors contributing to academic success outcomes. *Review of Educational Research*, *20*(10), 1–26 doi:10.3102/0034654314551064

Dawson, P., van der Meer, J., Skalicky, J., & Cowley, K. (2014). On the effectiveness of supplemental instruction: A systematic review of supplemental instruction and peer-assisted study sessions literature between 2001 and 2010. *Review of Educational Research*, *84*(4), 609–639. doi:10.3102/0034654314540007

Diekman, A. B., Brown, E. R., Johnston, A. M., & Clark, E. K. (2010). Seeking congruity between goals and roles: A new look at why women opt out of science, technology, engineering, and mathematics careers. *Psychological Science*, *21*(8), 1051–1057. doi:10.1177/0956797610377342

Diekman, A. B., Clark, E. K., Johnston, A. M., Brown, E. R., & Steinberg, M. (2011). Malleability in communal goals and beliefs influences attraction to stem careers: Evidence for a goal congruity perspective. *Journal of Personality and Social Psychology*, *101*(5), 902–918. doi:10.1037/a0025199

Estrada, M., Eroy-Reveles, A., & Matsui, J. (2018). The influence of affirming kindness and community on broadening participation in STEM career pathways. *Social Issues and Policy Review, 12*(1), 258–297. doi:10.1111/sipr.12046

Ford, D. Y., & Grantham, T. C. (2003). Providing access for culturally diverse gifted students: From deficit to dynamic thinking. *Theory and Practice, 42*(3), 217–225. doi:10.1207/s15430421tip4203_8

Fullilove, R. E., & Treisman, P. U. (1990). Mathematics achievement among African American undergraduates at the University of California, Berkeley: An evaluation of the mathematics workshop program. *The Journal of Negro Education, 59*(3), 463–478. Retrieved from http://www.jstor.org/stable/2295577

Gasiewski, J. A., Eagan, M. K., Garcia, G. A., Hurtado, S., & Chang, M. J. (2012). From gatekeeping to engagement: A multicontextual, mixed method study of student academic engagement in introductory STEM courses. *Research in Higher Education, 53*(2), 229–261. doi:10.1007/s11162-011-9247-y

Hsu, E., Murphy, T. J., & Treisman, U. (2008). Supporting high achievement in introductory mathematics courses: What we have learned from 30 years of the Emerging Scholars Program. In M. Carlson & C. Rasmussen (Eds.), *Making the connection: Research and teaching in undergraduate mathematics*. Washington, DC: Mathematical Association of America.

Hurtado, S., Cerna, O. S., Chang, J. C., Saenz, V. B., Lopez, L. R., Mosqueda, C., ... Korn, W. S. (2006). *Aspiring scientist's research: Characteristics of college freshmen interested in the biomedical and behavioral scientists.* Los Angeles, CA: UCLA Higher Education Research Institute. Retrieved from http://heri.ucla.edu/PDFs/NIH/Summer%20Report.PDF

Isaac, J. D., Sansone, C., & Smith, J. L. (1999). Other people as a source of interest in an activity. *Journal of Experimental Social Psychology, 35*(3), 239–265. doi:10.1006/jesp.1999.1385

Lockspeiser, T. M., O'Sullivan, P., Teherani, A., & Muller, J. (2008). Understanding the experience of being taught by peers: The value of social and cognitive congruence. *Advances in Health Sciences Education, 13*(3), 361–372. doi:10.1007/s10459-006-9049-8

Martin, D., & Arendale, D. (1993). *Supplemental Instruction: Improving first-year student success in high-risk courses* (2nd ed.). Columbia: National Resource Center for the First-Year Experience and Students in Transition, University of South Carolina. Retrieved from https://files.eric.ed.gov/fulltext/ED354839.pdf

Moust, J. H., & Schmidt, H. G. (1995). Facilitating small-group learning: A comparison of student and staff tutors' behavior. *Instructional Science, 22*(4), 287–301. doi:10.1007/BF00891782

Nora, A., & Cabrera, A. F. (1996). The role of perceptions of prejudice and discrimination on the adjustment of minority students to college. *Journal of Higher Education, 67*, 119–148. doi:10.2307/2943977

Pérez, P. A., & Ceja, M. (2010). Building a Latina/o student transfer culture: Best practices and outcomes in transfer to universities. *Journal of Hispanic Higher Education, 9*(1), 6–21. doi:10.1177/1538192709350073

Peterfreund, A. R., Rath, K. A., Xenos, S. P., & Bayliss, F. (2008). The impact of Supplemental Instruction on students in STEM courses: Results from San Francisco State University. *Journal of College Student Retention: Research, Theory & Practice, 9*(4), 487–503. doi:10.2190/CS.9.4.e

Rath, K. A., Peterfreund, A., Bayliss, F., Runquist, E., & Simonis, U. (2012). Impact of Supplemental Instruction in entry-level chemistry courses at a midsized public university. *Journal of Chemical Education, 89*(4), 449–455. doi:10.1021/ed100337a

Rath, K. A., Peterfreund, A. R., Xenos, S. P., Bayliss, F., & Carnal, N. (2007). Supplemental instruction in introductory biology I: Enhancing the performance and retention of underrepresented minority students. *CBE-Life Sciences Education, 6*(3), 203–216. doi:10.1187/cbe.06-10-0198

Rendón, L. I. (1994). Validating culturally diverse students: Toward a new model of learning and student development. *Innovative Higher Education, 19*(1), 33–51. doi:10.1007/BF01191156

Rendón Linares, L. I., & Munoz, S. M. (2011). Revisiting validation theory: Theoretical foundations, applications, and extensions. *Enrollment Management Journal, 2*(1), 12–33.

Sansone, C., Sachau, D. A., & Weir, C. (1989). Effects of instruction on intrinsic interest: The importance of context. *Journal of Personality and Social Psychology*, *57*(5), 819–829. doi:10.1037/0022-3514.57.5.819

Schuman, H., Steeh, C., Bobo, L., & Krysan, M. (1997). *Racial attitudes in America: Trends and interpretations*. Cambridge, MA: Harvard University Press.

Seymour, E., & Hewitt, N. M. (2000). *Talking about leaving: Why undergraduates leave the sciences*. Boulder, CO: Westview Press.

Smith, J. L., Cech, E., Metz, A., Huntoon, M., & Moyer, C. (2014). Giving back or giving up: Native American student experiences in science and engineering. *Cultural Diversity and Ethnic Minority Psychology*, *20*(3), 413. doi:10.1037/a0036945

Stephens, N. M., Fryberg, S. A., Markus, H. R., Johnson, C. S., & Covarrubias, R. (2012). Unseen disadvantage: How American universities' focus on independence undermines the academic performance of first-generation college students. *Journal of Personality and Social Psychology*, *102*(6), 1178–1197. doi:10.1037/a0027143

Tinto, V. (1997). Classrooms as communities: Exploring the educational character of student persistence. *Journal of Higher Education*, *68*(6), 599–623. doi:10.2307/2959965

Topping, K. J. (2005). Trends in peer learning. *Educational Psychology*, *25*(6), 631–645. doi:10.1080/01443410500345172

Treisman, U. (1992). Studying students studying calculus: A look at the lives of minority mathematics students in college. *The College Mathematics Journal*, *23*(5), 362–372. doi:10.2307/2686410

Trujillo, G., & Tanner, K. D. (2014). Considering the role of affect in learning: Monitoring students' self-efficacy, sense of belonging, and science identity. *CBE-Life Sciences Education*, *13*(1), 6–15. doi:10.1187/cbe.13-12-0241

Vygotsky, L. S. (1962). *Thought and language*. Cambridge, MA: MIT Press.

Vygotsky, L. S. (1980). *Mind in society: The development of higher psychological processes*. Cambridge, MA: Harvard University Press.

MATH EMPORIUM INSTRUCTIONAL COURSE DESIGN: ALGEBRA COURSE EVOLUTION AT AN HBCU

Kathy Cousins-Cooper, Dominic P. Clemence-Mkhope, Thomas C. Redd, Nicholas S. Luke and Seong-Tae Kim

ABSTRACT

Before 2011, student performance rates in college algebra and trigonometry at North Carolina A&T State University (NCA&TSU) were consistently below 50%. To remedy this situation, the Mathematics Department implemented the math emporium model (MEM) instructional method. The underlying principle behind MEM is that students learn math by doing math (Twigg, 2011). The MEM requires students to work on math problems and spend more time on material that they do not understand while allowing them to spend less time on material that they do understand. Also, students receive immediate feedback on problems from teaching assistants as they work through their online assignments. After implementing the MEM, student pass rates improved for both the MEM and traditional sections. Data to date also show that female students outperform male students in both instructional models. Further study is needed to determine the factors that have caused improvement in pass rates in addition to the implementation of the MEM. Some important lessons learned by the NCA&TSU math faculty from implementing the MEM into the college algebra and trigonometry courses are that successful implementation requires a long-term commitment, internal and

Broadening Participation in STEM: Effective Methods, Practices, and Programs
Diversity in Higher Education, Volume 22, 237–263

ISSN: 1479-3644/doi:10.1108/S1479-364420190000022011

external collaborations, and the collective ability to determine what works for the local setting.

Keywords: Math Emporium; algebra; HBCU; active learning; student engagement

INTRODUCTION

The consensus among educators and researchers is that algebra is extremely hard to teach and to learn and has been singled out as the subject most responsible for low college retention rates. Researchers further agree that, due to current trends in the secondary school systems, students are not adequately prepared to fix the high school-to-college algebra gap by themselves from the perspective of study habits, time spent on task, and metacognitive development (Noble & Sawyer, 2002; Roderick, Nagaoka, & Coca, 2009). Consequently, first-year students generally do not perform well in entry-level college algebra courses, resulting in high failure rates and significant course repetition; first-year students at North Carolina A&T State University (NCA&TSU) fell into this category. Before 2011, pass rates in the entry-level college algebra course sequence at NCA&TSU were consistently under 50%. Since success in algebra is critically important to students' overall success (Rakes, Valentine, McGatha, & Ronau, 2010), such high failure rates, in a course sequence with an annual enrollment of about 3,500 students, were a serious threat to the University retention goals.

To address this unsatisfactory algebra gap and its negative impact on retention, NCA&TSU has undertaken an extensive algebra course and learning space redesign project, with the course redesign modeled after the math emporium model (MEM) (Twigg, 2011). The MEM eliminates lecture and uses commercially available interactive computer software combined with personalized on-demand assistance and mandatory student participation to improve study habits, increase time spent on task, and enhance metacognitive development. Originated at Virginia Tech in 1997, the MEM has now been replicated with modifications and used extensively for instruction in courses outside of mathematics, at hundreds of institutions worldwide. Robust demonstrations of the model's success are evident across all student population demographics and institution sizes, and indeed, Twigg (2011) dubbed it the "silver bullet of education."

While the lack of preparation from an academic view is well documented (HERI-CIRP, 2010), what has received less emphasis is the lack of preparation relative to study habits and time management, which plays a large role in the preparation of students and their successes. To address these issues, aggressive steps must be taken to help to modify behaviors learned over the previous 12 years of schooling. Among the most important obstacles to increasing the success rate of students is possibly helping them to understand the learning process and the roles and expectations of students and those who are educating them, as well as time management and studying outside of class. The MEM not only addresses the specific math skills learning needs of college students but also incorporates student-driven mechanisms to address problems specific to populations traditionally underrepresented in the STEM workforce. The MEM systematically addresses students' lack of algebra skills and intentionally increases student time on task, improves student academic support, and closes student–faculty perception gaps, all of which are expected to eliminate the identified skills gaps and lead to improved student performance.

It is widely recognized that improvements that impact student retention and long-term progression to graduation require an institutional commitment by senior administrators, faculty, students, and other stakeholder groups (Kuh, Kinzie, Buckley, Bridges, & Hayek, 2006; Tinto, 1987). Effective programs need to create an academic culture that promotes engaged student learning, rather than "plug and play" retention programs culled from symbols of "best practices." (Kuh et al., 2006; Tinto, 1987). In his study of interactive interventions, Arendale (2004) observes that their success depends on a high level of sustained institutional support while concluding such methods have the highest potential of improving student performance and retention. The institution must be willing to modify course policies and expectations, and the curriculum needs to be well developed to fully integrate all course components (Arendale, 1997; Fullilove & Treisman, 1990). This is exactly the strategic direction that NCA&TSU has set as its course of action in addressing the low algebra course pass rates. NCA&TSU administrators and faculty recognize the importance and urgency, even with difficult economic constraints, of continuing to maintain support for the critical tutorial, mentoring, and supplemental instruction processes needed or necessary to enhance student learning. The project is part of an institutional effort to renovate STEM gatekeeper course instruction and builds upon a substantial institutional commitment to the necessary capital resources, with extensive course structure and learning space redesign.

Twigg (2011) attributes the lack of student success in math courses to the fact that students "do not actually do the problems [...] or spend enough time with the material." The MEM is designed to enhance students "time on task" to more effectively facilitate learning. The MEM enables faculty to engage students in an innovative learning paradigm with new expectations and motivations. The model is supported by nationally developed software and made possible by the unique organization of space, complemented by high-end technology. The project addresses the needs of those students who meet the academic criteria for admission to University of North Carolina colleges, yet fail to exhibit college-level proficiency to master the collegiate curricula related to math and science. While MEM does not help all students equally, overall performance in algebra courses at NCA&TSU has improved, with pass rates now consistently in the 60–75% range.

THEORETICAL UNDERPINNINGS OF THE MATHEMATICS EMPORIUM MODEL

The MEM is an extensively researched, effective form of active learning for gatekeeper mathematics course instruction. Active learning is described as the use of learning activities in which the student is given a marked degree of ownership and control over the learning activities, and the student can actively participate and shape the learning experience (Kyriacou, 1992).

The MEM incorporates learning activities designed to include peer and instructor discussion, which have been documented to increase interactive student engagement and learning (Hake, 1998). It also develops and encourages peer-facilitated learning, which has been shown to promote communication and cooperative learning (Shechtman, Haertel, Roschelle, Knudsen, & Singleton, 2010; Slavin, 2010). This methodology provides a learning context that is social, interactive, and highly engaging. These structures have been proven effective in enhancing important cognitive and affective constructs that contribute to learning and motivation for a range of diverse students (Antil, Jenkins, & Wayne,1998; Fullilove & Treisman, 1990; Kuh et al., 2006; Zakaria, Chin, & Daud, 2010). The National Council of Teachers of Mathematics (NCTM) has determined that this sort of interactive pedagogy is critical for math learning for *all* students (NCTM, 2006). The shared social contexts provide support for students to construct mental models, solve problems, extend mathematics conceptual understandings and build higher-order thinking skills (Nebesniak, 2012; Slavin, 2010; Zakaria et al., 2010). The underlying principle of the MEM is that *students learn math by doing math, not by listening to someone talk about doing math*, while instructors serve as facilitators who guide students, pull concepts together, and help them avoid common pitfalls (Twigg, 2011).

The consensus in the literature is that algebra, both in high school and in college, is not only important for students' success in other courses but is critically important to students' overall success (Rakes et al., 2010). Unfortunately, algebra is recognized by educators and researchers as a subject that is very difficult both to teach and to learn (Drijvers, Doorman, Kirschner, Hoogveld, & Boon, 2014; Furner & Gonzales-DeHaas, 2011; Gordon, 2008; Herriott & Dunbar, 2009; Stacey, Chick, & Kendal, 2004). In fact, failure rates in college algebra courses are often higher than in other courses (Herriott & Dunbar, 2009), and algebra has in fact been singled out as the subject most responsible for low retention rates (Gordon, 2008; Roderick et al., 2009). The pervasiveness of dismal performance has led to the questioning by some authorities of the effectiveness of the traditional method of instruction in algebra (NMAP, 2008; Rakes et al., 2010).

It is well understood that the teaching method can affect students' attitudes, while anxiety and negative attitudes have been found to have negative impacts on college-level mathematics courses (Hegeman, 2015). Further, students' beliefs and expectations about the difficulties of mathematics, their levels of engagement, and their likelihood or perceived value of success can profoundly influence their achievement in mathematics (Lipnevich, MacCann, Krumm, Burrus, & Roberts, 2011; Singh, Granville, & Dika, 2002; Stevenson & Newman, 1986). This has led to recommendations (Furner & Gonzales-DeHaas, 2011) that educators devise methods to create a class climate that emphasizes thinking, encourages the asking of questions, and allows students to actively engage in activities that allow them to learn from and correct their mistakes. The MEM accomplishes all three of these recommendations. The MEM is designed to increase interactive student engagement by building community through properly structured group work, computer-assisted learning, and instructor discussion.

Interaction in an inclusive community, which is premised on the fact that the more time students spend interacting with classmates, the higher the level of satisfaction, has been suggested as an effective practice to enhance learning (Blimling & Whitt, 1999; Chickering & Gamson, 1987). However, while it is widely supported by experts as essential to successful mathematics instruction (NCTM, 2006; NMAP, 2008), cooperative learning fails without properly structured group work, with little individual accountability and no group goals (Fullilove & Treisman, 1990; Slavin, 2010). In fact, the long-term effects of lack of structure to promote proper interaction inhibit mathematics learning, where students turn group work into simply sharing answers rather than ideas (Gillies & Ashman, 2003). Many students erroneously expect and attempt to learn all of the information that is needed for a course as passive listeners in the classroom and do not put in enough of their own effort and time to master material (Antil et al., 1998; Gillies & Ashman, 2003). Unfortunately, for many, this learning model is the only one that they have ever known, so they do not even know that they need to look for help. There is a need, therefore, to change the culture of the students and their understanding of the learning process, particularly in their foundational math courses, to indoctrinate them into college expectations and time management expectations.

Peer tutoring is an intervention in which one student provides academic assistance to another (Robinson, Schofield, & Steers-Wentzell1, 2005). The MEM involves peer tutoring in that students are encouraged to help each other when an instructor cannot assist them immediately. Sturdivant and Souhan (2011) described a project designed to facilitate peer-to-peer teaching implemented in differential calculus. They noted that it was a very inefficient way to present new material and that it was very challenging for the students, but it is worthwhile as a developmental technique. Robinson et al. (2005) reviewed the literature and concluded that mathematical peer and cross-age tutoring have positive effects on math achievement for African American and other minority students as well as White students. The increase in math achievement extends to the tutor as well as to the tutee. Other school-related outcomes to peer tutoring include increased time-on-task and improved classroom behavior (Robinson et al., 2005). This project will develop and evaluate specific guided instructional tools for structuring effective peer interaction in math to meet the social and cognitive developmental needs of the students.

The use of computers and online environments in the mathematics classroom has mixed results in the literature. While Burch and Kuo (2010) reported that those students given online homework performed better on exams and had a higher retention rate than those given traditional paper homework, Mathai and Olsen (2013) reported that online homework beneficial to the incoming higher-skilled group and of no benefit to the lower-skilled students. In a study of developmental math classes, Spradlin and Ackerman (2010) found that while there was no statistical difference in the posttest scores of the computer-assisted and the traditional groups, females outperformed males in both modes of instruction. In algebra studies, Drijvers et al. (2014) reported that computer-assisted students performed below the level of the traditional students; Tienken and Maher (2008) reported that computer-assisted instruction did not improve student performance and the computer-assisted group underperformed the traditional group; Vilardi and Rice (2014) reported that the technology-assisted students did not perform at the same level as those given the traditional instruction. It is noteworthy, however, that these results are for students who were not native to online environments; similar studies may well produce different results for students of the Internet generation, who should start being the majority of college entrants starting in the academic year 2018–2019.

The MEM incorporates learning activities designed to include peer and instructor discussion, which has been documented to increase interactive student engagement and learning (Hake, 1998). It also encourages peer-facilitated learning, which has been shown to promote communication cooperative learning (Shechtman et al., 2010; Slavin, 2010) and to improve mathematics achievement among both African Americans and White students (Robinson et al., 2005). Robinson et al. (2005) also found that both those tutoring and those being tutored not only improved in academic achievement but also experienced positive and attitudinal socioemotional outcomes, such as students' attitudes toward school and their sense of academic efficacy. DePree (1998) reported that, while small group peer-learning did not improve achievement in algebra, it did have a positive impact on course completion rates. In short, while the increased peer-to-peer student interaction may not directly impact algebra courses, it does impact overall student learning and retention rates. Twigg (2011) of the National Center of Academic Transformation (NCAT), which has many years of experience in redesigning math courses involving institutions including Virginia Tech, the University of Alabama, and the University of Idaho, highlights the following features as the primary drivers of the success of the MEM:

- Students spend most of their time actually solving math problems instead of listening to a lecture about solving problems.
- Students spend more time on areas in which they have challenges and less time on areas they have already mastered.
- Students get immediate assistance when they have problems and often have more human interaction with faculty.
- Students are required to do the math through scheduled learning activities.

Twigg (2011) further observes that redesigns in mathematics at NCAT partner institutions have:

- increased the percentage of students successfully completing a *developmental* math course by 51% on average (ranging from 10% to 135%) while reducing the cost of instruction by 30% on average (from 12% to 52%) and

- increased the percentage of students successfully completing a *college-level* math course by 25% on average (from 7% to 63%) while reducing the cost of instruction by 37% on average (from 15% to 77%).

INSTITUTIONAL CONTEXT AND IMPLEMENTATION PLAN FOR THE STRATAGEM

The purpose of the project described here is to improve performance and pass rates in four College Algebra and Trigonometry (CAT) courses at NCA&TSU. At NCA&TSU, these courses fill remedial and general education (GE) requirements or serve as prerequisites for a wide range of higher-level courses in various subjects and are therefore gatekeeper courses for a vast range of majors – and in particular those in the high-demand science, technology, engineering, and mathematics (STEM) fields. The persistently low pass rates have turned these courses into educational goal stoppers for an unacceptably high percentage of the approximately 3,500 students enrolled in them annually.

NCA&TSU is a historically Black college or university (HBCU) that is nationally top-ranked in graduating African Americans with STEM degrees. The student population of about 11,000 is over 87% African American at NCA&TSU, with over 50% being the first generation to attend college in their families, and approximately 80% qualified for need-based financial aid (including a high percentage who are Pell eligible; Oldham, Mers, & Burns, 2002). It is well documented that among those factors contributing the most to the risk of failure are poor academic preparation, social integration, and study skills and that among the leading student characteristics fueling these risks are the first generation to attend college, and coming from low-income families, which almost always correlates to attending low-performing high schools and/or being in the bottom half of one's high school class (Adelman, 1999; Burkum, Habley, McClanahan, & Valiga, 2010; Kuh et al., 2006). About 75% of incoming first-year students at NCA&TSU come from the lowest 25% of their graduating classes (NCA&TSU, Factbook; Oldham et al., 2002); the majority of NCA&TSU students thus have some or all of the characteristics that label them as being at risk. The algebra course redesign project therefore addresses not only the University's retention problems but also the nation's underrepresentation problems by targeting courses most likely to be taken by those most at risk.

Educators, researchers, and policymakers widely agree that one of the primary reasons for unsatisfactory student performance in college mathematics courses is underdeveloped algebra skills (Roderick et al., 2009) among high school graduates. The gap between race and ethnicity only exacerbates the problem for HBCUs. For example, according to ACT's "benchmarks of college readiness," which compare students' scores on subject-matter ACT exams with their grades in introductory college classes, 69% of all test-takers meet ACT benchmarks in English, but only 43% do so in mathematics (Noble & Sawyer, 2002). In mathematics, while 49% of whites meet the benchmarks, only 25% of Latinos and 12% of African Americans do so. While most first-year students at NCA&TSU start their math coursework in college algebra, many students are unable to make the transition to college-level math; the gaps in their skills are just too large. Unfortunately, there is not enough time in a student's curriculum to fix this issue and hence the high failure rates in algebra. Students entering NCA&TSU are placed in math courses according to their SAT scores or placement tests, with placements into Math 099 (a developmental course), Math 101, Math 103, Math 110 or 111 (pre-calculus), or Math 131 (Calculus 1).

Before the 2011–2012 academic year, the College Algebra sequence consisted of two courses: Math 101 (Fundamentals of College Algebra) and Math 102 (Algebra and Trigonometry). Starting in the Fall semester 2011, the College Algebra sequence was re-purposed into four courses, making up two distinct sequences: Math 101–102 (Fundamentals of Algebra and Trigonometry I and II) for non-STEM majors, and Math 103–104 (College Algebra and Trigonometry I and II) for STEM majors. No student would be expected to complete both sequences. All courses in each sequence were taught in the traditional lecture style, with three 50-minute lectures per week, and normal course section size of 45 students. The poor pass rates in Table 1 show that the traditional instructional program was not meeting the diverse needs of students at their levels of preparation.

Prompted by this persistently poor performance, the Mathematics Department conducted, with University support, a self-study of gatekeeper courses in 2010–2011, as part of the University's Wabash 2 Study (Blaich & Wise, 2011). Besides looking at student performance, the department worked with the NCA&TSU Academy of Teaching and Learning (ATL) to collect student input through focus groups. Student feedback showed that the traditional lecture model, with three 50-minute lectures per week, just was not working. Faculty had no additional contact with students outside of the class, office hours went mostly unutilized, and varying instructional techniques and course agendas and student's needs were simply not aligned. The study identified or confirmed several skill gaps beside poor preparation, which inhibit success in gatekeeper math courses, and prominent among them were as follows:

- time on task (60% of first-year students spend less than 10 hours/week studying) (Blaich & Wise, 2011; NCA&TSU Mathematics Department Annual Report, 2010);
- low student satisfaction with academic support (tutoring) (NCA&TSU sophomore Survey, Spring, 2011);
- low academic self-concept (40% new first-year students) (HERI-CIRP, 2010);
- gaps in student–faculty perceptions, in particular regarding prompt feedback, emphasis on memorization, and emphasis on time studying and academic work (NSSE, 2012); and
- low critical thinking, writing, and reasoning skills (Blaich & Wise, 2011; NSSE, 2012).

Moreover, technology use was at the time not well integrated into departmental instructional methods, and students reported that online assignments more often overwhelmed them than helping them (Blaich & Wise, 2011;

Table 1. 2008–2012 Math 101, 102, 103, and 104 Enrolment and Pass Rates

	2008–2009		2009–2010		2010–2011		2011–2012		Fall 2012	
	#Enrolled	%Passed	#Enrolled	%Passed	#Enrolled	%Passed	#Enrolled	%Passed	#Enrolled	%Passed
M101	1,241	39.5	1,249	35.9	1,681	38.1	743	65.0	–	–
M102	655	48.1	759	44.6	733	48.5	348	63.0	–	–
M103	0	0	0	0	0	0	303	53.0	332	67.8
M104	0	0	0	0	0	0	0	0	113	78.8

Source: Mathematics Department.
Note: Fall 2012 is left blank for M101 and M102 because this is the semester that the emporium was implemented for these courses.

NCA&TSU Math Dept & ATL, 2011). Besides highlighting the unsustainably low pass rates and some of the responsible factors, the self-study recommended an immediate, comprehensive instructional overhaul, beginning with the creation of Math 103–104, and the systematic adoption of proven instructional techniques that integrated the use of technology.

While the Math 101–102 sequence, the only algebra sequence that existed prior to 2011, fulfills a University General Education requirement, it is designed as a terminal math course and does not prepare students for further math study. However, no less than 30% of the students enrolled in the sequence pre-2011 were STEM majors. The STEM majors then needed to enroll in the four-credit hour Math 110 Pre-calculus course after Math 102, causing them to spend as many as two years in college before taking Calculus, the first required math course for most of them. Math 103–104 was created in 2011 as an enhanced Math 101–102 to serve as an extended two-semester Pre-calculus course with remediation for high school algebra skill deficiencies for STEM majors.

The NCA&TSU Mathematics Department thus in 2011 undertook a major project to completely redesign the four CAT courses: Math 101, Math 102, Math 103, and Math 104, to eliminate identified success inhibitors through proven instructional techniques. The project was designed to turn around the consistently low pass rates by creating an effective environment for engaged student learning and increasing overall math proficiency in the CAT courses. It would transition students from passive listeners to active participants in knowledge acquisition, while faculty transitioned from mere information providers to skilled facilitators of knowledge gathering. This was to be achieved through the development of course materials and full implementation of the MEM in the two CAT sequences, and the evaluation of student learning in the new course sequences. The project aligned well with the University and the then College of Arts and Sciences' strategic goal of promoting excellence in teaching and learning as well as ensuring student success and was initiated with a UNC Performance Funding grant.

The two new sequences are autonomous relative to each other. Math 101–102 is an Algebra and Trigonometry course that covers many of the same topics as Math 103–104 but in a less intensive manner and is intended to be taken by non-STEM majors; it is designed to be a terminal course sequence. The material is covered at a slower pace in the new set of courses. The modular format of the course enables the material in the corresponding first chapter of the original textbook to be split into two separate modules (Module 1 and Module 2), which allows students to master and be assessed on smaller quantities of information.

The Math 103–104 sequence is introductory in nature, serving as the basis for higher-level math courses. It is designed to provide the foundations for STEM majors who will subsequently enroll in Calculus courses. The sequence covers the same topics as Math 101–102, though more problems and higher difficulties in the problems are included; a typical student would not have cause to complete both sequences.

STRATAGEM GOALS, STRUCTURE AND COMPONENTS, AND IMPLEMENTATION, ADAPTATION DUE TO INSTITUTIONAL CONTEXT

This project was conceived with the overarching purpose of completely transforming the way that the two CAT course sequences are taught and the way students learn. It would produce exemplary materials that incorporate online delivery and eliminate duplication of faculty efforts, a delivery process model that enhances student learning, and a fully developed MEM learning environment led by a competent and highly engaged faculty. Actual course delivery would take place in specially designed emporium learning spaces with workstations, which facilitate active and collaborative learning among students. To achieve this complete course redesign and transformation of the learning process, the project had the following six distinct broad component objectives to be coordinated:

(1) redesign learning spaces;
(2) redesign course structure;
(3) develop course materials;
(4) develop faculty competence;
(5) implement the MEM; and
(6) assess the MEM.

(1) *Learning space redesign:* The MEM requires specially designed spaces, which enable active student–faculty collaboration and computer-assisted learning, and none existed at NCA&TSU in 2011. This required the re-purposing of existing spaces in three buildings, and hence strong University support and substantial external funding, to create a series of emporium laboratories. A UNC Performance Funding grant in 2011 enabled the re-purposing of two classrooms and a suite of faculty offices in the Academic Classroom Building and the conversion of a suite of study rooms in the Bluford Library, into the initial Emporium Testing Center; a major Glaxo Smith-Kline grant in 2014 established the GSK STEM Center of Excellence for Active Learning, which enabled the re-purposing of a suite of chemistry laboratories and faculty offices into a new Emporium Lab in Marteena Hall (the home of the Math Department).

(2) *Course structure redesign:* The MEM CAT sequence courses have modular content with each course consisting of 10 modules. Students still register for the individual courses in sections, which define their mandatory attendance times and venue and assign them a designated instructor. A common syllabus for each sequence is used, specifying the individual course objectives and competencies; the modular designs will allow the students to work at their own pace. This means that the stronger students continue to be challenged as they work through the course and the weaker students do not get discouraged because they complete work on their own schedules as long as they stay ahead of due dates. Common faculty notes, homework, quizzes, and tests are incorporated into the customized workbook and *MyMathLab* system.

In addition to the three weekly meetings in the emporium lab per week, students are required to meet with the assigned graduate assistant one hour per week in a traditional classroom setting. The purpose of the weekly class meeting is to focus students' attention on the week's upcoming tasks, recap the previous weeks' materials and their relationship with new material to be learned, as well as discuss study strategies and develop individualized study plans highlighting which areas need practice. Students are encouraged to keep a notebook demonstrating their work on their homework and practice tests and/or turn-in work on tests. This enables instructors and assistants to review all student work and recommend appropriate remediation whenever needed.

(3) *Course materials development:* A common online course manual has been developed to orient students to the course, including essential information about emporium model and the course structure, such as the course content and syllabus, the required *MyMathLab* homework as well as all other course resources, testing, and assessment materials.

Modular course packs: The traditional mode of instruction is centered on the textbook, with each faculty member developing her/his own instructional strategies and procedures to achieve common syllabus course objectives. In the MEM course redesign, the printed textbook is replaced by a combination of printed workbooks and online materials in modular format; each of the two CAT course sequences is packaged into 10 modules. The modules use current course objectives to determine the content of each module and use the course learning outcomes to select pretest, homework problems, quizzes, and posttest questions from problem banks to assess student mastery.

Module format: Each module has a similar format: module pretest, topic-oriented interactive online tutorials, homework assignments, a quiz, a practice test, and a module test. A student enters a module by taking the module pretest first. If the student achieves a predetermined score on the pretest, then the student is exempt from the homework assignments and quizzes, and the student's score on the pretest serves as his/her score on the test for that module. If a student scores below that threshold, then the student must work through the tutorials and assignments within the module. To work through the module, the student is directed to watch a video lecture and take notes and is supposed to read and study the corresponding sections in the workbook. The student can ask an instructor or student tutor for assistance after watching the video lectures and studying the workbook. The student works homework problems built in *MyMathLab* software as well as in their workbooks and takes a quiz halfway through the module. The student then takes the practice test to prepare for the module test. The student is provided three attempts to take the quiz, the practice test, and the module test. The procedure described above is designed to assure that if the student completes the modules that s/he is forced to prepare sufficiently until s/he has a greater chance of passing the class.

Another feature of the modular format that is appealing to students is that for instance, if a student finishes the first of the two-course sequence before the semester ends, then that student may start on the second immediately and not have to wait until the next semester begins. Also, it is possible for students to complete the material for two courses in one semester, with variable credit awarded for module completion as follows. If a student registered in the first of the two-course sequence does complete both courses in a sequence in one semester, then they can register for and get credit for the second in the following semester. While stronger students can move ahead of the class in a course, weaker students who do not pass the course in a semester can pick up where they left off the previous semester. When students complete the course, they have the option to begin work on the next course during the current semester to reduce lost time.

Workbooks: One of the deviations (inspired by GTCC) from the original Virginia Tech model that was adopted from the start in the NCA&TSU implementation is that, in addition to the course modules, students must purchase a workbook to work through exercises that go along with the content of the modules. The workbooks used in the project, which have already undergone several revisions (Cousins-Cooper, Luke, Redd, & Casterlow, 2016, 2015, 2013; Cousins-Cooper, Luke, & Redd, 2016; Cousins-Cooper, Luke, & Noble, 2013, 2012), are written by the project team and aligned with textbook chapters, though the content is created by the faculty, in particular, supported by algebra and trigonometry textbooks published by Pearson Publishing. The workbooks are a compilation of summary notes and problem worksheets, all in modular format synchronized with the course modules. While the workbooks are designed for skills practice and concept synthesis, the accompanying note-taking templates are designed to improve study skills and mathematics writing: they have prompters for what to look for, and write down, as students read assigned materials before class, and the resulting student notes are also assessed as part of the student's workbook grade.

MyMathLab Software: Central to the current NCA&TSU MEM implementation is commercially available software used in the labs. All the targeted FCAT courses are currently taught using textbooks published by Pearson Publishing. Therefore, the NCA&TSU project works with Pearson Publishing through their online homework and testing system, *MyLabsPlus*. *MyMathLab* provides immediate feedback and assistance to students and through its "Show Me An Example," "Help Me Solve This," and "Work A Similar Exercise" functions. Also, online quizzes and tests are graded immediately by *MyMathLab* upon submission, with hard copies of all student work available for review. We note that, while this project uses *MyMathLab*, it is also developing original online reading materials and problem banks, instead of simply selecting existing Pearson textbook materials.

(4) *Faculty development:* Although content knowledge is an important contributor to teacher effectiveness, it does not ensure effective pedagogy (Shechtman et al., 2010). It is well documented that professional development that integrates pedagogical understanding and knowledge of content increases implementation of standards-based teaching (Slavin, 2010). Therefore, to ensure successful implementation of the proposed course redesign project, it has been essential to providing sufficient professional development and training for instructors and student assistants who develop, teach, or facilitate redesigned

courses for the first time to understand the new philosophy of teaching that is required in the MEM setting. The professional development engages instructors in structuring their curriculum and instructional strategies to provide a learning context that is social, interactive, and highly engaging.

Professional development and training include visiting institutions that have successfully implemented MEM course redesign, providing focused development workshops, involving faculty in curricular decision-making, and holding regular meetings with all instructors and student assistants to offer ongoing training. The project has held on-campus faculty development workshops, in collaboration with facilitators from GTCC and the NCA&TSU Academy for Teaching and Learning; the redesign team has also visited the Virginia Tech Math Emporium several times. The team works with Pearson Publishing curriculum and assessment specialists for expert guidance and efficacy testing of course materials.

(5) *MEM implementation:* The NCA&TSU redesign aims to create an instructional program that will prepare each student for her/his educational goal whether it involves beginning a program of study in a field that requires advanced mathematics, or completing a general education mathematics course. To accommodate the widely varying learning styles, levels of preparation, and career goals, the following components of the emporium model are highlighted in the implementation:

- modularization of course content, with 10 modules for each of Math 101–102 and Math 103–104:
- teaching of multiple courses in the Math Emporium;
- inclusion of an array of learning opportunities: instructional software, lectures, team/group collaboration, small group study sessions, focus groups for students with same educational goals, videos, and skills building assignments with immediate feedback;
- customization of the learning environment for each student based on background and learning preference, creating individualized paths for all to reach the same learning outcomes;
- elimination of duplication of efforts by faculty and use a staffing model that utilizes a combination of faculty, graduate students, and undergraduate peer tutors; and
- mandatory attendance at the Math Emporium, with mandatory extra weekly group meetings according to the performance.

(6) *MEM assessment:* The assessment phase of the project is ongoing and currently has one primary goal: to compare course performance measures between the MEM and the traditional methods for the two CAT course sequences. Besides the comparison of final grades between the two instructional designs, comparative studies on pretest versus posttest learning gains, comparison of various measures of learning, and course effectiveness are being performed.

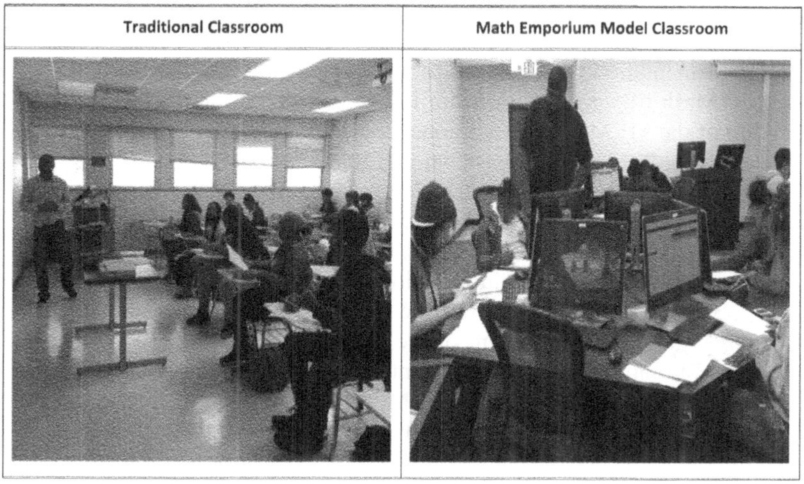

Fig. 1. Traditional and Emporium Classrooms.

STRATAGEM OUTCOMES AND EVOLUTION

A collection of four learning spaces, collectively known as the emporium laboratories, have been created to date as the result of this project: one 20-seat lab in Bluford Library, two 36-seat labs in the Academic Classroom Building, and one 60-seat lab in Marteena Hall. Each lab is equipped with thin-client servers for students, as well as outfitted with one instructor workstation and various other types of audio-visual presentation equipment. Fig. 1 below shows a traditional classroom and a MEM classroom at NCA&TSU.

In the pre-2012 course configuration, 10 full-time faculty and five part-time adjuncts formerly taught 2,200 each year in 20 traditional (non-MEM) 45-student sections, availing each student to only three contact hours per week besides office hours. With the MEM, each faculty member is assigned a group of students (either 36 or 60, based on classroom size) to counsel, monitor, and guide through the entire term. Through three required meetings per week and need-based open labs, students are guided in choosing what types of learning materials to use depending on their needs and in determining a timeline to work through their prescribed modules. Courses are intended to have sufficiently many instructors and teaching assistants to ensure a student to instructor/tutor ratio of no more than 12:1 so that expert help is always available for immediate one-on-one assistance. Frequent practice, immediate feedback, and reinforcement of concepts are provided by *MyMathLab*, while the faculty-written workbook provides mathematical writing and further practice and concept synthesis. Instead of spending time writing concepts on the board, instructor and assistants spend class time explaining and reinforcing concepts the students do not know. As shown in Table 2, the numbers of MEM versus traditional sections of the respective courses have evolved through the years, as have those of instructors and assistants per course.

The implementation was designed to increase pass rates in the two CAT course sequences (Math 101–Math 102, Math 103–Math 104) using the MEM. While the evaluation of the MEM versus traditional course performance is ongoing, the overall pass rates for 2012–2017 are shown in Table 3 for comparison. It is noted that during the study, the university switched from standard ABCDF grading to a plus/minus grading scale during the Fall 2014. No additional evaluation was performed based on the change in the grading scales. Occurring concurrently with the improving passing rates for the emporium sections was the improvement of the passing rates in the traditional sections. While the root cause is still being investigated, it is hypothesized that the development of the Math 103 and Math 104 sequence and resulting partitioning of the students contributed to the increase in performance. In separating the STEM versus non-STEM majors and making appropriate adjustments to the depth of content and delivery, it is believed that conditions were created that provided better avenues to success for all students in all sections.

As part of the redesign of the Math 101–102 and Math 103–104 courses, the redesign team authored two workbooks for use in the MEM courses and designed the online homework and online quizzes for the modules. The workbooks, homework, and quizzes are available for use in the traditional courses of the Math 101–102 sequence. Some faculty have taught in both formats and often chose to use the same Web-based assignments for assessments. While all of the tests in the MEM courses were present at the beginning of the semester, instructors of the traditional courses generated their own paper tests for assessment.

A comparison of the passing rates prior to the emporium redesign (see Table 1) with the passing rates after the implementation of the emporium (Table 3) shows that the passing rates for the courses in the college algebra sequence improved from below 50% to above 60%. Specifically, Math 101 passing rates improved from an average of 44% passing per year to 70% passing per year, and Math 102 passing rates increased from an average of 51% to 67%. There is a clear improvement in passing rates for these college algebra courses, but it is difficult to determine whether this improvement can be attributed to the implementation of the emporium, the separation of the STEM and non-STEM majors, or some combination of the two interventions.

There are a few things to keep in mind when viewing the performance of students enrolled in the traditional sections and the MEM sections of the CAT courses. As the MEM was being implemented, the faculty began to meet more about how to improve their students' performance in the CAT courses. There was much discussion among the faculty of the Math Department about which teaching strategy may be most effective or which topics in the CAT courses needed more coverage. There was much collaborative effort among faculty members to improve the learning experience of the students enrolled in these courses. While one can speculate as why the traditional classes performance also improved, this is an area in which further study is needed.

Table 4 summarizes statistical tests comparing the two instructional methods for 2012–2017 without covariate adjustment, which is a limitation.

Table 2. Evolution of the Algebra Course Structure.

Course	Type	Fall 2011				Fall 2014				Fall 2017			
		#Sect	#Instr	#TAs	#Hrs	#Sect	#Instr	#TAs	#Hrs	#Sect	#Instr	#TAs	#Hrs
101	Trad	17	9	0	51	7	7	0	21	6	3	0	18
	MEM	0	0	0	0	11	7	12	44	9	6	17	36
102	Trad	10	7	0	30	8	7	0	24	6	5	0	18
	MEM	0	0	0	0	6	4	7	24	6	5	10	24
103	MEM	0	0	0	0	12	9	10	48	16	10	21	64
104	MEM	0	0	0	0	4	3	7	16	5	3	10	20

Source: Mathematics Department.

Table 3. 2012–2017 Math 101, 102, 103, and 104 Enrollment and Pass Rates.

		2012–2013		2013–2014		2014–2015		2015–2016		2016–2017	
Course	Type	#Enrolled	%Passed	#Enrolled	%Passed	#Enrolled	%Passed	#Enrolled	%Passed	#Enrolled	%Passed
M101	Trad	787	70.5	805	71.9	454	68.1	390	61.5	375	76.0
	MEM	423	71.2	493	71.2	731	69.9	817	71.2	879	71.8
M102	Trad	412	70.2	758	72.4	516	60.3	483	61.5	403	70.2
	MEM	428	54.7	433	68.6	686	69.8	633	64.1	569	78.0
M103	MEM	–	–	548	64.6	526	56.3	569	56.8	597	61.0
M104	MEM	–	–	399	59.2	350	59.7	375	64.5	391	69.3

Source: Institutional Research.
Note: Math 103 and Math 104 were not taught using the MEM until the 2013–2014 academic year, which explains why there is a blank in those columns for 2012–2013 in Table 3.

Table 4. *p*-Values of a Two-sample Proportion Test.

	2012–2013	2013–2014	2014–2015	2015–2016	2016–2017
M101	0.868	0.8264	0.5464	0.0009	0.1415
M102	5.27E-06	0.1811	0.0007	0.3979	0.0072

A *p*-value of less than 0.05 indicates that there is a statistically significant difference between the two instructional methods. For Math 101, the only year with a statistically significant difference is in 2015–2016. Therefore, the Math 101 MEM sections outperformed the Math 101 traditional sections with statistical significance during 2015–2016. For Math 102, there are *p*-values less than 0.05 during 2012–2013, 2014–2015, and 2016–2017. During 2012–2013, the traditional sections of Math 102 outperformed the MEM sections, and during both 2014–2015 and 2016–2017, the MEM sections of Math 102 outperformed the traditional sections all with statistical significance.

A notable by-product of the MEM implementation is the difference in the performance of female students versus male students as shown in Table 5. There is a notable decrease in pass rates during most spring semesters, which may be attributed to the course being "off sequence." On a semester-by-semester basis, female students received a grade between "A" and "C−" at higher rates than male students in Math 103. While the reasons for this are unclear at present, it is hypothesized that the individualized nature of (1) the student–teacher interactions and (2) the student–content interactions during class provided a supportive environment for confidence-building and success. Further qualitative survey data are needed to truly isolate the primary factors.

In the Math 103–Math 104 sequence taken by STEM majors, female students also outperformed their male counterparts as shown in Table 6. Female students received more grades of "A," "B+," and "B" than male students. Prior to intervention, DFW rates for the courses could reach values of 50% and higher. After the intervention, the rates have reduced to closer to 40%. If only "on sequence" courses are considered (Math 103 courses occurring in the Fall and Math 104 courses that occur during the Spring), the rate improves as well.

Table 7 shows the pass rates by gender for 2016–2017. Table 7 does not strongly indicate the difference between female and male. Only the MEM Math101 and Math 104 have a statistically significant difference between female and male at 5% significance level. In Math 101, however, female students' pass rate in the MEM compared to the traditional was significantly higher than male students with a Cochran -Mantel -Haenszel test p-value < 0.0001. Math 102 has a similar pattern, but it is not significant.

LESSONS LEARNED AND RECOMMENDATIONS

When the MEM model of instruction was first implemented at NCA&TSU in the Fall 2012 semester, the redesign team decided that there would be no formal deadlines for assignments or tests. This was the policy used in the math

Table 5. 2012–2016 Mat 103 Pass Rates By Gender.

Mat 103	Fall 2012	Spring 2013	Fall 2013	Spring 2014	Fall 2014	Spring 2015	Fall 2015	Spring 2016
Male (%)	103/179 (57.5%)	61/144 (42.4%)	128/204 (62.7%)	50/92 (54.3%)	95/182 (52.2%)	61/120 (50.8%)	90/152 (59.2%)	38/107 (35.5%)
Female (%)	123/154 (79.8%)	58/111 (52.2%)	117/171 (68.4%)	59/81 (72.8%)	87/148 (58.8%)	53/76 (69.7%)	142/202 (70.3%)	53/108 (49.1%)

Source: Institutional Research.

Table 6. 2012–2016 Grade Distributions for the Math 103–104 Sequence By Gender.

Math 103 and 104 (2012–2016)												
	A	A−	B+	B	B−	C+	C	C−	D+	D	F	W
Male ($N = 855$)	79	16	8	152	21	24	201	16	9	81	187	61
	9.2%	1.9%	1.0%	17.8%	2.5%	2.8%	23.5%	1.9%	1.1%	9.5%	21.9%	7.1%
Female ($N = 1,642$)	272	29	35	363	39	28	364	29	17	130	243	93
	16.6%	1.8%	2.1%	22.1%	2.4%	1.7%	22.2%	1.8%	1.0%	7.9%	14.8%	5.7%

Source: Institutional Research.

Table 7. Pass Rates by Gender Using 2016–2017 Data.

		Female		Male	
		#Enrolled	%Passed	#Enrolled	%Passed
M101	Trad	248	77.8	127	72.4
	MEM	576	76.9	303	62.0
M102	Trad	264	70.0	139	70.5
	MEM	388	79.1	181	75.6
M103	MEM	315	63.2	282	58.5
M104	MEM	215	77.2	176	59.7

Source: Institutional Research.

emporium at other universities, and the redesign team felt that it would give the students enrolled in the MEM the ability to spend extra time to master concepts that they found particularly challenging. By the middle of the semester, it had quickly become obvious that the "no deadline" policy was not appropriate for the MEM classes at NCA&TSU. Many of the students that were enrolled in the MEM that semester would go days, even weeks, without completing any assignments in their course. As the end of the semester approached, they found themselves in the nearly impossible position of needing to complete a large portion of the course in a very short time frame.

From this experience, the MEM redesign team learned that there is no "one-size-fits-all" approach to the mathematics emporium. The policies that are successful for one campus may not be appropriate for another. One of the greatest benefits of the MEM is the ability to customize its policies to best suit an institution's student population. After that first semester, strict deadlines were introduced for all assignments and tests. Furthermore, the faculty involved in teaching MEM courses at NCA&TSU meet at the end of each semester to assess the policies of the MEM. The evolution of the MEM at NCA&TSU is an ongoing process, with minor modifications being implemented each year.

From the implementation and expansion of the MEM at NCA&TSU, the redesign team has learned the importance of organization, communication, and proper training. During the initial implementation of the MEM, only four faculty members were involved in teaching courses using the new methodology. Those four faculty members had access to shared office space near the emporium classroom, which facilitated communication among the group. As the emporium model has expanded, the number of faculty members involved has significantly increased. One of the perceived benefits of the MEM is the uniformity across all sections of a course (the use of the same policies, a common syllabus, common assignments, etc.). With a larger number of faculty members involved in the MEM, it becomes more difficult to communicate within the group to ensure that the desired uniformity across the sections is maintained. In an effort to foster communication, the emporium faculty at NCA&TSU held multiple meetings each semester to discuss any problems that may have arisen, to compare strategies that seem to be working well and to share any information or announcements that may be pertinent to the faculty involved.

The four faculty members that taught the initial MEM courses were involved in the planning of the course and worked very closely together during that first semester. They had a vision of how an emporium classroom would run, and they developed some "best practices" while teaching their courses. As the MEM expanded and additional faculty were asked to teach courses using the emporium model, those new faculty members needed to learn the ins and outs of the emporium. Some basic, relatively informal training sessions were offered, but some of the faculty members who were new to the emporium classroom felt underprepared and learned how the emporium operates while they were teaching in it. The redesign team has realized that the MEM would benefit from a more formal training for both new faculty members and teaching assistants. Plans for formal training sessions, including an emporium classroom handbook, are currently being developed.

The success of the NCA&TSU project is the product of very strong and tangible institutional support of a team of faculty members with a long-term commitment to become skilled in the use of the MEM for mathematics instruction and engaged in the national Emporium community. It is of note that the NCA&TSU Mathematics Department first considered the MEM mode of instruction for the CAT courses in 1998: two of the redesign team members were among the participants of the first Virginia Tech Math Emporium Faculty Workshop in 1998. At the time, the only MEM model was the Virginia Tech model, a 500-student continuously open laboratory devoted to a single course. While the administration was supportive, scaled down multi-course laboratories were never considered, large-volume spaces were not available, and existing space conversions were considered too expensive. It took long-term faculty commitment to self-development and creativity as well as continuing administrative commitment to securing resources for existing space conversions to overcome the original obstacles.

In its present form, the NCA&TSU MEM adaption for the CAT courses seems to be working and has improved learning over traditional instruction for all the algebra and trigonometry courses and all semesters. However, we have discovered, in agreement with other studies that MEM does not work the same for everyone. Webel, Krupa, and McManus (2017) report that "the emporium appeared to best serve students with higher math achievement, who enjoyed mathematics, and who spent more time taking their exams. Regarding mathematical learning, the emporium appeared to improve students' ability to recall and use formulas for familiar problem types but had limited impact regarding developing meaning for symbols or flexibility in solving unfamiliar tasks. Also, students expressed mixed feelings about the autonomy provided by the structure of the emporium."

While Cousins-Cooper, Staley, Kim, and Luke (2017) report significantly higher performance for MEM over the traditional classroom, a further comparison of the methods is starting to reveal non-uniform impacts. Clemence and Cousins-Cooper, & Clemence (2019) report significant difference in the pretest to posttest gain of the students in the MEM class as compared to those students in the traditional lecture class, no significant difference between the course effectiveness of the MEM classes and that of the traditional lecture college algebra class as measured by final course grade, and significant course effectiveness difference in the MEM as well as traditional lecture college classes between female students and male students (in favor of female students) as measured by final course grade. While there were increased gains for students in the MEM courses, the performance of all students increased relative to passing rates regardless of delivery method. Future work for the project will investigate to determine whether there are any causal factors from the emporium implementation that drove the improvement in the traditional courses. A more in-depth study of what factors (tangible and intangible) are driving the increased performance of females versus males in the courses is also planned. While it is hypothesized that separating the courses into two, distinct tracks (STEM versus non-STEM) contributed to improved performance for students by avoiding the one-size-fits-all mentality, further work is needed to determine what can be attributed to the course delivery style and what part can be attributed to the shift in course focus.

Overall, the current MEM for algebra at NCA&TSU is not perfect but has the potential for a major positive impact on student learning and performance in CAT courses.

ACKNOWLEDGMENTS

This material is based upon work supported by the following:

- National Science Foundation under Grant No. 1610482;
- Glaxo Smith Kline Foundation Grant; and
- University of North Carolina General Administration through a UNC Performance Funding Grant and a UNC Course Redesign Grant.

REFERENCES

Adelman, C. (1999). *Answers in the tool box: Academic intensity, attendance patterns, and bachelor's degree attainment*. Washington, DC: US Department of Education.

Antil, L. R., Jenkins, J. R., & Wayne, S. K. (1998). Cooperative learning: Prevalence, conceptualizations, and the relation between research and practice. *American Education Research Journal*, *35*(3), 419–454. doi:10.3102/00028312035003419

Arendale, D. (1997). *Supplemental instruction (SI): Review of research concerning the effectiveness of SI from the University of Missouri-Kansas City and other institutions from across the United States.* Proceedings of the 17th and 18th Annual Institutes for Learning Assistance Professionals: 1996 and 1997.

Arendale, D. R. (2004). Pathways of persistence: A review of postsecondary peer cooperative learning programs. In I. Duranczyk, L. Higbee, & D. B. Lundell (Eds.), *Best Practices for Access and Retention in Higher Education* (pp.27–40). Minneapolis: University of Minnesota, General College, Center for Research on Developmental Education and Urban Literacy.

Blaich, C., & Wise, K. (2011). *The Wabash National Study – The impact of teaching practices and institutional conditions on student growth.* 2011 AERA Annual Meeting. Retrieved from http://www.liberalarts.wabash.edu/storage/Wabash-Study-Student-Growth_BlaichWise_AERA-2011.pdf

Blimling, G. S., Whitt, E. J., & Associates. (1999). *Good practice in student affairs: Principles to foster student learning*. San Francisco, CA: Jossey-Bass.

Burch, K. J., & Kuo, Y. (2010). Traditional vs. online homework in college algebra. *Mathematics and Computer Education, 44*(1), 53–63.

Burkum, K., Habley, W., McClanahan, R., & Valiga, M. (2010). *What works in student retention (fourth national survey): Four-year colleges and universities with twenty percent or more Black students enrolled,* ACT 2010.

Chickering, A. W., & Gamson, Z. F. (1987). Seven principles for good practice in undergraduate education. *AAHE Bulletin, 39*(7), 3–7

Clemence-Mkhope, D. P., Cousins-Cooper, K. & Clemence-Mkhope, B. (2019). *Math emporium college algebra: Comparison of performance measures. International Journal of Instruction* (under revision).

Cousins-Cooper, K. Luke, N.S. Redd, T.C. (2016). *Fundamentals of algebra & trigonometry workbook.* New York, NY: Pearson.

Cousins-Cooper, K. Luke, N.S. Noble, R. (2013, 2012). *Fundamentals of algebra & trigonometry workbook.* New York, NY: Pearson.

Cousins-Cooper, K., Luke, N. S., Redd, T. C., & Casterlow, S. (2016, 2015, 2013). *College algebra & trigonometry for engineers and scientists workbook.* New York, NY: Pearson.

Cousins-Cooper, K., Staley, K. N., Kim, S., & Luke, N. S. (2017). The effect of the math emporium instructional method on students' performance in College Algebra. *European Journal of Science and Mathematics Education, 5*(1), 1–13.

DePree, J. (1998). Small-group instruction: Impact on basic algebra students. *Journal of Developmental Education, 22*(1), 2–6.

Drijvers, P., Doorman, M., Kirschner, P., Hoogveld, B., & Boon, P. (2014). The effect of online tasks for algebra on student achievement in grade 8. *Technology and Learning, 19*(1–2), 1–18.

Fullilove, R. E., & Treisman, P. U. (1990). Mathematics achievement among African American undergraduates at the University of California, Berkeley: An evaluation of the mathematics workshop program. *The Journal of Negro Education, 59*(3), 463–478.

Furner, J. M., & Gonzales-DeHaas, A. (2011). How do students' mastery and performance goals relate to math anxiety? *Eurasia Journal of Mathematics, Science, and Technology Education, 7*(4), 227–242.

Gillies, R. M., & Ashman, A. F. (Eds.). (2003). *Cooperative learning: The social and intellectual outcomes of learning in groups.* London: Routledge Falmer. (an imprint of the Taylor & Francis Group).

Gordon, S. P. (2008). What's wrong with college algebra? *Primus: Problems, Resources, and Issues in Mathematics Undergraduate Studies, 18*(6), 516–541.

Hake, R. R. (1998). Interactive-engagement versus traditional methods: A six-thousand-student survey of mechanics test data for introductory physics courses. *American Journal of Physics, 66*(1), 64.

Hegeman, J. S. (2015). Using instructor-generated video lectures in online mathematics courses improves learning. *Online Learning 19*(3), 70–87.

Herriott, S. R., & Dunbar, S. R. (2009). Who takes college algebra? *Primus, Resources, and Issues in Mathematics Undergraduate Studies, 19*(1), 74–87.

Higher Educational Research Institute-Cooperative Institutional Research Program. (2010). *Higher Educational Research Institute-Cooperative Institutional Research Program (CIRP) Freshman Survey, 2010.*

Kuh, G. D., Kinzie, J., Buckley, J. A., Bridges, B. K., and Hayek, J. C. (2006). *What matters student success: A review of the literature* (Commissioned Report for the National Symposium on Postsecondary Student Success: Spearheading a Dialog on Student Success). Retrieved from http://nces.ed.gov/npec/pdf/kuh_team_report.pdf

Kyriacou, C. (1992). Active learning in secondary school mathematics. *British Educational Research Journal, 18*, 309–318. doi:10.1080/0141192920180308

Lipnevich, A. A., MacCann, C., Krumm, S., Burrus, J., & Roberts, R. D. (2011). Mathematics attitudes and mathematics outcomes of U.S. and Belarusian middle school students. *Journal of Educational Psychology, 103*(1), 105–118.

Mathai, E., & Olsen, D. (2013). Studying the effectiveness of online homework for different skills levels in a college algebra course. *Primus, Resources, and Issues in Mathematics Undergraduate Studies*, *23*(8), 671–682.

National Council of Teachers of Mathematics. (2006). *The NCTM curriculum and evaluation standards*. Retrieved from http://standards.nctm.org

National Mathematics Advisory Panel. (2008). *Foundations for success: The final report of the national mathematics advisory panel*. Washington, DC: US Department of Education. Retrieved from https://www2.ed.gov/about/bdscomm/list/mathpanel/report/final-report.pdf

National Survey of Student Engagement. (2012). *National survey of student engagement-faculty survey of student engagement, 2012 edition*. Retrieved from http://nsse.iub.edu/html/annual_results.cfm

Nebesniak, A. L. (2012). Learning to teach mathematics with reasoning and sense making. *Public Access Theses and Dissertations from the College of Education and Human Sciences, University of Nebraska-Lincoln*. Retrieved from http://digitalcommons.unl.edu/cgi/viewcontent.cgi?article=1139&context=cehsdiss

Noble, J., & Sawyer, R. (2002). *Predicting different levels of academic success in college using high school GPA and ACT composite score*. ACT 2002-4 Iowa City, IA: ACT Inc.

North Carolina Agricultural & Technical State University Interactive Self-Service Factbook. Retrieved from http://www.ncat.edu/provost/ospie/ir/factbook.html

North Carolina Agricultural & Technical State University Math Dept & ATL. (2011). *NCA&TSU math department and academy for teaching and learning internal report: Wabash 2 Study Math Student Focus Group*. Presented to the Mathematics Department on March 31, 2011.

North Carolina Agricultural & Technical State University sophomore survey. Administered Spring 2011.

Oldham, J. M., Mers, R., & Burns, S. D. (2002). Cooperative efforts among HBCU mathematics departments in the production of African American Ph.D.'s in the mathematical sciences. *Proceedings of The Ronald E. McNair Symposium on Science Frontiers: The Role of HBCU's in 21st Century Higher Education*.

Rakes, C. R., Valentine, J. C., McGatha, M. B., & Ronau, R. N. (2010). Methods of instructional improvements in algebra: A systematic review and meta-analysis. *Review of Educational Research*, *80*(3), 372–400.

Robinson, D. R., Schofield, J. W., & Steers-Wentzell1, K. L. (2005). Peer and cross-age tutoring in math: Outcomes and their design implications. *Educational Psychology Review*, *17*(4), 327–362. doi:10.1007/s10648-005-8137-2.

Roderick, M., Nagaoka, J., & Coca, V. (2009). College readiness for all: The challenge for urban high schools. *The Future of Children*, *19*(1), 185–210.

Shechtman, N., Haertel, G., Roschelle, J., Knudsen, K., & Singleton, C. (2010). *Design and development of the student and teacher mathematical assessments*. Menlo Park, CA: SRI International.

Singh, K., Granville, M., & Dika, S. (2002). Mathematics and science achievement: Effects of motivation, interest, and academic engagement. *Journal of Educational Research*, *95*, 323–332. doi:10.1080/00220670209596607.

Slavin, R. (2010). Co-operative learning: What makes group-work work? In H. Dumont, D. Istance, & F. Benavides (Eds.), *The nature of learning: Using research to inspire practice*. OECD Publishing.

Spradlin, K., & Ackerman, B. (2010). The effectiveness of computer-assisted instruction in developmental mathematics. *Journal of Developmental Education*, *34*(2), 12–14,16,18,42. Retrieved from http://digitalcommons.liberty.edu/cgi/viewcontent.cgi?article=1217&context=educ_fac_pubs

Stacey, K., Chick, H., & Kendal, M. (Eds.). (2004). *The future of the teaching and learning of algebra: The 12th ICMI study*. Dordrecht: Kluwer.

Stevenson, H. W., & Newman, R. S. (1986). Long-term prediction of achievement and attitudes in mathematics and reading. *Child Development*, *57*, 646–659.

Sturdivant, R. X., & Souhan, B. E. (2011). Peer-to-peer teaching using multi-disciplinary applications as topics. *Primus: Problems, Resources, and Issues in Mathematics Undergraduate Studies, 21*(3), 283. doi:10.1080/10511970903039621

Tienken, C. H., & Maher, J. A. (2008). The influence of computer-assisted instruction on eighth grade mathematics achievement. *RAMLE Online, 32*(3), 1–13.

Tinto, V. (1987). *Leaving college: Rethinking the causes and cures of student attrition*. Chicago, IL: University of Chicago Press.

Twigg, C. (2011). The math emporium: Higher education's silver bullet. *Change: The Magazine of Higher Learning, 43*(3), 25–34. Retrieved from http://www.changemag.org/Achives/Back Issues/2011/May-June-2011/math- emporium-full.html

Vilardi, R., & Rice, M. L. (2014). Mathematics achievement: Traditional instruction and technology-assisted course delivery methods. *Journal of Interactive Online Learning, 13*(1), 16–28. Retrieved from https://www.learntechlib.org/p/153509/

Webel, C., Krupa, E., & McManus, J. (2017). The math emporium: Effective for whom, and for what? *International Journal of Research in Undergraduate Mathematics Education, 3*(2), 355–380. doi:10.1007/s40753-016-0046-x

Zakaria, E., Chin, L. C., & Daud, Md.Y. (2010). The effects of cooperative learning on students' mathematics achievement and attitude towards mathematics. *Journal of Social Sciences, 6*(2), 272–275.

PROCESS-ORIENTED GUIDED-INQUIRY LEARNING AT JACKSON STATE UNIVERSITY AND TUSKEGEE UNIVERSITY

Naomi F. Campbell, Melissa S. Reeves, Marilyn Tourné and M. Francis Bridges

ABSTRACT

Process-oriented guided-inquiry learning (POGIL) is a student-centered instructional strategy to actively engage students in the classroom in promoting content mastery, critical thinking, and process skills. The students organize into groups of three to four, and each group member works collaboratively to construct their understanding as they proceed through the embedded learning cycle in the POGIL activity. Each group member has a specific role and actively engages in the learning process. The roles rotate periodically, and each student has the opportunity to develop essential process skills, such as leadership skills, oral and written communication skills, team-building skills, and information-processing skills. The student groups are self-managed, and the instructor serves as a facilitator of student learning. A POGIL activity typically contains a model that the students deconstruct using a series of guided, exploratory questions. The students develop concepts (concept invention) as the group members reach a valid, consensus conclusion. The students apply their concepts to new problems completing the learning cycle. The authors implemented POGIL instruction in several chemistry courses at Jackson State University and Tuskegee University. They share

Broadening Participation in STEM: Effective Methods, Practices, and Programs
Diversity in Higher Education, Volume 22, 265–289

ISSN: 1479-3644/doi:10.1108/S1479-364420190000022012

their initial findings, experiences, and insights gained using a new instructional strategy.

Keywords: Active learning; process skills; guided-inquiry learning; STEM teaching; group learning; student-centered learning

INTRODUCTION

A perennial professorial dilemma is spending increasing time and effort to deliver the best content, only for the students' mean exam score to be disappointingly low. Students matriculate in the university with the goal and desire to perform well in college, with aspirations of attending graduate or professional school and, eventually, to have a great career. The students expect that professors will provide them with the necessary content, tools, and skills to be successful upon graduation. They also have the expectations that they will graduate on time and well prepared in their discipline for continued education or to enter the workforce. However, there is a divide at the undergraduate level between students' expectations and reality, especially in science, technology, engineering, and mathematics (STEM) disciplines. The divide between students' expectations and reality is especially true for underrepresented minority and female students (National Science Foundation, 2017). The purpose of writing this chapter is to provide evidence-based information on an instructional strategy that has been shown to improve students' critical thinking and process skills.

The most common instructional strategy is the lecture style of delivering content to the students (Stains et al., 2018). Lecture-based learning is typically used in the K-12 setting, which is reinforced by the plethora of standardized testing to measure the student's achievement in mathematics and sciences (Hart et al., 2015). This approach instructs students on what to think or memorize to be successful on exams. The students are successful by merely recalling facts or concepts with little understanding of the fundamental principles. These same students matriculate into college with the expectation that they will continue to have good performance on exams and perform well in college because their past success is a good predictor of future success. Some students may have had critical thinking and information-processing in high school or can adapt without any outside intervention, but the vast majority of students struggle in college-level science and mathematics courses (Malcom & Feder, 2016).

THEORETICAL UNDERPINNING

Active Learning

Learning is not the direct transfer of information from instructor to student. The student must process the information to construct their understanding of the concepts. Bloom taxonomy is a model that organizes learning into various levels, beginning with knowledge as the lowest level of learning and ascending to comprehension, application, synthesis, and evaluation at the highest level of learning (Bloom, 1956). Herron (1996) describes the lower level of Bloom's taxonomy as declarative knowledge or merely factual information. Declarative knowledge can be memorized and is stored in the brain as unrelated pieces of information without meaningful connection to other stored knowledge. Meaningful learning occurs when related facts are stored together in the brain (National Research Council, 2000). Declarative knowledge is essential because students need to know basic facts and concepts. Students may not understand how to move from the bottom of Bloom's taxonomy to application or synthesis without guided instruction. Active learning instructional strategies improve student's performance in science, engineering, and mathematics compared to lecture-based instructional strategies (Freeman et al., 2014; Wieman, 2014). The goal is to move beyond declarative knowledge to higher order thinking skills by actively engaging the students in the classroom. Students show improvement in STEM learning when instructors move from lecture-based instruction to student-centered instruction (American Association for the Advancement of Science, 2012).

Inquiry-based Learning

Inquiry-based learning is a student-centered instructional method that uses a process similar to the scientific method to make students involve in their learning. The students learn content through the same process that scientists use to conduct research (Gormally, Brickman, Hallar, & Armstrong, 2009; McNew-Birren & van den Kieboom, 2017). In contrast to rote memorization of facts, students construct knowledge by engaging in a discovery process. There is an increase in students' learning outcomes with inquiry-based learning as compared to those with the lecture-based instruction (Eberlein et al., 2008). Pedaste et al. (2015) reviewed 35 journal articles on inquiry-based learning to develop an inquiry-based learning framework with four main phases: orientation, conceptualization, investigation, and conclusion. The implementation of inquiry-based learning has several pathways, and the authors illustrate three of the possible pathways in their model. For example, a questioning and exploration loop leads to data interpretation and a conclusion, which include opportunities for discussion and self-reflection of learning (Mbamalu, 2001). In Kolb's Experiential Learning Theory, the learner has an encounter with a new experience (Kolb, Boyatzis, & Mainemelis, 2011; Konak, Clark, & Nasereddin, 2014). The learner observes, thinks, and processes information about the experience to develop or modify a concept, and applies the concept to a new situation to generate new experiences. These two models are similar to the scientific method, in which a scientist makes observations, generates a hypothesis, performs experiments to test the hypothesis, and makes observations to support or disprove the hypothesis (Pérez & Torregrosa, 1983).

Teaching Process Skills

Universities may not explicitly state this in their mission statements, but all universities want to produce students with excellent content knowledge and process skills, which include critical thinking, problem-solving, leadership, and teamwork skills. Employers also want students who can be a part of a team, collaborate with others, solve problems, and exhibit excellent oral and written communication skills. Students need content knowledge, critical thinking and problem-solving skills, oral and written communication skills, and information-processing skills to be successful in the workplace (Hobin, Fuhrmann, Lindstaedt, & Clifford, 2012). While the lecture-based instruction is typically used to deliver content to the students, students do not learn process skills by listening to the content (Aktamis & Ergin, 2008). Some would suggest that students do not learn content by listening to a lecture. Kirschner, Sweller, and Clark (2010) state that learning occurs when there is a change in long-term memory. Long-term memory storage is enormous, but the working memory has a limited capacity and stores around three to seven elements. The information stored in the working memory is lost after about 30 seconds if it is not rehearsed. In lecture-based instruction, new information is continually presented to the student, and unless the student makes some permanent record to review later, most of the content from the lecture will be lost (Hmelo-Silver, Ducan, & Chinn, 2007; Ozden & Gultekin, 2008). The science laboratory course associated with a lecture course was initially designed to enable inquiry-based instruction and to coordinate with the content taught in the lecture course (Hofstein & Lunetta, 1982). The premise was that the students would receive content in the lecture course and develop critical thinking and other process skills in the laboratory component of the course. However, most science laboratory courses, especially at the introductory level, often employ the "cookbook" style instructional format where the students are told "what to think" and "how to think" without requiring the student to process information or asking the students employ scientific and higher order thinking skills. The scientific process has been removed from the lecture and laboratory courses and replaced with vast amounts of content knowledge. This contrast can be seen as a difference in the actual size of inquiry-based textbooks and content-based textbooks. In a knowledge-based science laboratory, the student works with another student to follow the procedures as written, collect the data as instructed, and perform the calculations as shown without an initial inquiry question or model that would lead to hypothesis generation and experimental design to test the hypothesis.

In the desire to deliver content, the inquiry and information-processing has been taken out of science and replaced with mechanical experimentation and rote memorization of content. However, students are still expected to analyze, interpret, draw a conclusion and evaluate information when they have not received the necessary tools and opportunities to strengthen these skills. Lecture-based instruction delivers content to the students, but student assessment measures information processing, critical thinking, and reasoning skills. The difference in expectation between instruction and assessment is frustrating to students because they cannot understand their poor performance on exams when they have memorized, read, listened, and taken notes on the content. Most students are unable to make the connections between content-driven instruction and critical thinking and reasoning skills needed for exams. Students have the content knowledge, but they are unable to apply, analyze, draw conclusions, and make inferences using that knowledge. Therefore, they often leave the STEM pipeline before they have the opportunity to develop inquiry skills and strengthen information processing and reasoning skills (National Science Foundation, 2017).

From the student viewpoint, science is not exciting, but just a vast amount of unrelated facts in different courses. The pre-med biology major cannot understand why she/he needs to take two years of chemistry courses for medical school. While students think science classes and labs are boring, most students love their undergraduate research experiences (Chopin, 2002; Russell, Hancock, & McCullough, 2007). A student who cannot remember what she/he did last week in the lab can explain their summer research project after several months. Early research experiences for STEM majors have been shown to increase retention of women and minority students in STEM disciplines (Russell et al., 2007; Stanford, Rocheleau, Smith, & Mohan, 2017). One reason is the development and strengthening of the critical thinking and process skills needed to be successful in upper-level STEM courses (Aktamis & Ergin, 2008; Kruea-In & Thongperm, 2014).

The Motivation for Pursuing POGIL Instruction

POGIL is an instructional strategy that is student-centered and encourages the use of the scientific process to actively engage students in learning by building their knowledge through hands-on, investigative activities (Farrell, Moog, & Spencer, 1999; POGIL, n.d.a). Students work in groups of three to four and are assigned specific roles. Konak et al. (2014) showed group work as a factor that impacted students' learning outcomes and experiences, and enhanced students' reflective observation and abstract conceptualization. Students working in groups to solve a problem can enhance their critical thinking skills by allowing themselves to collaborate while discussing possible approaches with the group members and learning from each other. The students can foster forethought, self-reflection, and comprehensive learning. The students become aware of how they learn and what gaps are missing in their knowledge; they can then evaluate and adapt these skills to become increasingly active learners (Karplus, 1977; Springer, Stanne, & Donovan, 1999).

INSTITUTIONAL CONTEXT AND IMPLEMENTATION PLAN

POGIL

Process-oriented guided-inquiry learning is an evidence-based teaching strategy that has been shown to improve both process skills and content knowledge (Farrell et al., 1999; Moog, 2014; POGIL, n.d.a). POGIL differs from other student-centered instructional strategies by using distinctive classroom materials and the development of process skills (Eberlein et al., 2008). Designed for the students to learn content and to strengthen one or more process skills, the

POGIL activity usually begins with a model and exploratory questions about the model to help students improve their information-processing skills as they work to interpret the model (Farrell et al., 1999; Hanson, 2006; Moog, 2014).

Organization of a POGIL Classroom

The students organize into small self-managed groups (three to four students per group). The group members depend on each other to complete the tasks within a specific time frame. Each member of the group has a specific role and actively participates in managing group activity to complete an inquiry-based lesson. The various roles rotate among the group members throughout the semester so that every student will get the opportunity to develop different process skills as they work through the guided-inquiry learning activities. The assignment of specific roles to the group members keeps every student actively engaged in the learning process; helps with effective group management; teamwork; and strengthens other embedded process skills. The typical student roles are manager, recorder, presenter, and reflector. The manager is responsible for the overall management of the group, and the recorder summarizes, in written form, the main points of group discussions, observations, and insights. The presenter communicates observations and conclusions with others, and the reflector thinks about the group dynamics during the learning process. Each group works on a specific task in a designated time frame and reports their findings to the entire class. The reporting gives students the opportunity to develop communication skills, assess their performance, and reflect on what they have learned (POGIL, n.d.b).

Design of POGIL Activities

The design of POGIL activities is discussed in detail in materials from the POGIL Project provided at their website (http://www.pogil.org), including guidelines for authors (POGIL, n.d.c). There are two basic types of POGIL classroom activities: learning cycle activities and application activities (Eberlein et al., 2008). To summarize, learning cycle activities have three stages: exploration, concept invention, and application. Ideally, an activity takes one class period. In the first phase, students explore the features of a model, diagram, or information through a series of exploratory questions. In the second stage, a second model or modification of the original model is used to guide students through concept formation or invention. The scientific term for the concept may not be provided until the end of this section. In the third phase, students use the developed concept in applications. Some topics are not suitable for concept invention. In this case, the model may be the concept and students examine the concept through exploration and application.

A unique feature of POGIL activities is targeting specific process skills in addition to content. For example, an activity-targeting information-processing would require the group members to sketch a diagram from a description or develop a mathematical relationship from data. POGIL classroom activities are available in several disciplines, particularly in chemistry due to the origins of POGIL (Farrell et al., 1999). Semester or year-long collections of POGIL activities are available as workbooks for students (POGIL, n.d.d). The POGIL website also includes implementation and facilitation resources (POGIL, n.d.e) for the instructor and a reference list of published articles in various disciplines (POGIL, n.d.f). The NSF POGIL-PCL (Physical Chemistry Laboratory) grant funded in 2010 began the development of POGIL laboratory activities. Currently, laboratory activities are available for physical chemistry, and work is underway to develop POGIL activities for general chemistry, analytical chemistry, and biochemistry laboratories. The POGIL project is currently developing an Activity Clearinghouse which will permit the members to download reviewed and edited activities as well as provide a formal method of having individual activities reviewed and approved as official POGIL activities.

INSTITUTIONAL CONTEXT AND IMPLEMENTATION PLAN

African American students are underrepresented in STEM disciplines (National Science Foundation, 2017). It has been shown that student-centered instructional methods increase students' learning outcomes by engaging the students in activities that strengthen students' information-processing and critical thinking skills (Chase, Pakhira, & Stains, 2013; Farrell et al., 1999; POGIL, n.d.f). Two Historically Black Colleges and Universities (HBCUs) implemented pilot projects of POGIL instruction into the chemistry curriculum. Jackson State University is a public university in the State of Mississippi and Tuskegee University is a private university in the State of Alabama. The K-12 systems in both the states rank in the lower-third for academic achievement, and many of the students in these states enter college underprepared for the rigor of STEM disciplines (Education Week, 2018; Nation's Report Card, 2017).

Jackson State University

Jackson State University (JSU) is classified as a Carnegie research-intensive university and is the largest HBCUs in the state of Mississippi and the fourth largest HBCUs in the US. JSU is located in Jackson, the capital and largest city in Mississippi and is the only Urban University in Mississippi. Recent statistics (Jackson State University, 2017) for the Fall 2017 semester showed 8,558 enrolled students – 76% undergraduate and 24% graduate students. Approximately 90% of the enrolled students are African American, 6% Caucasian, and 4% Other. The College of Science, Engineering, and Technology is the largest college at JSU with 2,225 (26%) students (Jackson State University, 2017). The Department of Chemistry & Biochemistry recently merged with the Department of Physics and Atmospheric Sciences. The initial implementation of POGIL instruction occurred before the merging of the chemistry and physics departments. The chemistry department has sixteen full-time faculties, two full-time instructors, and several adjunct faculties; it confers degrees at the bachelor (American Chemical Society-approved program), master, and doctoral levels with approximately 120 undergraduate and 40 graduate students. Approximately 90% of the undergraduate chemistry majors and 50% of the graduate students are African Americans. On average, 15–20 undergraduate students graduates with a bachelor degree in chemistry per year.

Implementation of POGIL Activities in the Biochemistry Laboratory

Before the implementation of POGIL activities in the biochemistry laboratory, students followed the laboratory procedures as written without thinking about the experiments. Usually, the laboratory experiment included a pre-laboratory theory and post-laboratory questions to encourage critical thinking about the concepts embedded in the experiment. The traditional laboratory instruction is similar to the lecture-based instruction, in which students are told "what to think" and are not actively engaged in the problem-solving process using the scientific method. POGIL lab instruction requires the students to evaluate or process the data or information (declarative knowledge) to transform the current information (information-processing) to reach conclusions (concept invention). As indicated previously, development of POGIL laboratory activities has lagged behind the development of exemplary POGIL lecture activities. The guided-inquiry activities implemented in the biochemistry laboratory were developed at JSU and have gone through one cycle of editing by a POGIL authoring coach (POGIL, n.d.c). POGIL lab instruction starts with an exploratory question, scenario, or a prediction to initiate the process of inductive reasoning. The students discuss the pre-lab prompt in groups of four students and generate a testable hypothesis. The different groups "report out" to the entire class and engage in a guided discussion. The entire class in the lab come to a consensus on one or two central questions to answer. The students work in pairs to perform the experiment, and the group of four students comes together to discuss their results and reflect on what worked well, what did not work, and the possible modifications to their experimental design. The group members "report out" to the whole class and discuss results during the last 30 minutes of lab.

POGIL Lab Activities

Using this model, POGIL laboratory activities were developed for micropipettes, buffers, protein assay, and column chromatography experiments that incorporated guided-inquiry activities to enhance students' learning of important concepts in the biochemistry laboratory, to promote critical thinking skills and to elevate the level of student's engagement (Campbell, 2014a, 2014b). The developed lab activities have not received the official POGIL endorsement by the POGIL Project, but the term POGIL activity will be used to distinguish the POGIL instruction from that of the traditional lab.

The students were familiar with measuring large volumes of liquids, but only a few students had prior experience using micropipettes. The pre-lab activity was the same for both the traditional and POGIL labs, which was a review of simple statistics (mean, % error, mean deviation) and unit conversion (mL to μL). In the traditional lab activity, the students had detailed procedures to determine the accuracy and precision of pipetting water using a micropipette, analytical balance, and density conversion table and compare the stated pipette volume with the actual volume of the solution delivered. The micropipette activity was a simple experiment that students could complete in about 30 minutes as part of the orientation to the biochemistry laboratory.

Facilitation of Lab Activity

In the POGIL lab activity, the students had a question prompt and no procedures (Table 1). The students were given a set time to discuss the question prompt in their groups prior to reporting out their hypothesis. For the micropipette experiment, three to five minutes were sufficient time for the discussion.

After the reporting out, the three most frequent hypotheses to test were the effect of temperature, density, and viscosity of the solution on the accuracy of liquid transfer using a micropipette. An example of a student hypothesis would be, " we hypothesize that the accuracy of liquid transfer will decrease as the viscosity of the solution increase." After additional discussion among the POGIL groups, the students concluded that viscosity is a function of temperature. To quickly summarize without going through all of the iterations, the various groups developed experimental procedures to test their hypothesis. Initially, they used water and glycerol but then realized that the glycerol was too viscous to

Table 1. Learning Objectives for the Micropipette POGIL Laboratory Activity.

Item	Item Description
Micropipette Question Prompt	What properties of a solution affect the accuracy and precision of liquid transfer using a micropipette?
Content Learning Objectives	• To properly use micropipettes for liquid transfer • To properly use an analytical balance for mass measurements • To calculate % error and absolute mean deviation • To differentiate between accuracy and precision • To generate a hypothesis based on scientific knowledge • To develop an experimental design to test a hypothesis
Process Skills Objectives	• To strengthen teamwork skills by interacting with others to work towards a common goal, building consensus, and sharing ideas • To develop oral and written communication skills through the exchange of information, understanding through speaking, listening, writing key discussion points, recording experimental data in a laboratory notebook, and non-verbal behaviors • To develop problem-solving skills by identifying, planning and executing a strategy to find a solution to a question • To develop critical thinking skills by analyzing and evaluating relevant information to reach a conclusion supported with evidence • To develop information-processing ability by evaluating, interpreting, manipulating, or transforming information

pipet using a micropipette. Interestingly, since the instructor did not provide answers during facilitation, the groups began to collaborate and compare the results. Finally, the students prepared several different concentrations of water–glycerol solutions, obtained their results, retrieved information on the density and viscosity of water–glycerol solutions along with the temperature, which they realized was important, completed calculations, and reported out their results. This was a quick overview of one experiment to illustrate that the students were actively engaged during the activity because they had to gather information, process it, develop a hypothesis to test, design an experiment to test their hypothesis, determine control and variable samples, share information, collaborate with other groups, and come to a consensus conclusion. Additionally, the students developed concepts about the relationship between temperature, density, and viscosity, which was not a component of the traditional lab experiment.

Student Outcomes for POGIL Activities

The biochemistry laboratory had two sections that met on different days for the Fall 2012 and Fall 2013 semesters. The same laboratory techniques were taught in both laboratory sessions, and no modifications were made to the typical chemistry laboratory classroom. The implementation of POGIL into the biochemistry laboratory was funded by a grant from NSF and had the full support of the department chair and college dean. Each laboratory section on average had about sixteen undergraduate students (mostly seniors), in which about 50% were chemistry majors and 50% were biology majors. The Control group had the traditional laboratory instruction, and the POGIL group received the POGIL lab activity. As a side note, the students in the different laboratory sections were all in the same biochemistry lecture. Therefore, there were ample opportunities for student interactions outside the laboratory setting. As the semester progressed, some students in the Control group wanted to switch laboratory sections because they felt the students in the POGIL group received "better" instruction as one student stated, "they are learning more than me." The same techniques were taught in both the laboratory sections but the students in the Control group "sensed" that they did not learn what they needed to learn from the traditional lab experiments.

A summative assessment, that embedded questions on the final exam, was used to measure students' learning outcomes of important concepts. The assessment data showed an increase in student learning of important concepts in biochemistry (Fig. 1). In the Fall 2011 semester, there was a slight increase in the POGIL group mean final exam score. A significant problem with the first implementation of POGIL instruction was the length of the activities; that is, the activities were too long for the students to complete within the three-hour laboratory period. Student frustration was also high due to unknown expectations. POGIL instruction was the first experience that most students have had with guided-inquiry labs, and the lack of experimental procedures caused significant stress to the students.

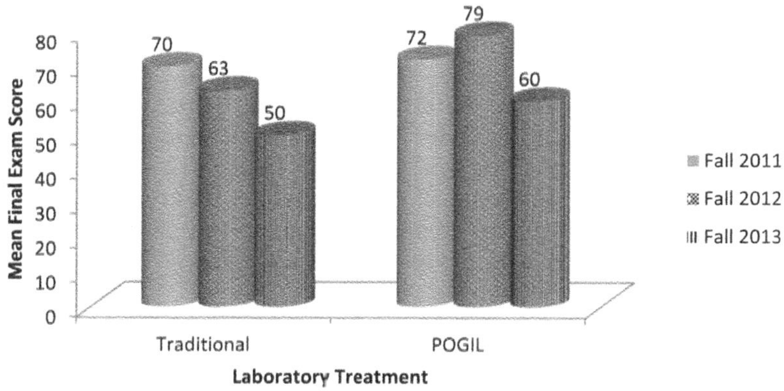

Fig. 1. The control group received traditional laboratory experiments, and the POGIL group received POGIL laboratory activities for micropipette, buffers, protein assay, and protein purification experiments. The same final exam was used in the Fall 2011 and Fall 2012 semesters. The data showed that the Fall 2011 mean final exam score was 70 ($n = 14$) for the control group and 72 ($n = 13$) for the POGIL group. The data for the Fall 2012 semester showed a mean final exam score of 63 ($n = 18$) for the control group and 79 ($n = 13$) for the POGIL group. In the Fall 2013 semester, the final exam was changed to embed more questions that required critical thinking, data analysis, and reasoning skills. The mean final exam score was 50 ($n = 14$) for the control group and 60 ($n = 14$) for the POGIL group.

For the second implementation in the Fall 2012 semester, most students had heard from their peers about the biochemistry lab. Also, the length of the lab activities was decreased, allowing sufficient time for information-processing, analysis, discussion, and reporting out. The amount of time spent for the active learning activities was determined from the first implementation. Also, the professor learned that is was important to orient the students to the guided-inquiry lab by providing expectations for group collaboration, and explaining the assignment of roles for and the responsibility of each group member. The professor also explained the importance of strengthening critical thinking, reasoning, and process skills for their continued success in academia or the workforce. The difference in the mean final exam score was sixteen points higher for the POGIL group as compared to that of the control group with the second implementation of POGIL instruction. However, an uncontrollable variable in the Fall 2012 semester was the number of students in each section of the biochemistry lab. The control group had about 30% more students than the POGIL group, and the size of the classroom probably played a role in students' learning outcomes.

In the Fall 2013 semester, the number of students in each section of the biochemistry lab was adjusted to fourteen. The final exam was changed by increasing the number of critical thinking, reasoning, and data analysis question types in the exam. The embedded questions on the final exam were application- and analysis-oriented questions as compared to those of the prior-year final exam, which contained more knowledge and comprehension questions. The overall mean final exam scores were lower for the Fall 2013 semester as compared to those of the two previous semesters (Fig. 1). However, the mean final exam score was ten points higher for the POGIL group as compared to those of the control group. These results indicate that the classroom size may have been a variable in the sixteen-point difference seen in the Fall 2012 semester. Overall, students in the POGIL group for all three years had on average higher final exam scores. Because of the improvement in the mean final exam scores for the POGIL group, POGIL lab activities were used in all sections of the biochemistry lab after the Fall 2013 semester.

POGIL Lab Project

A guided-inquiry biofuel laboratory module was developed and implemented for the Fall 2014 semester that incorporated research-based activities to enhance student learning of important concepts in enzyme kinetics, to increase student critical thinking and reasoning skills, and to promote student's interest in energy-related research. This project took four laboratory sessions to complete, and POGIL lab projects are not an official type of endorsed POGIL activity. The students received background information on enzyme kinetics and performed a literature search on biofuel production in the United States. The POGIL groups developed a series of experiments using acid-digestion of switchgrass (*Panicum virgatum*, a perennial grass used in biofuel research), followed by enzymatic digestion of the neutralized switchgrass extract and detection of glucose using a colorimetric assay.

Collaboration among the student groups was essential for completion of the biofuel project. An example of the collaborative effort involved the characterization of cellobiase, an enzyme used to hydrolyze β-glycosidic bonds in cellobiose. One student group performed the effect of temperature on enzyme activity; another student group determined the optimal temperature for the enzyme; and several groups completed experiments and combined results to determine the maximal velocity (V_{max}) and the Michaelis–Menten constant (K_M) for cellobiase. The student groups shared their data, and they were able to characterize the enzyme in one laboratory session.

Student Outcomes for the POGIL Lab Project

The students were assessed on their learning of essential concepts in biochemistry using embedded questions in the final exam. The same final exam was used for the Fall 2013 and Fall 2014 semesters. The Fall 2014 semester data were combined with the Fall 2013 semester data to examine the effect of adding a guided-inquiry project into the biochemistry laboratory curriculum. Both sections of the Fall 2014 semester utilized the same guided-inquiry with the addition of a biofuel guided-inquiry project. The mean final exam scores for the Fall 2014 lab sections were 72.7 ($n=15$) and 73.5 ($n=16$), which was higher than the mean final exam scores for the POGIL instruction without the biofuel project.

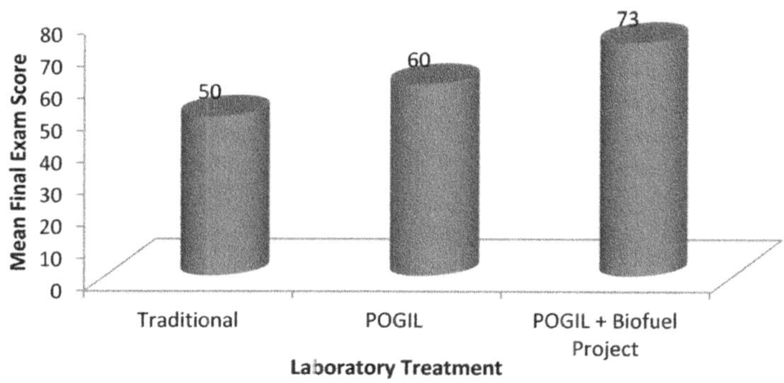

Fig. 2. The control group received traditional laboratory experiments, and the experimental group received POGIL laboratory activities. The Traditional and POGIL data is from the Fall 2013 semester. The POGIL + Biofuel Project data is from the Fall 2014 semester. The same final exam was used in the Fall 2013 and Fall 2014 semesters. The data showed a mean final exam score of 50 (n = 14) for the traditional instruction, 60 (n = 14) for the POGIL instruction, and 73 (n = 31) for the POGIL + Biofuel Project instruction. In the Fall 2014 semester, both sections of the biochemistry lab utilized POGIL + Biofuel Project instruction (the mean final exam scores for the two sections were 72.7 (n= 15) and 73.5 (n = 16)).

The results showed a positive correlation between student learning outcomes and the level of student engagement (Fig. 2).

One former student stated that she was able to discuss the biofuel laboratory experiment during a job interview and she felt that it contributed to her obtaining employment with a biofuel-manufacturing industry in Mississippi. She also credits the pipetting techniques, experimental design, standard curve preparation, and data analysis for her excellent job performance in her current position. Other students have commented that they have a better understanding of either a past summer research project or a current research project after completing the biochemistry laboratory curriculum. These results suggest that the curriculum changes had a positive impact on student learning and attitudes in the biochemistry laboratory.

To further examine the effect of the guided-inquiry project on students' learning outcomes, POGIL instruction was used in one section of the biochemistry lab, and POGIL instruction with a final POGIL biofuel project was used in the other lab section for the Fall 2015 semester. The previous analysis combined data from the Fall 2013 and Fall 2014 semesters, and during each implementation, modifications were made to either the activity or the facilitation. A new final exam was developed and used for the Fall 2015 semester. The mean final exam score was 60 (n = 22) for the POGIL instruction group and 65 (n = 21) for the POGIL with biofuel project group. These results indicate that POGIL lab projects may improve students' learning outcomes more than using single, stand-alone POGIL activities in the biochemistry lab.

Students' Perception of Their Learning

The student assessment of their learning gains (SALG) was used to assess students' perception of laboratory skills and reasoning ability. SALG is a free, online course-evaluation tool to obtain students' self-assessment of their learning (SALG, n.d). The results from the Fall 2013 semester showed a slight increase ($p = 0.10$) in student perception in the area of knowledge integration (Fig. 3). The inquiry-based laboratory experiments may have had a positive impact on student attitudes and perceived abilities (Fig. 3). Students stated that the lab required more "thinking," similar to their experiences in summer research experiences rather than an undergraduate instructional laboratory.

Focus Group Interview

The external evaluators conducted a focus group interview with about 25 students enrolled in the biochemistry lab. The lab instructor was not present during the focus group interview. The majority of the students were seniors, about half of the students were chemistry majors, and the remaining students were biology majors. The major conclusions from the focus group interview were as follows:

- The students stated that the lab experience was different from their previous chemistry lab experiences.
- The lab activities required them to use more critical thinking and inquiry skills similar to their experiences in research internships rather than an academic lab experience.

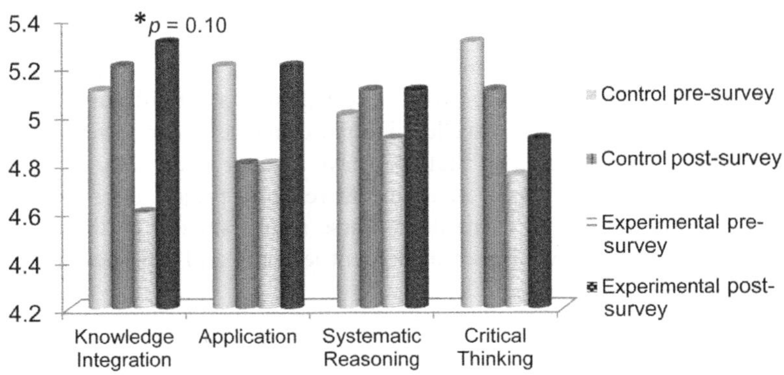

Fig. 3. The students in the control group (traditional labs) and the experimental group (POGIL labs) took a pre-survey at the beginning of the laboratory course and a post-survey after the last POGIL lab activity. The survey consisted of a series of questions that related to one of the four categories of knowledge integration, application, systematic reasoning, and critical thinking. The rating scale for the survey was as follows: 1 point (*not applicable*); 2 points (*not at all*); 3 points (*just a little*); 4 points (*somewhat*); 5 points (*a lot*); 6 points (*a great deal*). The experimental group showed a significant increase in knowledge integration ($p = 0.10$).

- The lab required group work.
- They felt that the review of background information was essential.
- The students appeared to have extensive content knowledge and were very curious about biofuels.

The external evaluators also noted that the students in the focus group were very motivated and engaged in their various disciplines. The students posed questions to the interviewers and appeared to be familiar with "active" participation in the learning process.

TUSKEGEE UNIVERSITY

Tuskegee University (TU) is an HBCU founded in 1881 by Booker T. Washington in southern Alabama. The current student population is approximately 3100 students, over 75% (2014) identify as African American. Approximately 60% of the student body is female. The majority of students are of science or engineering majors. TU implemented POGIL instruction in general chemistry (lecture only; selected sections) and physical chemistry (lecture and lab).

Implementation of POGIL Activities in Physical Chemistry Lecture Courses

The physical chemistry lecture course is the origin of POGIL instruction at TU. In the Spring 2007 semester, more than half of the students in the Physical Chemistry 2: Quantum Mechanics and Spectroscopy failed the first exam. Up to that point, the lecture-based instruction was used to deliver content including the use of PowerPoint slides, textbook, homework, and quizzes. After the failed exam, the instructional strategy changed to small-group, problem-based learning using self-written exercises. It was several years before that metamorphized into POGIL, and another few years before the implementation of POGIL grew in fidelity (particularly for the use of the learning cycle, facilitation, and process skills). This development of the course pedagogy was not funded but received full support and encouragement from the department head and no concerns from the administration An occasional student has been distressed by the lack of lecture, but the vast majority of students have ended the semester preferring the active learning environment versus the lecture-based learning.

During this decade of problem-based learning developing into POGIL, the number of students in the physical chemistry lecture has been as low as four and as high as sixteen. As is common in guided-inquiry interventions, the American Chemical Society (ACS) standardized exam score was relatively unaffected by the change in methods. While not the most effective measure of student learning from POGIL, the ACS standardized exam in Thermodynamics provides a relatively stable assessment from year to year.

As of Fall 2017, students in the physical chemistry course at Tuskegee work in groups of four with assigned roles. POGIL activities authored by Tricia Shepherd at St Edwards College in Austin, TX, with supplements authored by Reeves were used in Thermodynamics, and activities authored by Reeves were used in Quantum Mechanics. The students had homework, reading assignments, and every class began with a short quiz covering the previous class's material. The student groups performed daily self-assessment of content goals and process outcomes.

Implementation of POGIL Activities in General Chemistry Lecture Courses

The General Chemistry lecture course at TU had four different sections with POGIL instruction implemented in two of the sections. Tourne commenced POGIL instruction in the Fall 2014 semester after noticing students were not truly grasping concepts but were instead memorizing terms and "cramming" the night before an exam. Implementation of POGIL instruction in the general chemistry course has been slow and has been modified every semester to accommodate different requests and/or logistical issues (Libby, 2008). Each of the four sections had more than 60 students with an upward limit of 115 students. For the most part, students were accustomed to lecture-based instruction, and they were uncertain about the POGIL instruction. Therefore, it was important on the first day of class for the professor to explain the purpose of POGIL instruction.

IMPLEMENTATION OF POGIL ACTIVITIES IN PHYSICAL CHEMISTRY LAB COURSES

In August 2011, NSF funded the grant "POGIL-PCL and the Development and Implementation of Guided-inquiry Experiments for Physical Chemistry," award number 1044624, with PI Sally Hunnicutt and co-PIs Robert Whitnell and Alexander Grushow. The PIs held the first POGIL-PCL workshop on 4–6 January 2012, at Virginia Commonwealth University, which Reeves attended. Over the course of the next five years, Reeves attended all the POGIL-PCL workshops and helped develop at least three experiments (Reeves, Berghout, Perri, Singleton, & Whitnell, 2017; Reeves & Whitnell, 2014) within the POGIL-PCL framework (Hunnicutt, Grushow, & Whitnell, 2015).

Due to the POGIL-PCL project and the collaborations it fostered, Physical Chemistry lab at Tuskegee University is a fully POGIL lab. The number of experiments has decreased, but students make many more decisions in the lab. A significant amount of lab time is spent in advanced data analysis, especially since POGIL-PCL emphasizes data pooling among teams. Multiple students over the years have commented that this was the first time they understood everything they were doing in the lab and why.

Large Classroom Implementation

A typical POGIL session requires students to join a four-person team to complete an hour-long conceptually driven activity. To ensure that students have successfully grasped the primary objective of the activity, the facilitator will usually request them to report out their results (by either testing them individually using clicker remotes or allowing each team to give a summary of what they have learned). In a small classroom, these activities are easily managed and executed, but it can become challenging in larger classes. For one, the sheer number of groups a facilitator will need to manage can become exceedingly high if the classroom has over 100 students (Yezierski et al., 2008). Without proper facilitation, students may be left waiting for help before they can proceed to other questions in the activity, which will, in turn, hurt their chances of completing the activity. For the process to work successfully (for both the student and the facilitator), it is best to recruit other facilitators for the sessions. Teaching assistants (TAs) are vital to this type of teaching methodology for large classrooms; in effect, multiple facilitators are walking around the student groups ensuring a reasonable pace in completion of the activity. If a TA is not available, it might be a good idea to keep the sections of the activity short enough so that the facilitator can regroup with the entire classroom at set intervals within the class period. In this fashion, the facilitator can clarify any confusion students have, allowing the students to proceed to the next section. In the general chemistry course at Tuskegee, undergraduate TAs assist the professor.

Organization and clear expectations are the other necessities for a large POGIL classroom (these are also the necessities in small classrooms but are not as detrimental to time management). For the most part, a facilitator can still count on the fact that a significant portion of their students has never participated in a POGIL class activity and thus must be clear on their expectations. It is imperative to explain what POGIL is, how it works (the theory behind it), and what the students will be doing in a typical class period. Setting a routine is helpful for any classroom but especially for a large POGIL classroom. In other words, students are expected to do quite a bit before they can start the activity – they need to pick up materials, know which team to join, know their role in the team, know the time limit to complete the sections, and know how to report their findings. If all these requirements are clarified early in the semester, it will help save time and confusion for everyone. Depending on the classroom, it might also be helpful to state where (logistically) the groups will be placed and who will be responsible for setting up the chairs.

It is important to note that even though there might seem to be more challenges to adopting this type of pedagogy in a large classroom, there are many variations to POGIL. The facilitator can change or amend any part of the process to make it work best for their situation. For instance, the facilitator might utilize different types of reporting methods, or require different teams to report on different sections of the activity or even require students to upload their report after class to maximize the amount of time that students have to complete the activity in class (Baepler, Walker, & Driessen, 2014). The facilitator might also decide to utilize POGIL only for specific concept building and not for others, or even use POGIL a certain number of days a week and lecture traditionally on other days. No variation is wrong as long as the students are conceptually building their knowledge of the complex subject matter.

OUTCOMES AND EVOLUTION

There was an increase in STEM learning, students' perception of their STEM learning, and students' attitudes about chemistry for students using POGIL instructions. JSU and TU successfully implemented POGIL instruction in several courses and received external funding from NSF to broaden the implementation of POGIL instruction into the chemistry curriculum. At JSU, one full-time faculty and two adjunct faculties in the chemistry department use POGIL activities at some level in the general chemistry lecture, biochemistry lecture, and biochemistry laboratory. The full-time faculty received training in POGIL facilitation through The POGIL Project and serves as a resource person at JSU for other faculties interested in implementing POGIL instruction. Faculty in the biology, chemistry, and education departments at JSU received an NSF grant in June 2017 to implement POGIL instruction in the general chemistry and organic chemistry lecture courses to improve learning outcomes of biology majors. There are approximately 950 undergraduate biology majors at JSU. At TU two full-time faculty use POGIL instruction in the general chemistry lecture, physical chemistry lecture, and physical chemistry laboratory. An NSF HBCU-TIP grant awarded August 2017 is being used to develop guided-inquiry experiments in multiple courses in the chemistry curriculum. Both faculties have had POGIL training, and Reeves has facilitated POGIL workshops.

LESSONS LEARNED AND RECOMMENDATIONS

Student Skepticism Toward Group Learning

When students realize that group work is fundamental to POGIL instruction, the response is almost inevitably the same – a wave of hesitation "sighs" heard across the room. Hesitation toward group learning is especially apparent for lower level classes where the only experience they might have had with group assignments is where a teacher gives them a task and everything about that task (who is responsible for what, how often they will meet, by when should it be completed) gets decided by the group members. If all the participating members are dedicated and strive to do an equal amount of work, the partnership will be a success! Unfortunately, this is usually not the case, and groups end up being a spectrum of over- and under-achievers leaving the brunt of the work to only a portion of the members. One of the efficacies of POGIL is that all participating members have an assigned role; only that member is allowed to do their assigned role thereby requiring the participation of everyone. Once students realize every team member has a purpose and the activity cannot be completed without them, their skepticism or hesitation seems to subside. Also, the fact that the group activities take place in the classroom rather than outside the classroom makes students more amenable.

Hesitation to Conceptual Learning

In lower level courses, it is common to see significant student hesitation toward conceptual learning – students tend to memorize topics, equations, algorithms, and vocabulary rather than thinking through a problem. Most students will want to memorize an equation and simply "plug-n-chug" to get an arbitrary answer – not even thinking why they are using the equation, what the equation tells them, or whether the result is plausible. For this reason, it behooves the professor to explain the difference between the two approaches early on in the semester (first day of class) and provide examples of activities that require skills such as information-processing, critical thinking, problem-solving, and assessment. The professor might even want to go a step further by providing students with a handout of all equations and/or constants that might be utilized in class (if applicable) and let them know it will always be provided in their examinations – to reiterate that memorization is not necessary. Most students have never been confronted with the idea of conceptual learning and must be encouraged to be successful.

Time Constraint

Time constraints are always a real "threat" to the professor but even more so when utilizing POGIL. It takes longer to cover a concept employing guided inquiry as compared to lecturing on the concept. Students require time to explore the model, process information, and collaborate as they work through the POGIL activity. A POGIL worksheet will typically have only one or two concepts introduced in the activity. At TU, the first half of the general chemistry course covers 10 chapters in one semester and is a time crunch even while lecturing traditionally. The TU Chemistry Department is mandated to cover specific topics so students can successfully take the ACS standardized exam at the end of the semester. Similar to adapting one's style to larger classrooms, the professor can change/amend their curriculum to make it work for them (while still covering the mandated topics). It is best to decide what topics are most critical for students to have a fundamental understanding and leave those for POGIL, while other topics may be covered either in class – traditional lecturing, or outside of class – videos, as homework assignments, or in extra study sessions (Baepler et al., 2014).

Similarly, at JSU the General Chemistry course covers Chapters 1–4, 6–10 in the first semester. However, covering every chapter does not correlate with student learning outcomes. As more knowledge becomes available from scientific discovery, there seems to be an urgency to include all of these discoveries in the next edition of the instructional textbooks. The textbooks are getting larger, students are spending more time on rote memorization of content, professors spend more time trying to cover all of the content, and the central concepts are obscured (Mulford, 1996). There are core concept inventories for the main STEM disciplines and some sub-disciplines: for example, the ACS Exams Institute anchoring concepts content map (Holmes & Murphy, 2012).

Implementation Issues

JSU received initial funding from the National Science Foundation (NSF) for faculty development and implementation of POGIL instruction. Also, faculty at both TU and JSU received funds from The POGIL Project to attend the National Conference for Advanced POGIL Practitioners (NCAPP), a Facilitator Training Workshop, POGIL workshops, and the POGIL National Meeting. The POGIL Project also funded a half-day POGIL workshop at JSU and TU. In 2017, TU's chemistry department received an NSF grant to develop guided-inquiry experiments as well as to hold a two-day POGIL Lab Track workshop for faculty. The reason for stating the amount of faculty development in implementing POGIL instruction is because best practices in facilitation are as crucial as good POGIL activities (Daubenmire et al., 2015).

Fidelity in implementation takes time. Whether funded or non-funded, we suggest beginning with a small pilot cohort. Before implementing a POGIL classroom activity, one should consider attending a one-day workshop or preferably a three-day POGIL workshop. It is tempting and less time-consuming to use POGIL worksheets as a supplement to lecture-based instruction, but most of the process skills, critical thinking skills, and reasoning skills embedded in the POGIL activity are lost in students learning outcomes without collaborative group interaction and proper facilitation (Daubenmire et al., 2015). One learns best practices in the facilitation of a POGIL activity by participating in the POGIL workshops. In the POGIL workshop, the facilitator becomes the student and constructs knowledge about the various components of POGIL and best practices in facilitating activities by actively engaging in various POGIL activities.

Correct Answers

An essential aspect to facilitation is handling students' questions about the correct answer; small changes in facilitator behavior, such as requesting students to justify their answers, lead to richer student discussion (Stanford, Moon, Towns, & Cole, 2016). Usually, students want the "correct" answers to the questions because they want to memorize them for the exam. Providing a direct answers to the students' questions stops student engagement in the activity, as they wait for "correct" answers. Once students realize that the facilitator provides probing questions rather than direct answers to their questions and that they are responsible for their learning, the students become more actively engaged in the learning cycle of exploratory questions, concept invention, application, and exploratory questions. Instead of a direct answer to students' questions, groups are encouraged to confer with other groups, or probing questions are asked to obtain group consensus. At the end of the activity, the students' main points and consensus conclusions are summarized. The facilitator guides the students to make valid conclusions, results, analyses, and inferences using their own words.

POGIL Teamwork

The POGIL activities are inquiry-based lessons, but without small groups and classroom facilitation, the embedded process skills in the activity are lost. Doing the group activity in class utilizes time spent on content but merely covering ten to twelve chapters in a semester does not mean students have learned the content. A teacher-focused approach must be changed into a student-focused approach. The goal should not be to cover as many chapters or content as possible, but rather how many learning outcomes and process skills students gain in a semester.

Complaints

There usually is some resistance with a change in instructional strategy. In particular, students in the JSU biochemistry laboratory were frustrated by the lack of detailed procedures. The first experiment in the biochemistry lab is a straightforward, 30 minutes of micropipette experiment with detailed procedures. In the three-hour lab, this gives the students about 30 minutes to realize that there are really no experimental procedures, about 30 minutes to generate a consensus hypothesis to test, another 30 minutes to decide to use water as the control liquid and glycerol as the test liquid, and finally 30 seconds to realize that they cannot micropipette glycerol. The students used the remaining ninety minutes to modify their protocol to make different percent glycerol solutions, perform an online search for the viscosity and density of the prepared solutions, conduct the experiments, record their observations, report out their results, and combine everyone's data. As the students became engaged in the activity, the complaints were ceased, and their interest piqued. There were discussions within each group, between groups, and also whole-class discussions. The students also saw the value of collaboration and sharing because their data set of $n = 3$ for three different percent glycerol concentrations was now $n = 12$ for six different percent glycerol concentrations. The students showed an interest in their results, the results of other groups in the laboratory, and were engaged in discussions about the relationship among temperature, viscosity, concentration, and density.

POGIL Instruction in Every Course

Once POGIL instruction in a lecture or laboratory course shows an increase in students' engagement, learning outcomes, and process skills, there is the desire to implement POGIL instruction in the entire department. As previously stated, best practices in facilitation is also an essential component of POGIL instruction. Usually, it is best if the professor shows an interest in learning about POGIL instruction, and attends POGIL workshops to learn about the design of POGIL activities and how to facilitate the activities. It is possible that not everyone in the department will be interested in POGIL instruction, and that is okay. Remember, there is a large POGIL community that is available for support, implementation assistance, facilitation training, and authoring POGIL activities.

Assessment and Education Research

Effecting Institutional Change

Once an early adopter of POGIL realizes the benefits of guided inquiry and active learning, the next step is to expand POGIL instruction to the department, college, or university. That active learning is effective is no longer in question (Freeman et al., 2014). As previously stated, best practices in facilitation is also an essential component of POGIL instruction. Training additional faculty through POGIL workshops is the most efficient way to expand the cohort of practitioners at an institution. Both JSU and TU held on-campus training in POGIL through grants obtained by the authors. Reeves and Tourné were invited at TU by the college dean to present POGIL methods in a faculty Lunch and Learn setting. Campbell facilitated a one-half day POGIL workshop for faculty at JSU. Administrative buy-in is invaluable, but even in an ideal situation, not everyone will adopt these methods. The large and growing POGIL community can offer support in implementation, facilitation, and authoring of POGIL activities.

Assessment and Education Research

Education research involves human subjects; hence, it is necessary to obtain approval from the local Office of Research's Institutional Review Board. Time and effort spent on improving students' learning outcomes by developing and adopting new instructional methods in the classroom should be documented and findings disseminated to the scientific and general public about improvement in learning outcomes and best practices in instructional methods.

CONCLUSIONS

The experience of the authors is that guided inquiry, active learning pedagogies in general and POGIL specifically are successful in improving students' learning outcomes in chemistry at HBCUs. The improvement in students' learning outcomes occurred in both lecture and laboratory courses and in small and large classrooms, with initial hesitance but eventual strong student support. The early adopters of successful pedagogies can become catalysts for institutional change through example and through work to provide training for other faculties.

ACKNOWLEDGMENTS

Authors Reeves and Tourné would like to acknowledge their students and the support of the Department of Chemistry and the College of Arts and Sciences at Tuskegee University, particularly Department Head Albert Russell and Dean Channapatna Prakash. This material is based upon work supported by the National Science Foundation under Grant No. 1719359. Authors Campbell and Bridges would like to acknowledge collaborators Drs Jaehwa Choi, Mercer University School of Medicine; Macon, Georgia and Fengxiang X. Han, Jackson State University; and the project evaluators, Drs Eugene Kennedy and Su-Seng Pang, Louisiana State University, Baton Rouge, Louisiana. We also acknowledge the students participated in the study and the support of the Department of Chemistry, Physics, and Atmospheric Sciences at Jackson State University; support was received from The POGIL Project and the National Science Foundation NSF DUE 0941955 and 1712477 awards.

REFERENCES

Aktamis, H., & Ergin, O. (2008). The effect of scientific process skills education on students' scientific creativity, science attitudes, and academic achievements. *Asia-Pacific Forum on Science Learning and Teaching*, *9*(1), Article 4. Retrieved from https://www.eduhk.hk/apfslt/v9_issue1/aktamis/index.htm

American Association for the Advancement of Science. (2012). Describing & measuring undergraduate STEM teaching practices. Retrieved from http://ccliconference.org/files/2013/11/Measuring-STEM-Teaching-Practices.pdf

Baepler, P., Walker, J. D., & Driessen, M. (2014). It is not about seat time: blending, flipping, and efficiency in active learning classrooms. *Computers & Education*, *78*, 227–236.

Bloom, B. S. (Ed.). (1956). *Taxonomy of educational objectives. Handbook 1: Cognitive domain* (2nd. ed.). White Plains, NY: Addison-Wesley Longman.

Campbell, N. F. (2014a). Process-oriented guided inquiry learning to enhance learning of concepts in a biochemistry laboratory. FASEB J. 28, (1_supplement), 618.9. Retrieved from https://www.fasebj.org/doi/10.1096/fasebj.28.1_supplement.618.9. Accessed on June 4, 2018.

Campbell, N. F. (2014b). Assessment of process-oriented guided inquiry learning on learning outcomes for minority students. Retrieved from http://acselb-529643017.us-west2.elb.amazonaws.com/chem/248nm/ program/view.php. Accessed on June 4, 2018.

Chase, A., Pakhira, D., & Stains, M. (2013). Implementing process-oriented, guided-inquiry learning for the first time: Adaptations and short-term impacts on students' attitude and performance. *Journal of Chemical Education*, *90*, 409–416.

Chopin, S. F. (2002). Undergraduate research experiences: The translation of science education from reading to doing. *The Anatomic Record*, *269*, 3–10. Retrieved from https://onlinelibrary.wiley.com/doi/epdf/10.1002/ar.10058. Accessed on June 10, 2018.

Daubenmire, P. L., Bunce, D. M., Draus, C., Frazier, M., Gessell, A., & van Opstal, M. T. (2015). During POGIL implementation the professor still makes a difference. *Journal of College Science Teaching*, *44*(5), 72–81.

Eberlein, T., Kampmeier, J., Minderhout, V., Moog, R. S., Platt, T., Varma-Nelson, P., & White, H. B. (2008). Pedagogies of engagement in science: A comparison of PBL, POGIL, and PLTL. *Biochemistry and Molecular Biology Education*, *36*(4), 262–273. doi:10.1002/bmb.20204

Education Week. (2018). State grades on K-12 education: Map and rankings. Retrieved from https://www.edweek.org/ew/collections/quality-counts-2018-state-grades/report-card-map-rankings.html. Accessed on June 6, 2018.

Farrell, J. J., Moog, R. S., & Spencer, J. N. (1999). A guided inquiry general chemistry course. *Journal of Chemical Education*, *76*, 570–574.

Freeman, S., Eddy, S. L., McDonough, M., Smith, M. K., Okoroafor, N., Jordt, H., & Wenderoth, M. P. (2014). Active learning increases student performance in science, engineering, and mathematics. *Proceedings of the National Academy of Sciences*, *111*(23), 8410–8415. doi:10.1073/pnas.1319030111

Gormally, C., Brickman, P., Hallar, B., & Armstrong, N. (2009). Effects of inquiry-based learning on students' science literacy skills and confidence. *International Journal for the Scholarship of Teaching and Learning*, *3*(2), Article 16. Retrieved from https://doi.org/10.20429/ijsotl.2009.030216

Hanson, D. M. (2006). *Instructor's guide to process-oriented guided-inquiry learning*. Stony Brook, NY: Pacific Crest.

Hart, R., Casserly, M., Uzzell, R., Palacios, M., Corcoran, A., & Spurgeon, L. (2015). Student testing in America's great city schools: An inventory and preliminary analysis. Retrieved from https://www.cgcs.org/cms/lib/DC00001581/Centricity/Domain/87/Testing%20Report.pdf

Herron, J. D. (1996). A framework for discussion. In I. D. Eubanks (Ed.), *The chemistry classroom: Formulas for successful teaching* (pp. 7–15). Washington, DC: American Chemical Society.

Hmelo-Silver, C. E., Ducan, R. G., & Chinn, C. A. (2007). Scaffolding and achievement in problem-based and inquiry learning: A response to Kirschner, Sweller, and Clark (2006). *Educational Psychologist*, *24*(2), 99–107. doi:10.1080/00461520701263368

Hobin, J. A., Fuhrmann, C. N., Lindstaedt, B., & Clifford, P. S. (2012). So you think you have skills. Retrieved from http://www.sciencemag.org/careers/2012/09/so-you-think-you-have-skills. doi:10.1126/science.caredit.a1200107.

Hofstein, A., & Lunetta, V. N. (1982). The role of the laboratory in science teaching: Neglected aspects of research. *Review of Educational Research*, *52*(2), 201–217.

Holmes, T., & Murphy, K. (2012). The ACS Exams Institute undergraduate chemistry anchoring concepts content map I: General Chemistry. *Journal of Chemical Education*, *89*(6), 721–723. doi:10.1021/ed300050q

Hunnicutt, S. S., Grushow, A., & Whitnell, R. (2015). Guided-inquiry experiments for physical chemistry: The POGIL-PCL model. *Journal of Chemical Education*, *92*(2), 262–268. doi:10.1021/ed5003916

Jackson State University. (2017). Institutional research, planning, and assessment. Retrieved from http://www.jsums.edu/institutionalresearch/2017-summer-2017-fall-2018-spring/

Karplus, R. (1977). Science teaching and the development of reasoning. *Journal of Research in Science Teaching*, *14*(2), 169–175.

Kirschner, P. A., Sweller, J., & Clark, R. E. (2010). Why minimal guidance during instruction does not work: An analysis of the failure of constructivist, discovery, problem-based, experiential, and inquiry-based teaching. *Educational Psychologist*, *41*(2), 75–86. doi:10.1207/515326985ep4102_1

Kolb, D. A., Boyatzis, R. E., & Mainemelis, C. (2011). Chapter 9: Experiential learning theory: Previous research and new directions. In R. J. Sternberg & L. Zhang (Eds.), *Perspectives on thinking, learning, and cognitive styles* (pp. 227–247). New York, NY: Routledge.

Konak, A., Clark, T. K., & Nasereddin, M. (2014). Using Kolb's experiential learning cycle to improve student learning in virtual computer laboratories. *Computers & Education*, *72*, 11–22.

Kruea-In, N., & Thongperm, O. (2014). Teaching of science process skills in Thai contexts: Status, supports, and obstacles. *Procedia – Social and Behavioral Sciences*, *141*, 1324–1329.

Libby, D. R. (2008). Phasing into POGIL. In R. S. Moog, & J. N. Spencer (Eds.), Process oriented guided inquiry learning (POGIL), *ACS Symposium Series* (Vol. 994, pp. 49–59). Washington, DC: American Chemical Society.

Malcom, S., & Feder, M. (Eds.). (2016). Barriers and opportunities for 2-year and 4-year STEM degrees: Systemic change to support diverse student pathways. Retrieved from https://doi.org/10.17226/21739

Mbamalu, G. E. (2001). Teaching science to academically underprepared students. *Journal of Science Education and Technology*, *10*(3), 267–272.

McNew-Birren, J., & van den Kieboom, L. A. (2017). Exploring the development of core teaching practices in the context of inquiry-based science instruction: An interpretive case study. *Teaching and Teacher Education*, *66*, 74–87.

Moog, R. S. (2014). Process oriented guided inquiry learning. In M. AMcDaniel, R. F. Frey, S. M. Fitzpatrick, & H. L. Roedinger (Eds.), *Integrating cognitive science with innovative teaching in STEM disciplines*. St Louis, MO: Washington University Libraries. doi:10.7936/K7PN93HC

Mulford, D. (1996). *An inventory for measuring college students' level of misconceptions in first semester chemistry* (Master's thesis). Retrieved from https://www.chemedx.org/JCEDLib/QBank/collection/CQandChP/CQs/ConceptsInventory/Concepts_Inventory.html. Accessed on June 11, 2018

Nation's Report Card. (2017). State performance compared to the nation. Retrieved from https://www.nationsreportcard.gov/profiles/stateprofile?chort=1&sub=MAT&sj=&sfj=NP&st=MN&year=2017R3. Accessed on June 6, 2018.

National Research Council. (2000). Chapter 1: Learning from speculation to science. In *How people learn: Brain, mind, experiences, and school*. Retrieved from https://www.nap.edu/read/9853/chapter/3

National Science Foundation, National Center for Science and Engineering Statistics. (2017). *Women, minorities, and persons with disabilities in science and engineering*. Special Report NSF 17–310. Retrieved from http://www.nsf.gov/statistics/wmpd/

Ozden, M., & Gultekin, M. (2008). The effects of brain-based learning on academic achievement and retention of knowledge in science course. *Electronic Journal of Science Education, 12*(1), 3–19.

Pedaste, M., Mäeots, M., Siiman, L. A., de Jong, T., van Riesen, S. A. N., Kamp, E. T., ... Tsourlidaki, E. (2015). Phases of inquiry-based learning: Definitions and the inquiry cycle. *Educational Research Review, 14*, 47–61.

Pérez, D. G., & Torregrosa, J. M. (1983). A model for problem-solving in accordance with scientific methodology. *European Journal of Science Education, 5*(4), 447–455.

POGIL. (n.d.a). What is POGIL? Retrieved from https://pogil.org/about-pogil/effectiveness-of-pogil. Accessed on December 11, 2017.

POGIL. (n.d.b). Basic POGIL classroom implementation. Retrieved from https://pogil.org/resources/writing-submitting-pogil-activities. Accessed on December 11, 2017.

POGIL. (n.d.c). Author guidelines for developing POGIL classroom activities. Retrieved from https://pogil.org/resources/writing-submitting-pogil-activities. Accessed on December 11, 2017.

POGIL. (n.d.d). Curricular materials: Classroom activities. Retrieved from https://pogil.org/educators/become-a-pogil-practitioner/curricular-materials. Accessed on June 6, 2018.

POGIL. (n.d.e). Implementing POGIL. Retrieved from https://pogil.org/educators/resources. Accessed on June 6, 2018.

POGIL. (n.d.f). References. Retrieved from https://pogil.org/educators/additional-resources/references. Accessed on June 6, 2018.

Reeves, M., Berghout, H., Perri, M., Singleton, S., & Whitnell, R. (2017). How can you measure a reaction enthalpy without going into the lab?: Using computational chemistry data to draw a conclusion. In Abstracts of Papers, 254th ACS National Meeting & Exposition, August 20–24 (p. CHED–86). Washington, DC: American Chemical Society.

Reeves, M. S., & Whitnell, R. M. (2014). New computational physical chemistry experiments: using POGIL techniques with ab initio and molecular dynamics calculations. *ACS Symposium Series* (Vol. 1180, pp.71–90). Washington, DC: American Chemical Society. Retrieved from https://doi.org/10.1021/bk-2014-1180.ch006

Russell, S. H., Hancock, M. P., & McCullough, J. (2007). The pipeline: Benefits of undergraduate research experience. *Science, 316*, 548–549.

SALG. (n.d.). Student assessment of their learning gains. Retrieved from https://salgsite.net. Accessed on June 8, 2018.

Springer, L., Stanne, M. E., & Donovan, S. S. (1999). Effects of small-group learning on undergraduates in science, mathematics, engineering, and technology: A meta-analysis. *Review of Educational Research, 69*(1), 21–51. doi:10.3102/00346543069001021.

Stains, M., Harshman, J., Barker, M. K., Chasteen, S. V., Cole, R., DeChenne-Peters, S. E., ..., Young, A. M. (2018). Anatomy of STEM teaching in North American universities. *Science, 359*(6383), 1468–1470.

Stanford, C., Moon, A., Towns, M., & Cole, R. (2016). Analysis of instructor facilitation strategies and their influences on student argumentation: A case study of a process-oriented guided inquiry learning physical chemistry classroom. *Journal of Chemical Education, 93*(9), 1501–1513. doi:10.1021/acs.jchemed.5b00993

Stanford, J. S., Rocheleau, S. E., Smith, K. P. W., Mohan, J. (2017). Early undergraduate research experiences lead to similar learning gains for STEM and non-STEM undergraduates. *Studies in Higher Education, 42*(1), 115–129.

Wieman, C. E. (2014). Large-scale comparison of science teaching methods sends clear message. *Proceedings of the National Academy of Sciences, 111*(23), 8319–8320.

Yezierski, J. E., Bauer, C. F., Hunnicutt, S. S., Hanson, D. M., Amaral, K. E., & Schneider, J. P. (2008). POGIL implementation in large classes: Strategies for planning, teaching, and management. In R. S. Moog & J. N. Spencer (Eds.), *Process-oriented guided-inquiry learning (POGIL): ACS Symposium Series* Vol. 994 (pp. 60–71). Washington, DC: American Chemical Society.

A CULTURAL SHIFT: A TRANSFORMATIVE APPROACH TO ADVISING STEM STUDENTS AT AN HBCU

Mary A. Smith, Angela M. White, Kelsie M. Bernot, Cailisha L. Petty, C. Dinitra White, Grace E. Byfield, Robert H. Newman, Roy J. Coomans and Checo J. Rorie

ABSTRACT

As the US transitions to a majority–minority population, the underrepresentation in the science, technology, engineering, and mathematics (STEM) workforce must be resolved to ensure that our nation maintains its competitiveness and global economic advantage. The persistent problem of retaining underrepresented minority (URM) students in STEM continues to be a national priority after several decades of attention. The role of historically black colleges and universities (HBCUs) in addressing this challenge cannot be overstated, given their history in producing African American STEM graduates. As the largest HBCU in the country, North Carolina A&T State University (NC A&T) serves a combined undergraduate and graduate population of 11,877 students, 78% of which self-identify as African American. To overcome the multiple challenges that impede retention and persistence to degree completion in biology, the Department of Biology at NC A&T has adopted a major cultural shift in its advising strategy. The new approach encompasses a Life Mapping and Advising Model that builds faculty–student relationships and engages both parties effectively in the process. The

Broadening Participation in STEM: Effective Methods, Practices, and Programs
Diversity in Higher Education, Volume 22, 291–316

ISSN: 1479-3644/doi:10.1108/S1479-364420190000022013

model includes six important pillars to drive student success: (1) dedicated advising space, the Life Mapping and Advising Center (LMAC), (2) effective advisors, (3) integrated peer mentor and peer tutoring programs, (4) an intrusive advising strategy, (5) integration with first-year student success courses, and (6) life coaching. Although the program is in its infancy, based on the first-year assessment data, we have observed many promising trends that, together, point toward successful retention and persistence of our students in the major.

Keywords: Intrusive advising; life coaching; peer tutoring; peer mentoring; retention; STEM

INTRODUCTION

Over the last two decades, our nation has experienced an ethnoracial shift in that we have moved closer to becoming a majority–minority nation. In fact, the Census Bureau projects that ethnic minorities will make up roughly 57% of the US population by 2060. Interestingly, a National Research Council report (NRC, 2011) stated that the number of underrepresented minorities (URMs) in science and engineering would need to nearly triple in order to equal their proportion of society. For this reason, the persistent problem of retaining URMs in science, technology, engineering, and mathematics (STEM) has become a national concern. For the US to compete globally, we must not only encourage more URMs to pursue STEM degrees, but we must also retain them within the pipeline. For instance, a national report from the Higher Education Research Institute (2010) found that over 50% of first-year students who declared a STEM major at the beginning of their college matriculation left those fields before they reached graduation. While positions within science and engineering are increasing, too few African American students are persisting through STEM programs to earn a Bachelor of Science (BS) degree and entering STEM careers (National Science Board, 2014, 2016; Russell & Atwater, 2005). Increasing the retention, persistence, and graduation rates of African Americans and other minorities is essential to sustaining the nation's competitiveness, economic development, and global leadership in science and technology (National Science Board, 2012).

Resolving the issue of underrepresentation in STEM requires recognizing and addressing the causes for attrition. The varied reasons for students leaving STEM are well documented and include academic, institutional, individual, social, as well as other external obstacles that are not as easily generalized. Academic issues include under preparation for college; underperformance in "gatekeeper courses" in mathematics, biology, and chemistry; frustration with the overwhelming pace and load of science curricula; and fear of asking for help (Seymour & Hewitt, 1997). When high school students matriculate into college, they often lack adequate study skills and do not understand the need to quickly adopt new strategies that are often unfamiliar to them. In addition, these students often delay adoption of new study strategies and behaviors for success until it is too late to recover.

URMs have high aspirations to succeed in charting their path to a career of interest as a STEM major, but some drop out of STEM after recognizing the rigor of the training required for the careers that they planned to pursue. They may give up because of low expectations by others or because they lack the grit, motivation, emotional support, self-efficacy, or self-regulation as a learner to either persist or select an alternative STEM goal. Along with these issues is the inability to build academic and social connections (Anderson & Kim, 2006; Tinto, 1987). Additionally, inadequate financial support forces some students to balance their academic goals with working long hours at external jobs that have little to no educational value. For some, obstacles also include issues associated with single parenthood, meeting demands to help their families financially, or time spent resolving family conflicts. Finally, students' perceptions of the institutional or departmental climate play a significant role in STEM attrition. Such perceptions may center on the lack of role models who look like them, poor academic advising, institutional support, and a sense of belonging/isolation within the STEM community (Blickenstaff, 2005; Chang, Eagen, Lin, & Hurtado, 2011; Chang, Sharkness, Hurtado, & Newman, 2014).

Considering all the factors that influence student attrition, engaging students inside and outside of the classroom is critical for fostering student success. Failure to consider the diverse reasons for students leaving STEM programs early will prevent the US from sustaining its global leadership in science and technology. At the university level, the retention of freshmen and sophomores has received notable attention. Because of the need to recruit and retain students from all ethnic, cultural, and racial backgrounds, many higher education institutions have shifted their focus to enhancing their academic advising structures as a strategy to improve student retention (Earl, 1987). At many (if not most) institutions of higher learning, faculties have been assigned the responsibility for advising but have not been properly trained or held fully accountable for the best outcomes. For this reason, we have written this chapter to demonstrate the positive influence that a comprehensive advising model, such as ours, can have on the engagement and retention of URMs in STEM.

The Department of Biology faculty has designed a comprehensive advising program that enhances faculty–student relationships and integrates best practices for fostering student retention, including intrusive advising and life coaching. We administer the program in a dedicated space called the Life Mapping and Advising Center (LMAC), which is housed in the biology building. To provide insight into the structure and process of implementing our LMAC model for other STEM departments, we have divided this chapter into eight sections. In the first two sections, we provide an overview of the stages of transforming advising at North Carolina A&T State University (NC A&T) and the Institutional context, respectively. In the third section, we discuss the context and evolution of innovating advising and the most recent transformation within the Department of Biology. The fourth section highlights the theoretical underpinnings of the LMAC model. We discuss program goals and implementation in

Section 5, followed by the outcomes in Section 6. The last two sections cover the lessons we have learned since the implementation of the model, as well as recommendations and challenges.

INSTITUTIONAL CONTEXT AND STRATEGY FOR TRANSFORMING ADVISING

NC A&T is a public, historically black, doctoral/high research activity, land-grant institution located in Greensboro, NC. Throughout its history, NC A&T has maintained a rich tradition in academics, research, and social justice outreach. Its strategic plan, *Preeminence 2020*, is designed to establish NC A&T as a preeminent institution of choice for high-achieving students who are inspired by outstanding teaching and learning, civic engagement, transformative research, and creative scholarship to solve societal challenges. As the largest Historically Black College and University (HBCU) in the country, NC A&T serves a combined undergraduate and graduate population of 11,877 students, 78% of which self-identify as African American, a population highly underrepresented in STEM. NC A&T faces many challenges pertaining to retention and persistence in STEM. Importantly, these challenges also represent opportunities to be a leader in addressing the demographic disparities in the STEM pipeline, graduate and professional programs in science, and the STEM workforce. Already, the College of Engineering is recognized for its success, ranking first in the nation for awarding BS degrees in engineering to African Americans. Indeed, *Preeminence 2020* outlines the strategies designed to help the university achieve preeminence in all disciplinary areas of STEM. To achieve this goal, the institution is focusing its attention on student success and raising the retention rate for the university to 85% across all disciplines.

TRANSFORMATION OF ADVISING AT THE INSTITUTIONAL LEVEL

NC A&T has adopted a centralized advising model at the university level through its Center for Academic Excellence (CAE; http://www.ncat.edu/divisions/academic-affairs/cae/). The CAE is well structured and resourced to address the needs of students in the diverse disciplines across our campus. The core functions of CAE are to provide (1) academic coaching/advising, (2) academic monitoring, (3) supplemental instruction (tutoring, skill building and mentoring), and (4) student success events. Similarly, the College of Science and Technology (CoST; http://www.ncat.edu/cost/current-students/index.html) specifically advises students in science and technology. In addition to these two levels of advising, all faculties at the University are required to serve as advisors to students at the departmental level. Thus, NC A&T supports hybrid models of advising to reach all students. The Department of Biology provides a third level of advising for its students that is intrusive and student-learning centered. We also collaborate with the CAE and CoST Office of Student Success to take advantage of the resources that they offer to facilitate student success.

DEPARTMENTAL CONTEXT FOR EVOLVING ADVISING OVER TIME

The Department of Biology is the largest STEM department on the campus, the largest science department in the CoST, and one of the most popular majors among all disciplines at the University. Therefore, it is critical that the department is successful in increasing its retention because of the impact it will have on the university reaching its student success goals. The undergraduate population in the Department of Biology represents 22% of the undergraduate STEM students at the University and 39.2% of the students in the CoST. Undergraduate enrollment in the department has steadily increased over the past three years, pushing the total full-time population to 473 in fall 2017 (Fig. 1). Most of the undergraduates in the program are underrepresented minorities in STEM; 87% are African American and 5% are non-White Hispanic. The gender distribution in biology is 84% female and 16% male.

Enrollment growth has presented the faculty of the Department of Biology with both challenges and opportunities: challenges associated with accommodating larger numbers of students and the opportunity to achieve preeminence in student retention. The ongoing growth has inspired faculty members in the department to embrace change and reflect on the best strategies to engage all its students to excel in science. To that end, the department has achieved significant gains by adopting high impact, student-centered, and evidence-based teaching and learning practices (independent research, course-based research, active-learning, service learning) to prepare students for competitive futures in the discipline. Undergraduates in biology are being educated in a research-supportive curriculum and developing the skills needed to become competitive contemporary scientists. All seven required foundation courses offered during the first two years have been transformed to scaffold learning and reinforce core competencies for doing science. We have mapped our courses to emphasize critical thinking, analytical and quantitative reasoning, ethical reasoning, strong communication skills, interdisciplinary thinking, and problem-based learning.

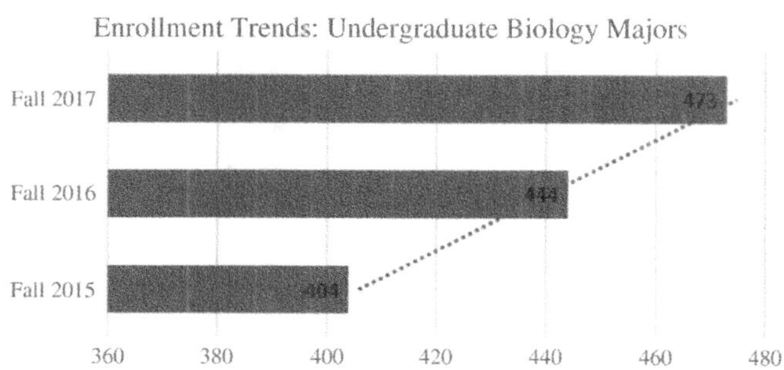

Fig. 1. Undergraduate Enrollment Trends for Biology. *Note*: The graph shows the total number of full-time undergraduate students enrolled as Biology majors over the past three years.

Pedagogical and curriculum reforms in biology have shown a positive impact on student learning but not on retention. The average first-year retention for all first-time, full-time students who entered as biology majors between 2009 and 2015 was 71.4% (Fig. 2). This retention rate was 4.9% lower than that for the university as a whole during that same period and far below the institutional goal of 85% retention. Of this 71.4%, 12.6% changed their major, transferring within the university to another department. Therefore, the average first-year retention for students who remained in biology was only 58.5% for this period (Fig. 3).

The retention data also show that the attrition rate is the most dramatic after the first year. Therefore, in 2011, we began to modify our advising strategy to increase student–faculty interactions to exemplify a caring environment for new students. Prior to 2011, all faculties were assigned a subset of biology advisees and were responsible for helping them to essentially meet the requirements for the major. This approach was not standardized or sufficiently effective for improving retention. As illustrated in Fig. 4a, the relationship between faculty advisors and advisees was weak. Since students were randomly assigned, by chance, they either ended up with an advisor who was knowledgeable and committed to student success or to one who was less interested and too often unavailable. This model, in too many cases, resorted to what Wilcox (2016) refers to as the "checklist" approach to advising. It is useful for tracking the academic progress of advisees but limited in recognizing, personalizing, and addressing specific goals and objectives of individual students. We concluded that our poorly planned and disparate advising approach at the departmental level was leaving too many capable students behind.

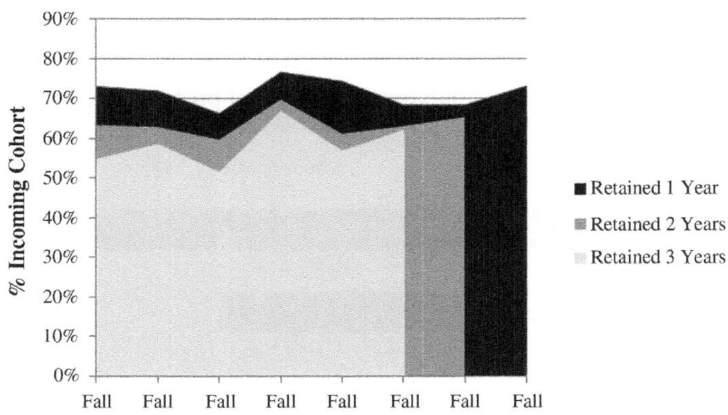

Fig. 2. Student Retention as Any Major. *Notes*: Each incoming cohort from Fall 2009 to Fall 2016 was followed for their retention at the university in any major. Historically, the largest gap in retention is between years one and two across all the academic years shown.

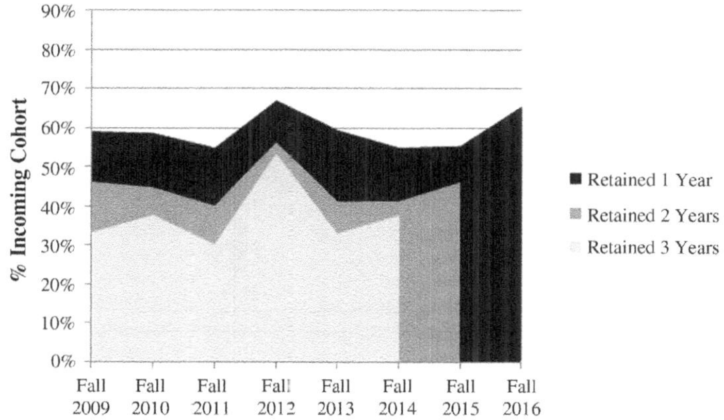

Fig. 3. Student Retention as Biology Majors. *Notes*: Each incoming cohort from fall 2009 to fall 2016 was followed for their retention at the university in the Biology major. From 2012, there has been a steady decline in retention rates, until the implementation of the LMAC in Fall 2016. Here we see a dramatic 10% increase in one-year retention rates from 55% to 65%. historically, from 2009 to 2015, we observed the largest change in retention was between year 1 and year 2.

Since the retention of first-year students is critical, we decided to implement a multi-pronged advising approach that encompasses different purposes and methods (Wilcox, 2016). Wilcox emphasizes that effective interactions with each individual student require the use of a wide range and unique set of approaches. Wilcox makes a very important point: "that [it is] only through open-ended dialog, reflection, and discussion that the educational experience is personalized and integrated, and as a result, becomes meaningful and transformational. Checklist-based interactions rarely lead to these important 'aha' moments."

THEORETICAL SUPPORT FOR THE LIFE MAPPING AND ADVISING CENTER MODEL

In 2015, we decided to standardize our advising practices to better meet the academic, social, and emotional needs of our students. Our advising practices are integrated through what we have defined as a "Life Mapping and Advising Center Model," per Dr Angela White. She suggested that we ground our practices in theory such that we aligned our advising practices with the Intrusive Advising Concept. According to Earl (1987), intrusive advising is a "deliberate, structured student intervention at the first indication of academic difficulty to motivate a student to seek help" (p. 28).

Furthermore, intrusive advising hones in on the motivational dimension of learning and success. Earl (1987) posits that intrusive advising tends to be more effective with underclassmen whose academic challenges are the result of

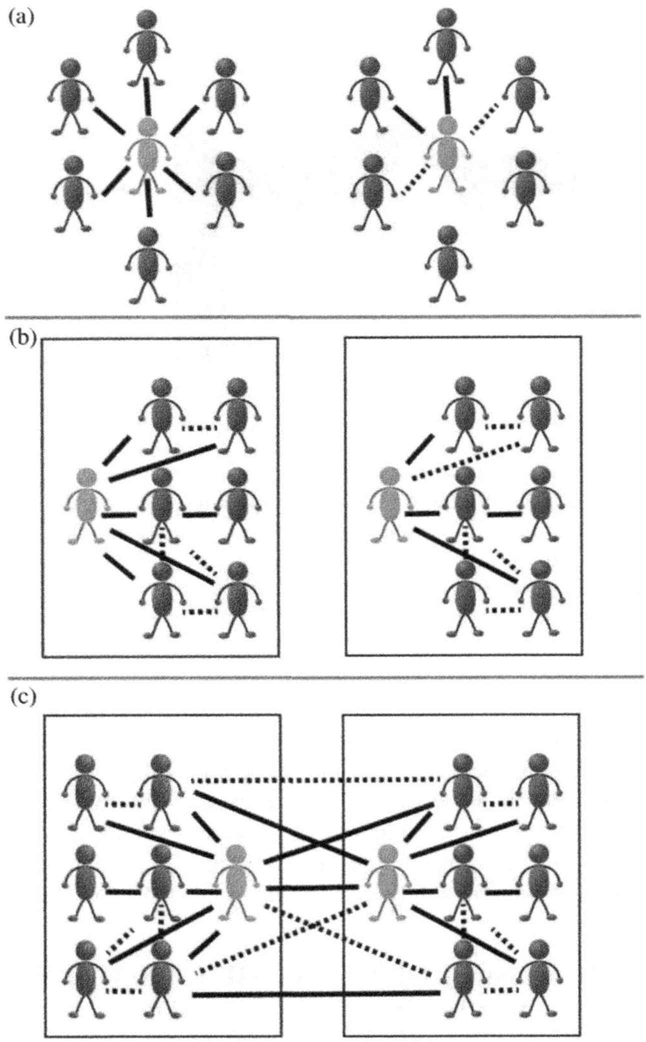

Fig. 4. Advising Model Transformations. *Notes*: (A) Original advising model shows two advisors in grey. The advisor on the left has a strong relationship with all six students (black) as designated by a solid line connecting the advisor to the advisees. However, the advisor on the right has strong relationships with only two advisees, weak relationships (dotted line) with two advisees and no relationship with two advisees. There is no interaction between advisors or between students. Quality of advising is disparate. (b) Advisor and advisees are paired in BIOL 105 Biology orientation course. This increases the quality of advising by ensuring that all students have a connection to an advisor in the context of a classroom. Also, students begin to have weak or strong relationships with other advisees in the classroom. (c) With the addition of the LMAC, advisors now interact with other

motivational rather than competency factors. Thus, we worked from the assumption that exposing our freshman and sophomore students to an intrusive advising model with the LMAC as our central site for engagement would result in a consistent increase in our retention and graduation rates. Most importantly, such an advising approach would improve our students' satisfaction with academic advising.

Multiple theories have shaped our perspective and helped to guide our approach. The theories have augmented our understanding of how to develop an advising structure that will impact the academic, social, and psychosocial development of our STEM majors for success. The approach implemented in our recently opened LMAC is situated within a transformative/participatory framework. One of the primary goals of transformative-participatory research is to gain greater insight and transform a situation based on the specific needs of the participants involved in the research (Creswell, 2007). As we investigated the efficacy of our model, this worldview was appropriate because we sought to understand the impact that the LMAC strategy would have on the academic performance and social development of our students; therefore, we actively engaged with the students who will be impacted by our LMAC approach. In addition to being guided by a transformative-participatory worldview, we are also guided by other relevant theories: the *social cognitive theory* (Bandura, 1977; Bandura, Adams, & Beyer, 1977; Bandura, 1986), which developed from the work of psychologist Albert Bandura; the intrusive advising approach, originally conceived by Earl (1987), Chickering's Theory of Identity Development, as well as Erickson's Stages of Development; and two student retention-focused theories, including the Theory of Departure (Tinto, 1987) and the Continuum of Community (Schlossberg,1989).

LMAC PROGRAM GOALS, STRUCTURE AND COMPONENTS, AND IMPLEMENTATION

The LMAC was developed with several primary objectives contributing to the framework of the model (see Table 1). All of the macro- and microcomponents of the LMAC were implemented in an effort to achieve each of the objectives within a three-year timeframe.

LMAC STRUCTURE AND COMPONENTS

The Department of Biology initiated the LMAC model for advising in the summer of 2016 and opened the Center in the fall of 2016. The model consists of six

advisors, increasing access to resources and strengthening advising skills. Additionally, advisors develop strong relationships both with advisees and with non-advisees, creating a larger faculty network for advisees. Advisees have stronger relationships with other students both inside and outside of their biology orientation class through the LMAC, peer tutoring, and the big brother/big sister peer mentoring program.

Table 1. Framework of the Life Mapping and Advising Center (LMAC) Model.

Measurable Academic Objective	**Measurable Performance Indicators**
1.0. To increase freshmen retention by 5% through biology faculty utilization of cognitive, metacognitive, and non-cognitive strategies in the classroom and in the LMAC, annually	1.1. Seventy percent of the new freshmen will earn good academic standing with a minimum GPA of 2.0 and 67% course completion by the end of the academic year because of the following specific LMAC sponsored activities • One hundred percent of new freshman will be enrolled in BIOL 105 & 205 for the development of metacognitive strategy use • Intrusive advising strategies in the LMAC will be used to reinforce metacognition and promote the development of non-cognitive factors 1.2. Approximately 70% of new freshman will learn to seek help and make use of one or more of the following specific LMAC sponsored activities during the academic year • Administered "Help-seeking" Questionnaire and use results during advising sessions to empower students to become help-seekers. • Faculty-led supplemental instruction and/or peer tutoring in Chemistry • Pairing new freshman with peer mentors (Big Brother or Big Sister) to the development of a sense of belonging • Life-coaching sessions in student success courses to help students develop as self-regulated learners through self-awareness, motivation, career development, and emotional intelligence • Individualize life coaching to promote the success of students who are at risk
Measurable Administrative Objective	**Measurable Performance Indicators**
1.0. To guarantee that 95% of freshmen receive consistent and effective intrusive advising by the biology faculty using LMAC-specific activities throughout the academic year, and summer session I	1.1. Ninety percent of the biology faculty will participate in the implementation of an innovative LMAC advising model throughout the academic year. Specific LMAC activities include the following • pre-matriculation planning for incoming first-year students by mid-August to include welcoming letter, academic placement based on standardized test scores by mid-August • performance tracking and data logging throughout the academic year to detect early alerts • identify students on probation at the end of each semester and develop action plans for retention by mid-January and again by mid-May • require all students on probation to report to the LMAC every two weeks to re-assess academic goals, interests, and skills with the assistance of faculty advisors, and/ or life coaching throughout the second semester

strategic pillars to address the attrition and retention of biology majors: (1) LMAC facility, (2) faculty advisors, (3) an engaged network of peers (e.g., Biology Big Brothers & Big Sister (BBBS) peer mentors and tutors), (4) intrusive advising, (5) life coaching, and (6) integration with the BIOL 105/205 "Biology Orientation" student success courses. Intrusive advising and integration with the student success courses will be discussed elsewhere in this chapter (see Sections 4 and 5). We are encouraged by the abundance of literature that supports the

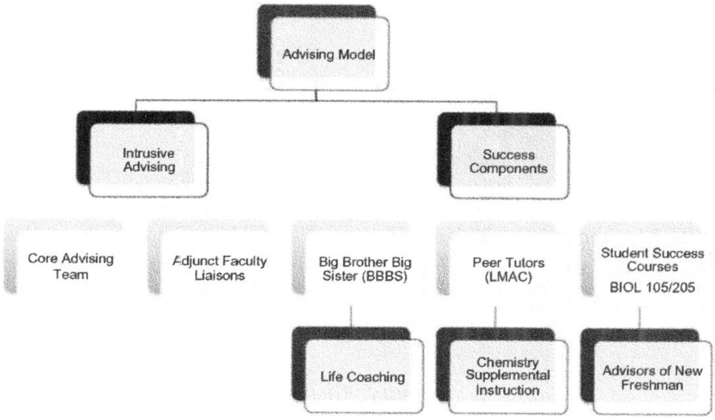

Fig. 5. Components of the LMAC Advising Model.

elements of our comprehensive and collaborative model over the traditional single faculty activity to meet diverse cognitive and non-cognitive issues that students present (Afshar & O'Hara, 2011; Erlich & Russ-Eft, 2011; Hemwall & Trachte, 2005; Mahoney & Schibik, 2004; Miller, 2010) (Fig. 5).

The LMAC Facility

The LMAC is designed to be a welcoming space where students receive an aggressive and integrated approach to advising. It is strategically located on the first floor of Barnes Hall, the main Biology building, to ensure visibility and accessibility. The space includes two private cubicles for advising and life coaching, a space for computer-based, self-guided career exploration, three computers, and an open area for peer tutoring with a white table (Fig. 6).

The hours of operation for the Center are Monday through Thursday from 10 a.m. to 12 p.m. and 1 p.m. to 4 p.m., providing flexibility to student schedules. We have critical personnel engaged, including faculty advisors and student leaders (e.g., Biology BBBS peer mentors and peer tutors). Dr Angela White, the Biology Student Success faculty in the Department of Biology, is the faculty leader of the LMAC and a member of the LMAC Leadership and Management Team. The team is responsible for programming, evaluation, assessment, and improvement. All members of the LMAC Leadership serve as advisors in the LMAC and as instructors in the first-year student success course, Biology Orientation I (BIOL 105). This arrangement serves to link the classroom and the LMAC metacognitive and non-cognitive strategies for student success (Fig. 4c). The strength of our LMAC Leadership and Management Team (the authors of this chapter) is in bringing expertise in science education, educational psychology, research, and student-centered learning, which is allowing us to bridge multiple gaps between student engagement within and outside of the classroom.

Fig. 6. Life Mapping and Advising Center. *Note*: These images show the setup of the LMAC (left), a student meeting with an advisor in one of the cubicles (center), and peer tutors working with students on the whiteboard table inside the LMAC and whiteboard (right).

Faculty Advisors

Ninety percent of the biology faculties serve two hours per week within the Center to meet with students for one-on-one advising. Specifically, the advisors meet with students to discuss their progress in their courses, personal or familial issues, financial issues, academic and career goals, or any other issue that the students may wish to discuss. The department's goal is to meet with each student at least three times every semester, either in the LMAC or with their assigned primary advisor in the faculty's office. Our advising with new first-year students begins during the summer during New Student Orientation, as we make our first point of contact to welcome the students to the department and assist them with registration. Additionally, we send the students a welcome email or letter in the mail. We have found that these early interactions with freshmen increase their motivation and sense of belonging in our biology community, evidenced by reflections they have written in their Biology Orientation course. Continuing students who are placed on academic probation are scheduled to meet with an advisor in the LMAC or a certified life coach on a bi-weekly basis. We intensively track the progress of students on academic probation during the regular school year and if they are enrolled in summer school.

Peer Engagement (Mentors and Tutors)

In addition to changing how faculties are engaged with advising, we also changed how students are engaged (Fig. 5). We recruited upper-level students to serve in two capacities: (1) as peer mentors through our BBBS program and (2) through our peer tutoring program in the LMAC. Peer tutoring focuses on helping students with course content and study strategies, whereas peer mentoring is broader, with peers providing advice, support, and knowledge gained from a more experienced upperclassman to a novice mentee (Kram & Isabella, 1985). In its strictest sense, peer mentoring matches mentors and mentees who are roughly equivalent regarding age, experience, and power (Angelique, Kyle, & Taylor, 2002) and can benefit both mentors and mentees (Colvin & Ashman, 2010). Peer and tiered mentoring have been shown to facilitate students' sense of belonging and contribute to more successful retention and graduation rates, particularly for URMs (Wilson et al., 2012). To that end, juniors and seniors are matched with freshmen and encouraged to communicate and to participate in BBBS activities, largely social events such as bowling or movie nights, the creation of a departmental haunted house, and field day. In addition to receiving correspondence from a faculty member during the summer, new freshmen are also contacted by their assigned Biology Big Brother or Big Sister.

Life Coaching

Comprehensively, life coaching seeks to develop an individual's self-concept, abilities, and personal interactions (Costello & Stone, 2012). There are several types of coaching that apply to college student success (Dalton & Crosby, 2014; Maher, 2009). According to Dalton and Crosby (2014), life coaching "involves a partnership with a knowledgeable guide who can provide an objective, outside perspective and who encourage clients to take action regarding important decisions." Although the practice of life coaching is fairly new, Costello and Stone (2012) propose it as an impactful practice that college support services should use to enhance various non-cognitive factors important to student success. Recently, controlled studies have shown that students who receive life coaching at the collegiate level are more likely to persist in college than students who do not receive life coaching (Bettinger & Baker, 2014). Moreover, Bettinger and Baker (2014) found that students who were assigned a life coach were more likely to persist during the period in which they participated in life coaching and were more likely to be enrolled one year after the cessation of coaching.

THE LMAC MODEL OUTCOMES AFTER THE FIRST YEAR OF IMPLEMENTATION

Impact on the First-year Retention

The LMAC management team evaluated the retention data from fall 2009 to fall 2016. Aside from one stellar cohort in 2012, there was a progressive decline in the retention of biology students prior to the implementation of the LMAC model in 2016. This decline occurred not only for students who remained biology majors but also for students who transferred to other majors. Our retention data for the first year of the LMAC (2016–2017) showed great promise in the model. For instance, there was an increase in retention for all students, regardless of whether they were in their first, second, or third year at NC A&T. On average, the ratio of the total retention (71.4%) for students who stayed in the major to students who left the major was 59:1 for cohorts 2009–2015.

In contrast, the ratio increased by 10 points (69:1) after the LMAC intervention for cohort 2016. Importantly, we observed larger increases in retention in students, who remained biology majors than we did in students who transferred to other majors, suggesting that this is not a university-wide retention increase and is specific to the Department of Biology (Fig. 4a and 4b). These data suggest that the LMAC model has positively impacted the retention of our population of biology students. These changes may be linked to our inclusion of intrusive advising and life coaching in the model.

Impact on Student Satisfaction for Advising

One of our primary objectives in instituting the LMAC model was to improve student satisfaction with academic advising. In order to evaluate the efficacy of our model in achieving this goal, we employed a mixed-methods approach; therefore, we collected quantitative and qualitative data. Students enrolled in the Department's Senior Project course (BIOL 496) were surveyed in spring and fall 2017, as well as spring 2018, to determine their advising experiences and their satisfaction with both the traditional advising and the LMAC advising systems. From the spring 2018 survey, in which 53 students responded, the level of satisfaction averaged 3.78/5 when asked to rate their satisfaction with their primary advisor. When asked to rate their satisfaction with the LMAC, the average rating was 4.11/5. These two questions were based on a five-point Likert scale (1 = *poor*; 5 = *excellent*). While statistical significance was not determined for these two questions, the surface difference in values revealed that students were more satisfied with the LMAC model than the traditional advising model. Moreover, student responses to the question "Which of the advising models do you prefer: traditional advising or LMAC advising?" revealed that 82.6% of students prefer the LMAC advising model.

To gain a richer and more in-depth understanding of the students' perceptions of both advising models, we conducted a focus group that consisted of 28 seniors and collected responses from the open-ended questions on our advising satisfaction survey. When asked to specifically describe their experiences with faculty advisors and the LMAC, students shared both positive and negative experiences. Ultimately, this qualitative data both confirmed and elaborated on our quantitative findings of increased satisfaction with the LMAC model. To maintain the confidentiality of the faculty and students, we assigned pseudonyms to the individuals mentioned in student remarks.

POSITIVE COMMENTS

I have always had a great advisement experience. Dr. Q is my advisor, and he has always helped me and was knowledgeable of the resources I needed. But I must say the LMAC has made advisement extremely easy.

The LMAC improved my experience. I don't think we should get rid of personal advisors for students, as a personal touch is sometimes needed and the LMAC cannot provide that. However, often a personal advisor is unavailable [when] you really need them, and the LMAC fills that gap. Additionally, I think it's a good way for faculty to keep up with the state of the student body.

The LMAC was helpful to my advisement experience because I was able to get different perspectives of information from the faculty.

The biology department is a place of supportive staff, professors, and mentors. Also, it is well organized. Dr. Y is very understanding and accommodat[ing] for the student's needs. [The] LMAC is the greatest idea yet for the department. The overall environment is welcoming.

The department has been improving with certain policies and programs to assist in the 'biology experience.'

NEGATIVE COMMENTS AND SUGGESTIONS FOR IMPROVEMENT

Some advisement sessions could have been better. There were sometimes when I had to go to other advisors outside of my primary advisor because I felt like I wasn't getting what I needed. I would also strongly suggest getting a set of advisors just for transfer students!

It was helpful. But I do think that some advisors need a little more training.

I wish it were bigger.

My advisor was pretty good, but the process to get my pin was always so tedious and unnecessary.

The LMAC is very beneficial. The only improvement could be making the hours more commonly known and if there are any special hours during certain weeks.

Based on data from an exit survey administered to students enrolled in the Department's Senior Project course (BIOL 496), students were moderate to extremely satisfied with advisement in the department. According to the students, the LMAC improved accessibility to a faculty advisor was more efficient and was pleasant.

Faculty Assessment of the LMAC Model after the First Year of Implementation

By the beginning of the second year of the LMAC model's existence, we reached our goal of 90% biology faculty participation. To determine the impact of the LMAC model from a faculty perspective, we administered a Faculty Satisfaction survey through Qualtrics. The survey consisted of Likert-format questions, as well as open-ended questions. The survey focused on six primary areas: Advantages and Disadvantages for Students, Faculty, and the Department.

PERCEIVED ADVANTAGES FOR STUDENTS

When faculties were asked about their perceived advantages of the LMAC model for students, the most salient responses were ease of access and flexibility. Because students are no longer tied to one advisor, they can receive assistance from any advisor serving in the LMAC as it fits their schedules; thereby, eliminating perceived challenges associated with getting in touch with their advisor. Additionally, the faculty described the diverse offerings available for the students in the LMAC, in that they can receive assistance from peer tutors, career advisement, and academic advisement. Some of the faculty comments regarding the advantages for students included:

Students have open access to an advisor as well as peer tutors throughout the week. In the traditional model, if a faculty member was busy, in class, out of town, etc. the student would be forced to wait. The LMAC model allows students' needs to be met more quickly. Students may also receive career advisement and academic advisement in one location. Finally, I feel like many students who would have avoided seeing an advisor now have more reasons to come to the advising center. I have seen them come in and request a particular advisor with whom they developed a connection. I have also seen students who are happy that they are not glued to a single advisor and can get help for another faculty in their own department.

Flexibility and ease of access.

Students no longer have an excuse that they can't get advising.

PERCEIVED DISADVANTAGES FOR STUDENTS

When faculty were asked about their perceived disadvantages of the LMAC model for students, the most salient responses were the effect the model has on the traditional advisor/advisee relationship and lack of continuity. If a student meets with a different advisor each time he/she comes into the LMAC, there is a lack of consistency that affects the opportunity for relationship building. Many of the faculty felt that their relationship with their primary advisees suffered or was not as strong due to the implementation of the LMAC model; however, at the same time, they noted that they were able to foster relationships with more of the students in the department. Consequently, the lack of continuity is also connected to meeting with advisors with different levels of advising experience/expertise, which at times induced frustration for the students. Many students may not have received consistent or fully accurate information due to meeting with a different advisor each time, or they were required to repeat their situation every time they visited the LMAC. We have resolved this issue of redundancy by arranging for each faculty advisor to have access to each of our students' records using the student success management system, GradesFirst. Therefore, when a student meets with a faculty advisor in the LMAC, a report is created by the advisor in GradesFirst that may be readily retrieved by a different faculty advisor in the future. Comments from faculty included:

There is a tendency to get advised by different faculty members, thus destroying continuity.

They may not be able to meet with their advisor while their advisor is in the LMAC, they may have classes that prevent them from coming to LMAC hours at all (likely very rare).

Not having the same interaction with their primary advisors. I don't think I know them individually as well as the previous model. I know more students in general, but not necessarily my advisees in particular. Also, students sometimes meet with advisees of various levels of expertise. This could be a disadvantage regarding how helpful their visit will be.

PERCEIVED ADVANTAGES FOR THE FACULTY

When describing the advantages for faculty, the comments honed in on the benefit of advising collaboratively. Faculty members can learn from each other, which mitigated any anxiety felt by those with less advising experience or knowledge of policies/procedures. Given the fact that policies and procedures change from year to year, the LMAC model ensured that faculties were always aware of those changes. Moreover, there is a disparity in the levels of advising expertise; therefore, we intentionally ensured that a less experienced advisor was paired with a more experienced faculty advisor. This type of collaborative advising provided the time and space for less experienced faculty advisors to receive scaffolding through various situations experienced by students. Faculty also perceived that the model increased opportunities to build rapport with more students instead of solely having a positive relationship with their own advisees.

Furthermore, the model has also provided opportunities for faculty members to build relationships with one another during downtime in the LMAC. Very rarely do faculties have the time to simply have a conversation about themselves with one another; however, being located in the same space allowed for faculty to get to know each other on a more personal level. The faculty shared comments such as:

> *Advising as a team makes it so much easier to advise properly. In the old model, I was not always totally confident that I was aware of the most up-to-date policies and best-practices for advising. That caused advising to be somewhat fractured because individual faculty members were advising 'to the best of their ability.' The lack of 'training' was also a negative aspect. Now, I am out of my office with an advising partner who is there if I have questions. Also, having a leader for the management team is great. I feel very supported and properly informed. I am no longer afraid that I may improperly advise a student.*

> *The LMAC allows for collaborative advising for faculty, in that, it provides faculty with another team member to draw insight from.*

> *Becoming better acquainted with A&T's and Biology's procedures/policy. Greater opportunities to build rapport with students.*

PERCEIVED DISADVANTAGES FOR THE FACULTY

When describing the disadvantages for the faculty, there was a general concern for assigned, primary advisees. Despite the LMAC model, some faculty members have a higher number of assigned advisees. For example, in fall 2017, one faculty member was assigned about 80 advisees, while another had 30. Additionally, faculty buy-in was an ongoing challenge due to the fact that some faculty did not serve in the LMAC or voiced concerns about the value of the model. Many of the faculty members who saw the value of the model and consistently served in the LMAC became frustrated with the lack of accountability required of those who did not serve in the LMAC. The refusal to serve in the LMAC had the potential to destroy the relationships between faculty members or cause the "house" to become divided. For instance, many who served in the LMAC would see those who did not participate in the model and question why there were no repercussions for their lack of service. Faculty responses included:

> *One disadvantage for faculty is not spending as much time with their primary advisees. Also, some faculty members still have a higher caseload than others, because some faculty members have not always supported the LMAC.*

> *None (advantages); same thing can be done in Professor's office.*

> *There are some faculty who did not serve in the LMAC and therefore didn't share the load equally with those serving in the LMAC and on the management team. Also, not knowing their assigned advisees as well.*

There are usually more than several weeks in a semester where it is difficult to manage time in LMAC with a heavy teaching load, and also aim for high productivity in scholarship and responsiveness to other university duties.

PERCEIVED ADVANTAGES FOR THE DEPARTMENT

The faculty discussed the potential for increased retention rates, access to better resources, and a systematic model for advising in relation to the perceived advantages for the department. Several faculties believe that the employment of a "student-centered" approach to advising has resulted in a more positive student perception of the faculty and department overall. For example, one faculty member noted the importance of knowing students on all levels, not just academically. The faculty member stated, *"The Department is able to get a better understanding of the various factors that affect a student's well-being so that they can help the whole student be successful."* Prior to this model, many students perceived the faculty to be disinterested with who they are as a person and disengaged with the steps toward their overall success and development. However, students now view the department as a community in which they are a valued participant. Other comments from faculty included:

The department is now able to increase the quality of the interactions with students and gather appropriate data to provide a 'big picture' of the students' needs. This advising format may help the department retain students by doing a better job of helping them navigate the curriculum and deal with issues that may derail a student's success.

The LMAC model has great potential to improve departmental retention. provides for more efficient advising, and greater accountability. This model actually holds the faculty accountable for the advising responsibilities that we are expected to carry-out. This model shifts to a more student-centered approach for the department, and has improved the reputation of the department.

I will be stunned if we don't see increased retention. Building better culture because of integrating peer tutoring into LMAC. Providing better resources for faculty and students

The LMAC model has allowed the department to create a systematic model for advising, recording and collecting data, developing community, and providing information to students.

(The LMAC) Represents a tangible investment into student success

PERCEIVED DISADVANTAGES FOR THE DEPARTMENT

The faculty response in relation to the disadvantages of the LMAC for the department echoes some of the sentiments expressed in the disadvantages to faculty section, such as "lack of accountability" and "unequal advising." Additionally, there was a concern about having time to mentor students within the context of a research setting, as well as a lack of attention given to higher achieving students. An ongoing concern has been the sustainability of the LMAC, as funds are needed for services such as life coaching, summer faculty employment, professional development, data analysts, peer tutor stipends, supplemental instruction, BBBS programs, and a LMAC-specific administrator.

For this reason, the LMAC management team has tirelessly sought to secure internal and extramural funds for this model. Some of the faculty comments regarding the disadvantages for the department included the following:

> *It's different from all other departments at NC A&T. Adds work to faculty who are already advising too many students.*
>
> *Still have unequal advising among faculty members where some faculty are not interested in advising.*
>
> *The LMAC model requires the department to constantly seek funding in order to support some of the components or future components of the LMAC.*
>
> *Disengagement from research-related mentorship.*
>
> *We are putting a lot of resources into helping the students who are barely making it through the department which is leaving us less time to really spend with the top students who require attention in other facets. In this respect, while we are helping a lot of our students make it through our program, our top students may not be getting the help and support and mentorship that they are going to need to feel confident in their next steps.*

STUDENTS PERCEPTION OF THE TRADITIONAL VS THE LMAC MODEL FOR ADVISING

Effective advising is vital to the retention and overall success of STEM majors. Over the past several years, the Department of Biology has thoroughly examined not only the academic needs of students but also the outside barriers and influences that far too often threaten students' success. Moreover, the department has simultaneously listened to the needs and concerns of the faculty and has taken a unique approach to alter the overall atmosphere surrounding advisement in the department. Since deciding that advisement would be an important focal point for the success of our students, the faculties have devoted invaluable manpower, time, and resources to collectively exploring novel methods for addressing issues observed through traditional advising.

In the past, the Biology Department approached advising much like other departments. A group of students, which varied in size from one advisor to another, were assigned to a single faculty member who was expected to maintain a basic connection with the students in order to ensure successful matriculation through the curriculum. Unfortunately, this model did not take into account the various issues that can derail a students' success. Additionally, the rigidity of the traditional advising model does not allow for flexibility often needed by the faculty. For example, demands on the faculty member's time, including teaching, research, and scholarship expectations, often require faculty to be away from the office and therefore unavailable to assist students much of the day. Unfortunately, the challenging time constraints cause faculty to perceive advising as a burden. A survey to evaluate the effectiveness of the traditional advising model versus the LMAC model was given to seniors in the Biology Department. Results strongly support the need for a better, more balanced advising structure. Students described a wide range of experiences with their advisors. One student shared "*I wish my advisor was more engaged in my academic career. She, so far, has not been helpful at all. I've been advising myself. If I could get another advisor I would,*" while another student reported "*I would like to think that we have a very close relationship, the help that she has provided with my personal life has greatly improved my academic success, knowing that someone believes in me and having them critique my success is a big help.*" For many students, the idea that they must "figure it out" on their own with minimal attention from a knowledgeable advisor paves the way for those students to be left behind and to drop out of the STEM pipeline.

One of the most notable accomplishments of the LMAC model is that the Biology Department faculty now have a different collective perception of the students. Moving from the "checklist" model to a truly "student-centered" model has created new opportunities and experiences for both students and faculty in the department. The LMAC model takes into consideration findings reported by Seymour and Hewitt (1997), who noted that "the educational experience and the culture of the discipline (as reflected in the attitudes and practices of [STEM] faculty) make a much greater contribution to [STEM] attrition than the individual inadequacies of students or the appeal of other majors" (p. 392). Therefore, the idea is no longer to simply "get the students through the curriculum" as prescriptive advisors (Gordon, 1992), but for the faculty to recognize the students as individuals with the potential to be successful. In transforming our approach to advising, we hope to catalyze more authentic interactions between faculty and students; thereby, promoting concern for the "whole student" within each faculty advisor. This approach requires that faculty be given the time to learn about the individual student's talents, career goals, academic and personal needs, abilities, and challenges, which will give students a greater sense that the department cares about their development in every area. Ultimately, this shift in the culture of our discipline will improve overall performance and retention in the major. This cannot be easily achieved when the ratio of advisees to an advisor is large and unmanageable; therefore, the LMAC creates the balance needed to promote these perceptual and relational transformations.

Finally, several positive and unexpected outcomes have developed from the implementation of the LMAC model that were not possible under the traditional model. For example, the LMAC model relies on consistent faculty communication. Therefore, meaningful and intentional interactions between faculty members have greatly increased. The experiences of both students and faculty confirm recommendations given by Petress (1996), in that advisors must receive training and guidance to do a competent job. Faculty reported that "*the LMAC allows for collaborative advising for faculty, in that, it provides faculty with another team member to draw insight from.*" Another faculty member reported that advantages of the LMAC model include "*Shared workload, we become better advisors because it is so easy to get a second opinion from another faculty member in the room, more team discussions about how to handle challenging students, easy to train new faculty in advising, where there was prior no training at all.*" Borns (2002) stated that "developing effective training opportunities for academic advisors is important for student satisfaction and persistence" (p. 1). To this end, the LMAC has served as a space that regularly augments the advising competence of our faculty; therefore, this model has great potential to increase student satisfaction and retention. In addition to the improvement of advising through collaboration, tangible products have resulted from this team approach. When the LMAC management team determined the need for new tools to streamline the pre-registration process, one team member designed a highly functional Excel spreadsheet that allows students and advisors to easily determine and track students' progress through the curriculum. Two other members teamed up to create a YouTube video explaining to students and faculty how to complete the new "tracking tool" and use it to inform students of their current status, remaining courses, overall grade point average, and grade point average in their major. These innovations have made the registration process significantly more efficient and accurate.

SUGGESTIONS FOR IMPROVING THE LMAC MODEL

As we consider the accomplishments of the model thus far, we realize the areas that require attention. Our goal is to provide a model that benefits both the faculty and students, so we will use the recommendations provided by both groups of stakeholders to further enhance the model. In this section, we will take a closer look at the areas of convergence and divergence from recommendations made by both groups. Both the faculty and the students stated that we needed more advisor training and a larger space. Faculty also suggested that our focus spans across all groups of students, and not just low performing/probationary students. Faculty suggestions for improvement included the following:

> *I strongly request that this model remains in place in the biology department. I would also like to see more space provided for peer tutoring. A larger space would accommodate more students and provide more room for private discussions (extend the height of the cubicle walls, size of the cubicles, and possibly provide roundtables to promote discussion).*

> *I believe that we need to find a way to make sure that the LMAC is servicing more than just our 'at risk' students and find a way to make sure that we are providing resources to the rest of our student population to make sure that they know we are investing in them as well.*

> *Training faculty: We should have regular training perhaps at the beginning of each faculty meeting where we are updated on the system we are using and the data we are collecting regarding student satisfaction, retention, and graduation.*

> *Secure a multi-year grant to support the LMAC.*

LESSONS LEARNED AFTER THE INAUGURAL YEAR OF THE LMAC IMPLEMENTATION

Implementation of the LMAC model was a strategic method to address the cognitive, as well as non-cognitive issues, that affected students' success in the Biology Department. Low first-year and second-year retention rates among Biology majors required dramatic changes in the advisement structure. Several important lessons have emerged from the department's willingness to make a major change to how faculty and students interact through advisement. Some of the most important lessons are discussed in the following text.

Evolution of the advisement model from the traditional model to the LMAC model has provided incredible opportunities for the faculty to learn from the increased interactions with students. Perhaps most importantly, these interactions have heightened our awareness of student issues that are not cognitive. Traditionally, many advisors focused solely on assisting students with classes. Very little to no thought was given to factors that may prevent students from successfully transitioning through classes in the proper sequence or performing well academically. With the implementation of the LMAC model, advisors are now more likely to ask students about their adjustment to the university, how well they are balancing courses with positive social experiences and helping to connect students with proper resources to alleviate issues early. A survey given to the senior Biology majors helped to shed light on the students' advising experience pre- and post-LMAC.

Three major themes emerged from the survey data: (1) poor advisement is very frustrating, leads to dissatisfaction in the department, and may hinder retention and degree completion, (2) the LMAC model has helped to alleviate multiple points of student frustration, and (3) the importance of maintaining a personal touch in any advising model. Students reported that access to their advisor in the traditional model was challenging and sometimes led to them seeking advisement from a different advisor or attempting self-advisement. Under the LMAC model, students report a high degree of value in having daily access to a faculty advisor, study center, and tutoring assistance. Students also want to feel connected to the department through their advisor. Clearly, we show that the relationship with the advisor is very important and maintaining access to a primary advisor, in addition to having access to other advisors through the LMAC, is vital to building students' trust, retaining students in the major, or helping them to transition to a major more appropriate to their interests.

Through the LMAC model, we have also learned more about the significance of peer-to-peer interactions. Anecdotally, we can say that our BBBS and peer tutoring program have resulted in increased student engagement within the classroom and our biology community. This, in turn, has promoted a greater sense of belonging for our freshmen. During the fall 2017 semester, many first-year students submitted reflections in their Biology Orientation class that enthusiastically expressed how impactful the BBBS events and interactions with the big sibling was on their transition from high school to college. Additionally, preliminary studies suggest that such interactions promote the development of students' science self-efficacy and science identity. With regard to increased engagement, this is not limited to our freshmen majors, as we have witnessed our upperclassmen truly taking ownership over the planning of activities and tutoring sessions for their peers. We believe that the outcomes of the peer component of the LMAC confirm previous petitions to establish a stronger peer network among STEM majors (Espinosa, 2011). Therefore, we strongly recommend that other STEM departments that may consider adopting our model incorporate BBBS and peer tutoring, due to the potential for such peer-to-peer interactions to reduce attrition rates.

Even though the LMAC model was initiated to address student retention, the LMAC management team anticipated that the changes would also positively affect the faculty by providing advisement training, peer support, equal distribution of advisees among the faculty, and minimizing the amount of time dedicated to advising duties. Through a survey administered to the faculty, important successes, as well as some continued challenges, were highlighted. The following four major themes emerged: (1) novel collaborations among faculty, (2) enhanced knowledge of the university and departmental policies, (3) improved advising skills, and (4) improved interactions with students. Advisors reported that the on-the-job training with a more experienced advisor was highly beneficial. Likewise, advisors reported that *my comfort level with advising has greatly improved!* Other faculty members commented that the training for new advisors was highly beneficial and the ability to learn through observing a more seasoned advisor was invaluable. It should be noted that some faculty members who were initially resistant to the model have since recognized the benefits for both students and themselves. Finally, advisors also commented on their interactions with students in the LMAC, suggesting that they enjoyed *getting to meet and interact with a greater variety of students and to have central resources to help in advising*.

CHALLENGES AND RECOMMENDATIONS FOR THE FUTURE

Each new generation matriculating into colleges and universities presents unique characteristics and introduces additional challenges for effective communication between faculty and students. Several studies point to differences between Millennial students and those of previous generations (Ciocco & Holtzman, 2008; Keeling, 2003; Montag, Campo, Weissman, Walmsley, & Snell, 2012). Advisors need to understand such generational shifts and develop innovative approaches to help these students set – and find ways to achieve – realistic career goals. Christy Price responded to a gap between students' expectation for success and the effort they put forth in the classroom (Price, 2011). Her findings summarize five Rs as strategies for engaging Millennial students: (1) Research-based methods, (2) Relevance, (3) Rationale, (4) Relaxed, and (5) Rapport. These strategies, which all represent characteristics of the Millennials' ideal learning environments, emphasize the Millennial's preference for engagement, appreciation for faculty showing interest in them, and their willingness to pursue learning outcomes when faculty connects with them on a personal level.

Transitioning from a traditional advising model to a transformative/participatory model required a multi-level commitment from the leadership, faculty, and staff of the Biology Department at NC A&T. The department chairperson, Dr Mary Smith, led the charge for change and identified faculty who shared a common interest in improving student success in the department. Once a team was assembled, she encouraged the group to think broadly about the needs of the students, encouraged the conception of fresh ideas, and supported the innovative recommendations put forth by the faculty and staff. Securing funding, reallocating usable space, and designing an appropriate area to create a true advising center were all challenging, but pivotal components of the process. Perhaps the most challenging aspect of the process was developing a method to support and encourage faculty members who were uncomfortable with advising, particularly intrusive advising. Pairing advisors in the LMAC and providing opportunities for training sessions have greatly improved the skills and confidence of advisors in the department. However, continued implementation of these strategies will be very important to the success of the LMAC.

Future challenges will continue to revolve around the need for additional finances to support staff, summer salary for some faculty members, consumables, and possibly a larger space to support the growing number of students in the department. Based on evaluations of students and faculty, the LMAC model has improved the advisement experience for students and faculty in the Department of Biology. Continuous evaluation and data-driven changes will ensure that the LMAC will meet the needs of both students and faculty in the department for years to come.

REFERENCES

Afshar, T., & O'Hara, L. (2011). Pedagogical issues concerning academic advising. *Journal of College Teaching & Learning (TLC)*, *3*(11), 25–44. (2006). doi:10.19030/tlc.v3i11.1656

Anderson, E. L., & Kim, D. (2006). *Increasing the success of minority students in science and technology*. Washington, DC: American Council on Education.

Angelique, H., Kyle, K., & Taylor, E. (2002). Mentors and muses: New strategies for academic success. *Innovative Higher Education*, *26*(3), 195–209.

Bandura, A. (1977). Self-efficacy: Toward a unifying theory of behavioral change. *Psychological Review*, *84*(2), 191–215. doi:10.1037/0033-295X.84.2.191

Bandura, A. (1986). Prentice-Hall series in social learning theory. *Social foundations of thought and action: A social cognitive theory*. Englewood Cliffs, NJ: Prentice-Hall, Inc.

Bandura, A., Adams, N. E., & Beyer, J. (1977). Cognitive processes mediating behavioral change. *Journal of Personality and Social Psychology*, *35*(3), 125–139. doi:10.1037/0022-3514.35.3.125

Bettinger, E., & Baker, R. (2014). The effects of student coaching in college: An evaluation of a randomized experiment in student mentoring. *Education Evaluation and Policy Analysis*, *36*(1), 3–19. doi:10.3102/0162373713500523

Blickenstaff, J. (2005). Women and science careers: Leaky pipeline or gender filter. *Gender and Education*, *17*(4), 369–386. doi:10.1080/09540250500145072

Borns, R. F. (2002, May 6). Creating an academic advising training program on your campus. *The Mentor: An academic advising journal*. Retrieved from http://www.psu.edu/dus/mentor. Accessed on December 1, 2002.

Chang, M., Eagen, M., Lin, M., & Hurtado, S. (2011). Considering the impact of racial stigmas and science identity: Persistence among biomedical and behavioral science aspirants. *The Journal of Higher Education*, *82*(5), 564–596.

Chang, M., Sharkness, J., Hurtado, S., & Newman, C. (2014). What matters in college for retaining aspiring scientists and engineers from underrepresented racial groups. *Journal of Research in Science Teaching*, *51*(5), 555–580.

Ciocco, M. D., & Holtzman, D. (2008). Prior learning assessment: E-portfolios and the millennial students. *The Journal of Continuing Higher Education*, *56*(3), 69–71.

Colvin, J. W., & Ashman, M. (2010). Roles, risks, and benefits of peer mentoring relationships in higher education. *Mentoring & Tutoring: Partnership in Learning*, *18*(2), 121–134.

Costello, C. A., & Stone, S. L. M. (2012). Positive psychology and self-efficacy: Potential benefits for college students with attention deficit hyperactivity disorder and learning disabilities. *Journal of Postsecondary Education and Disability*, *25*(2), 119–129.

Creswell, J. W. (2007). *Research design: Qualitative, quantitative, and mixed methods approaches* (2nd ed.). Thousand Oaks, CA: Sage.

Dalton, J. C., & Crosby, P. C. (2014). The power of personal coaching: Helping first-year students to connect and commit in college. *Journal of College and Character*, *15*(2), 59–66.

Earl, W. R., (1987). Intrusive advising for freshmen. *Academic Advising News*, *9*(3).

Erlich, R. J., & Russ-Eft, D. (2011). Applying social cognitive theory to academic advising to assess student learning outcomes. *NACADA Journal*, *31*(2), 5–15.

Espinosa, L. (2011). Pipelines and pathways: Women of color in undergraduate STEM majors and the college experiences that contribute to persistence. *Harvard Educational Review*, *81*(2), 209–240.

Gordon, V. N. (1992). *Handbook of academic advising*. Westport, CT: Greenwood Press.

Hemwall, M. K., & Trachte, K. C. (2005). Academic advising as learning: 10 organizing principles. *Nacada Journal*, *25*(2), 74–83.

Higher Education Research Institute. (2010). *Degrees of success: Bachelor's degree completion rates among initial STEM majors*. Los Angeles, CA.

Keeling, S. (2003). Advising the millennial generation. *NACADA Journal: Spring-Fall*, *23*(1–2), 30–36. doi:10.12930/0271-9517-23.1-2.30

Kram, K. E., & Isabella, L. A. (1985). Mentoring alternatives: The role of peer relationships in career development. *Academy of management Journal*, *28*(1), 110–132.

Maher, J. (2009). *Spiritual life coaching: A new model of spiritual direction for collegians* (Doctoral dissertation). Retrieved from ProQuest Dissertations and Theses database. (UMI No. 304849534).

Mahoney, T., & Schibik, T. (2004). A new advising model using a non-traditional approach: Advising in a changing environment. *Proceedings of the Midwest Business Economics Association*, 117–122.

Miller, L. K. (2010). The impact of intrusive advising on academic self efficacy beliefs in first-year students in higher education. *Dissertation Abstracts International: Section A. Humanities and Social Sciences*, *72*(2–A), 491.

Montag, T.,Campo, J., Weissman, J., Walmsley, A., & Snell, A. (2012). In their own words: Best practices for advising millennial students about majors. *NACADAJournal*, *32*(2), 26–35.

National Research Council. (2011). *Expanding underrepresented minority participation: America's minority science and technology talent at the crossroads*. Washington, DC: The National Academies Press.

National Science Board. (2012). *Report to the president, engage to excel: Producing one million additional college graduates with degrees in science, technology, engineering, and mathematics*. Washington, D.C.: Executive Office of the President, President's Council of Advisors on Science and Technology.

National Science Board. (2014). *Science and engineering indicators 2014*. Arlington, VA: National Science Foundation (NSB 14–01).

National Science Board. (2016). *Science and engineering indicators 2016*. Arlington, VA: National Science Foundation (NSB-2016-1) | Digest (NSB-2016-2).

Petress, K. C. (1996). The multiple roles of an undergraduate academic advisor. *Education*, *117*, 91–92.

Price, C. (2011). *Five strategies to engage today's students, Magna Online Seminar*. 1 Nov. 2011. Madison, WI: Magna Publications.

Russell, M. L., & Atwater, M. M. (2005). Traveling the road to success: A discourse on persistence throughout the science pipeline with African American students at a 27 predominantly white institution. *Journal of Research on Science Teaching*, *42*, 691–715.

Schlossberg, N. K. (1989). Marginality and mattering: Key issues in building community. *New Directions for Student Services*, *1989*(48), 5–15.

Seymour, E., & Hewitt, N. M. (1997). *Talking about leaving: Why undergraduates leave the sciences*. Boulder, CO: Westview Press.

Tinto, V. (1987). *Leaving college: Rethinking the causes and cures of student attrition*. Chicago, IL: University of Chicago Press.

Wilcox, E. (2016). An end to checklist thinking: Learning-centered advising in practice. Retrieved from http://nacada.ksu.edu/tabid/3318/articleType/ArticleView/articleId/6101/article.aspx

Wilson, Z. S., Holmes, L., Sylvain, M. R., Batiste, L., Johnson, M., McGuire, S. Y., ... Warner, I. M. (2012). Hierarchical mentoring: A transformative strategy for improving diversity and retention in undergraduate STEM disciplines. *Journal of Science Education and Technology*, *21*(1), 148–156.

TRANSITIONING FROM A TRADITIONAL LECTURE STYLE ORGANIC CHEMISTRY CLASSROOM INTO A "FLIPPED" CLASSROOM

Angela Winstead and Liuli Huang

ABSTRACT

The transition from a traditional lecture style method of teaching to the flipped classroom in sophomore-level Organic Chemistry I and II courses at an Historically Black University (HBCU) is described. The process of implementation was explained and the students' performance was analyzed. The flipped teaching method made a much bigger positive impact to Organic I than Organic II Chemistry course. A higher percentage of A, B or better, and C or better were observed for Organic I Chemistry course. The DFW rate was also significantly lower for the Organic I Chemistry flipped classroom. However, Organic II results were very similar between the students from both teaching methods.

Keywords: Flipped classroom; active learning; organic chemistry; HBCU; STEM

INTRODUCTION

Organic Chemistry is one of the most challenging science courses. It is fast-paced, content heavy, and requires learning and application of new content in a short time period. The course is required for chemistry majors, biology majors, and students who aspire to attend medical schools and other health-related careers. Faculty are required to plow through massive amounts of topics each

Broadening Participation in STEM: Effective Methods, Practices, and Programs
Diversity in Higher Education, Volume 22, 317–339

ISSN: 1479-3644/doi:10.1108/S1479-364420190000022014

semester. Students are expected to master fundamental concepts such as resonance, acid-base theory, hybridization, and formal charge in three weeks and then apply them to mechanistic problems. It is like asking a student to learn the rudiments of Spanish in a month and requiring a Spanish novel at the end of the semester. By midterm, both sides are frustrated; the combination of future mechanisms and past unlearned concepts seem unsurmountable by the students, and though faculty are aware that the students have not grasped the concepts, they must continue to deliver the remaining content. In addition, the results of intermittent knowledge checkpoints in the form of quizzes and exams, etc. arrive after the concept is covered and provide neither faculty or students the opportunity to address deficiencies in a fast-paced course. In an attempt to acquire a real-time gauge of student understanding, increase student's awareness, and to facilitate the exchange of knowledge, the traditional sage-on-the-stage model was abandoned for a more student-centered approach. This chapter describes the transition from a traditional lecture style method of teaching to the flipped classroom in sophomore-level Organic Chemistry I and II courses at an Historically Black University (HBCU).

HBCUs and STEM Education

The attention of the nation is focused on diversifying Science, Technology, engineering, and math (STEM). The number of African American (AA) students who intended to enroll in STEM decreased from 39% to 36% from 1997 to 2012 (Science and Engineering Indicators Report, 2016). In 2014, the number reached 40% which is essentially the same as 17 years ago. With respect to degrees attained by AAs in STEM, the number has decreased for most of the STEM fields (Fig. 1). These data support the fact that the diversity in STEM is both an interest and persistence issue.

Studying approaches that can potentially improve the production of STEM undergraduate degrees at HBCUs are critical. HBCUs play a vital role in the production of AA STEM researchers (Jett, 2013; Owens, Shelton, Bloom, & Cavil, 2012). Even though HBCUs represent only 3% of the nation's four-year colleges and universities (U.S. Department of Education. Institute of Education

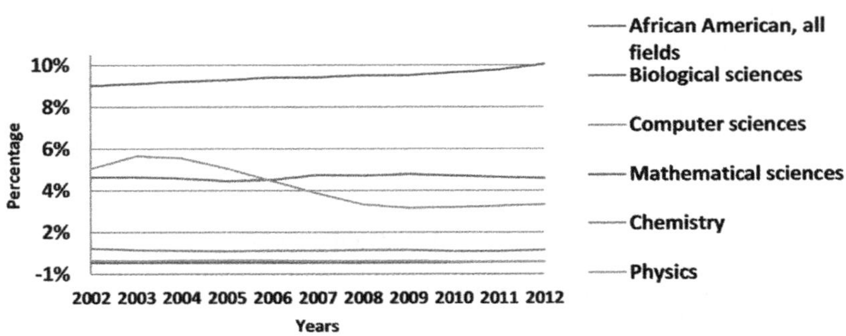

Fig. 1. Bachelor's Degrees Earned by African Americans.

Sciences), they have produced 22% of the bachelor's degrees granted to AAs (Science and Engineering Indicators Report, 2016). Moreover, out of the top 51 institutions that produce AA PhD recipients in STEM fields (2010–2014), HBCUs contributed 50.2% of these awardees. Currently, HBCUs are experiencing a surge in enrollment (for example, Shaw University +49%, Tuskegee University +32%, and Virginia State University +39%). This surge coupled with the effectiveness of HBCUs in the production of potential PhD scientists warrants a comprehensive study of the effectiveness of teaching strategies that enhance AA performance in STEM courses at HBCUs.

Morgan State University (MSU) is an HBCU designated as Maryland's public urban institution of higher learning. It serves a student population that is ethnically and culturally diverse and has recently been classified as a Research Intensive Urban Institution. The university has median 885 SAT, and 61.5% of its students are Pell Grant recipients. MSU was ranked number 14 in the top 50 baccalaureate institutions of black science doctorate recipients (Science and Engineering Indicators Report, 2016). The Department of Chemistry at MSU has two degree tracks: General and American Chemical Society (ACS) certified. From 2004 to 2017, 50% of the Morgan chemistry BS graduates attended graduate and/or professional schools.

Active Learning in STEM

A teaching strategy that has been successful in reducing the achievement gap for minorities in STEM is active learning (Haak, HilleRisLambers, Pitre, & Freeman, 2011). Benefits include enhanced course grades, increased self-efficacy, and reduced DFW rates (Laws, Sokoloff, & Thornton, 1999; Lorenzo, Crouch, & Mazur, 2006). An extensive meta-analysis of 225 studies of peer-reviewed, conference proceedings and unpublished dissertations on student performance in active learning versus traditional lecturing in STEM classes was conducted by Freeman et al. (Freeman et al., 2014). The authors initially examined 642 papers for their set criteria, and 225 were ultimately analyzed in the report. The study reported an increase by 6% in examination scores in active learning sections. Average failure rates were 21.8% for active learning and 33.8% in traditional lecturing classrooms (Fig. 2). This was observed across all STEM disciplines regardless of course level and class size. It should be mentioned that in classes smaller than 50, the benefit was greater with the increase on concept inventories. Freeman calls into question the continued use of traditional lecturing in light of the findings of this comprehensive quantitative study.

ACTIVE LEARNING – FLIPPED CLASSROOM

"Flipped Learning is a pedagogical approach in which direct instruction moves from the group learning space to the individual learning space, and the resulting group space is transformed into a dynamic, interactive learning environment where the educator guides students as they apply concepts and engage creatively in the subject matter"(FLN, 2014). A flipped classroom comprises two major

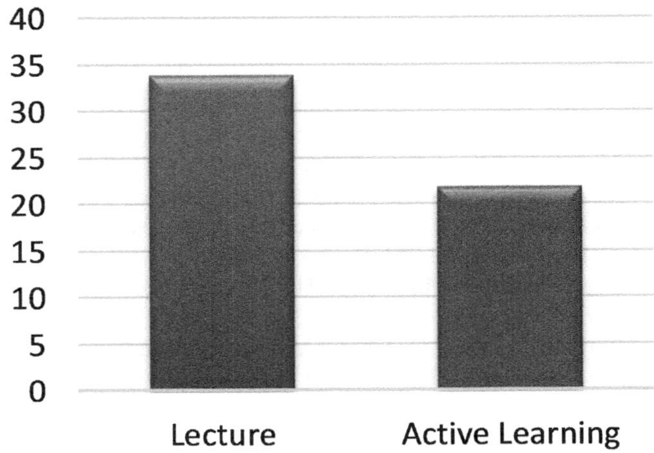

Fig. 2. Failure Rate.

components: (1) pre-class preparation that may include reading, practice quizzes, and watching videos and (2) in = class session consisting of active learning activities. In 2007, Bermann and Sams introduced the flipped classroom to their high school chemistry classes (Bergmann & Sams, 2012). Since then, this style has reached a variety of disciplines at the high school and college level. In 2015, Seery published a review article on flipped learning in higher education Chemistry courses (Seery, 2015). In the areas of general chemistry, organic chemistry (I & II), analytical chemistry, spectroscopy, physical chemistry, and biophysical chemistry related 12 peer-reviewed articles were extensively reviewed. The various approaches to the implementation of flipped classrooms was surveyed, and common approaches in pre-class preparation material, preparation assessment, in-class activities, and after-class assignments were reported. PowerPoint recordings were the prevailing delivery method for pre-classroom preparation. In-class quiz or pre-class quiz were used in 58% of the surveyed articles. Most studies used problem sets as the in-class activity, and 91% gave no after-class assignments. Overall, student performance was either the same or improved; students enjoyed the flipped classroom and, thereby, some students who may have otherwise dropped the course continued to persist. A study conducted at an all-women's 4-year HBCU examined the effectiveness of the flipped classroom in a General Chemistry 2-course sequence over a five-year period (Hibbard, Sung, & Wells, 2015). Students enrolled in the flipped classroom earned increased raw scores on the ACS standardized exam for both Fall and Spring semesters. Notably, students' scores were above the national norm for the Spring semester.

At MSU, the School of Computer Mathematical and Natural Sciences (SCMNS) has invested in the enhancement of teaching pedagogy. Specifically, the Department of Chemistry has implemented active learning into over 70% of its General and Organic Chemistry sections. The percentage of courses utilizing some form of active learning in 100 level chemistry SCMNS courses increased from 60% to 83% from Fall 2016 to Fall 2017.

THE TRANSITION

The effort to transition a traditional organic chemistry lecture classroom into an active learning classroom through the use of "flipped" pedagogy is described. The transition from traditional teaching strategy to active learning method at HBCUs can be challenging. (1) On average, faculty members teach 3–4 courses a semester in addition to the service, mentoring, and research requirements. The result is limited time to invest into changing their teaching pedagogy. (2) Facilities and/or funding to support renovations for an active learning classroom may be limited. (3) Technical support required for video recordings and training may not be readily available. Specifically, the following challenges were faced during the transition: classroom space, optimal class size concerns, student apprehension, and resistance and hesitation regarding department-wide implementation by faculty. How these challenges were addressed using our REACH (Read and Review, Examine and Explore, Apply and Achieve, Confer and Conform, and Hint and Help) framework coupled with classroom management software will be discussed.

Classroom Space

The department had six combined lecture-laboratory classrooms where most of its classes were taught. This type of classroom made it difficult to implement active learning activities and collaborative learning due to the laboratory benches and obstructions such as sink faucets, instrumentation, and elevated bench hoods. In 2014, the department transformed a small meeting space into an active learning classroom with minimal financial resources (Fig. 3). The space was outfitted with six round foldable tables, chairs, two additional 4×8 whiteboards, and a presentation stand for cost a under USS$3000. There are a total of five whiteboards and one screen in the classroom. Each table was positioned in front of a whiteboard and assigned to the four-person group seated at the table. The room can also be rearranged in presentation style to accommodate student and faculty presentations. The classroom accommodated small General and Organic Chemistry classes, as well as upper-level Chemistry classes. Since 2014, external funding supported three additional shared science classrooms to be transformed into active learning classrooms.

Course Content Development – REACH

A framework called REACH was developed to assist faculty and students with adjusting and succeeding in the flipped classroom. Each letter in REACH represents a key element in the successful implementation of flipped learning into the classroom. We sought out a term that would describe the transition from something very comfortable to the unfamiliar. Flipped learning requires one to reach beyond their current standard method of learning and teaching. REACH stands

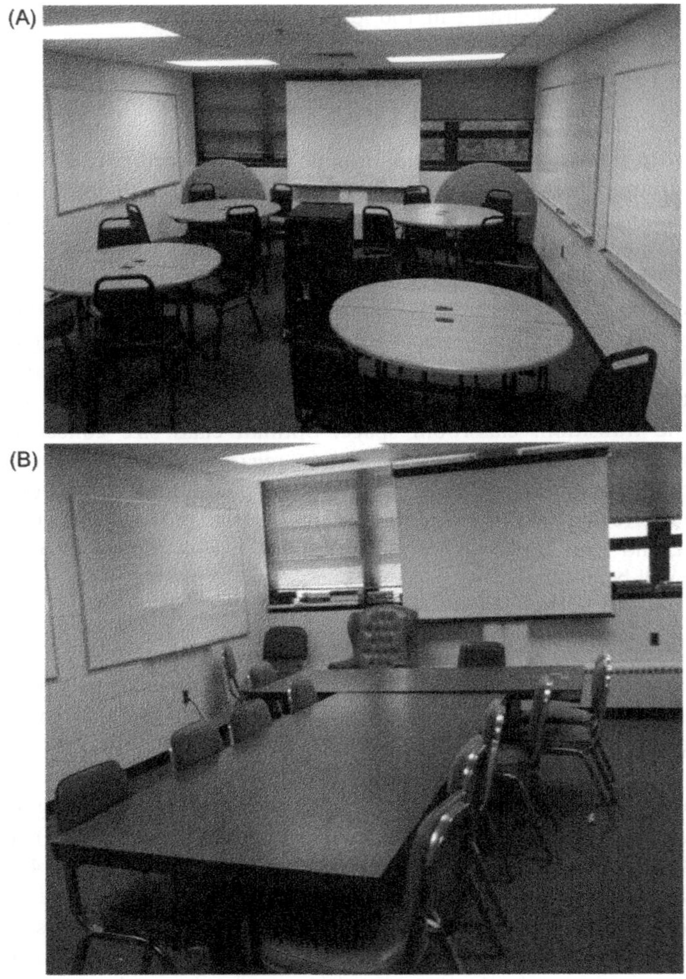

Fig. 3. Before and After Renovations of Meeting Space.

for Read and Review, Examine and Explore, Apply and Achieve, Confer and Conform, and Hint and Help (Fig. 4).

Read and Review

Read and Review describes the essence of the flipped classroom, the pre-lecture component. The concepts that students are expected to learn on their own are contained herein. It is essential to ensure that the content is delivered in a way that is accessible to different learning styles including visual, auditory, and verbal. It allows students the flexibility to use a combination of delivery styles that best support their learning. Read and Review is divided into three subcategories:

Read and review.	Examine and Explore.	Apply and Achieve.	Confer and Confirm.	Hint and Help.
READ Lecture notes and Text *LISTEN* Lecture, PowerPoint *WATCH* Youtube Videos Khan Academy	*PRACTICE* • Adaptive Learning homework Multimedia instruction • Molecular Models • Lewis Electron Dot Models	*APPLY* • End-of-Chapter homework	• QUIZZES • CLASSROOM PARTICIPATION Talk, learn, reinforce • SELF ASSESSMENT Self-tests at the midway point	*HELP OUTSIDE THE CLASSROOM* Tutoring Office hours

Fig. 4. REACH Diagram.

read, listen, and watch. Examples include assigned reading in the textbook, faculty-prepared lecture notes, voice-over PowerPoint lectures, publisher slides, and whiteboard screencasts such as YouTube videos and Kahn Academy instructional videos. Institutions that utilize Blackboard Learn have Readspeaker's text-to-speech feature preinstalled. While not as dynamic as an instructor's lecture, it could be used as a tool to provide faculty time to dedicate to the development of worksheets and other active learning instruments. At the completion of this category, students are expected to be able to recall the facts and basic concepts in the form of classroom conversations and pre-class quizzes (discussed in Section "Confer and Confirm").

Examine and Explore

Examine and Explore is the space that encourages students to test their proficiency in understanding the basic concepts with the opportunity to explore the concept further through simulations and visualizations. Students are recommended to examine and explore after the pre-class reading and in-class discussions on the topic. *Electronic resources*: Most organic chemistry textbooks provide instructors and students with simulation and/or visualization tools. For example, Cengage's OWLv2, an interactive online homework system has text-specific mastery and multimedia activities for most concepts. A useful free supplemental visualization resource is Chemtube3d at the University of Liverpool. Interactive animations for a large number of organic reactions are available at this site. Many of these aforementioned resources are interactive and provide students with either instant feedback (OWLv2) or a walk-through explanation of the visualization. *Hands-on:* The course utilized both Lewis Dot models (Flinn Scientific) and 3-D molecular models. The use of molecular models in organic chemistry classes to assist with stereochemistry and basic spatial visualizations is certainly not a new teaching tool (Minne, 1929). In 2016, Stull et al. proposed three benefits for using molecular modeling (Stull, Gainer, Padalkar, & Hegarty, 2016). One of these benefits proposes that cognitive off-loading onto these external models frees up students' working memory to process translation between different representations leading to improved learning

(Wu & Shah, 2004). The Lewis Dot Models were designed to assist with teaching Lewis electron dot structures. In this course, it is used to provide students the opportunity to physically move electrons when learning reaction mechanisms and drawing resonance structures. Examine and Explore enhances students' ability to understand concepts and display this through scientific explanation of mechanisms and concepts through in-class activities and discussion.

Apply and Achieve

Apply and Achieve challenges students to take the basic concepts and to apply them to a more difficult set of problems. This occurs during both in-class and post-class learning. During class, students work through worksheets created by the instructor. The hands-on resources are incorporated into the worksheets throughout the course. The post-class assignment (end-of-chapter questions) is regarded as review or practice for the exam. The end-of-chapter questions are a combination of all of the concepts covered in the chapter in one assignment. The interactive online homework system (OWLv2) provides the students with feedback, but less guidance than received during in-class problem-solving sessions. Apply and Achieve enables students to get a sense of the concepts that they have mastered and those that need to be refreshed through tutoring, office hours, or revisiting adaptive learning problems.

Confer and Confirm

Confer and Confirm guides students through the process of demonstrating a fundamental understanding of basic-level concepts and encourages the exchange of knowledge among their peers and the instructor. Five-question quizzes are given to confirm students' basic knowledge of the concepts after they have read and reviewed the material. They are labeled as a "Ticket In" to indicate to the students that this knowledge is required to actively participate in the classroom. The classroom is designed to facilitate small group discussions with round collaborative tables and each group is assigned a whiteboard. The in-class worksheets contain a mix of questions that are designed to be worked individually and as a group. However, all students are required to work through mechanisms on the whiteboard and provide oral scientific descriptions of their answers.

Hint and Help

Although the flipped classroom enhances the in-class teacher–student interaction, often students still need help outside of the classroom with tutors, study groups, teaching assistants, and instructor office hours. The Hint and Help on Blackboard Learn reminds students of the extra help that is available to assist with their student learning.

REACH – Running the Classroom

The transition from being the "sage" required a change not only in content delivery but also the manner in which the classroom was managed. REACH served as a guide to the four pillars of F-L-I-P ™ developed by FLN in 2014 (FLN, 2014):

(1) Flexible environment – educators are flexible in the design of their learning spaces. They allow support and flexibility in when and where students learn.
(2) Learning culture – educator shifts from "sage on the stage" to a student-centered model. Students are actively engaged during class time, and questions and conversation are encouraged.
(3) Intentional content – educators carefully determine the concepts that are suited for out-of-class learning and which concepts should be taught. The in-class activities reinforce the concepts presented in videos/readings, etc.
(4) Professional educator – the professional educator must be observant, attentive, and possess the ability to provide instant feedback to students working problems in class.

Similar to Section "Course Content Development," the REACH framework was also designed to provide faculty with clear succinct strategies to achieve successful classroom management of "organized chaos" (also known as "the flipped classroom"). This was important to the transition because faculty dedicate considerable time to course content development (Section "The Transition") and have little to no time to read how-to articles for classroom management of a flipped classroom. They need to read and review content that is suitable for outside learning and ensure that all videos are working properly at the beginning of each semeste; review the Learning Management System (LMS) statistics to see if students are utilizing the content; andexamine and explore assessment strategies for the flipped classroom. In a small class or with the assistance of a teaching assistant, each student must be observed as they work problems to identify their level for the concept and suggest review material for fundamental concepts related to the problem. Apply and Achieve assists students to achieving their full potential. Application problems with different levels of difficulty must be identified to challenge students at the appropriate level. This can prevent advanced students from getting bored and may reduce the frustration of struggling students. It must be confirmed whether students understand the basic concepts by reviewing the five-minute quizzes or other pre-class assignment. Hint and Help is available to instructors through conferences, campus professional development opportunities, and the community of other active learning instructors in their field.

CLASSROOM DESIGN

Organic Chemistry is a two-semester course that is required for Chemistry and Biology majors. The course makeup is sophomore Chemistry majors and sophomore and Junior Biology majors. Organic I and II are both offered both semesters during the academic year. There are two sections of both courses each semester. One section is taught using traditional lecture style format and the other section was transitioned from a lecture style to a flipped classroom style in 2014. The course presented here is considered the "on sequence" course of

Fall–Spring. Students enrolled in the course are required to be enrolled in the laboratory concurrently or should have passed the laboratory in a previous semester. The course meets three times a week for 50 minutes. The course enrollment is between 15 and 20 students. The classroom has five round tables that seat four students per table. There are five large whiteboards in the room.

Learning Management System Organization

The LMS utilized at MSU is Blackboard Learn. The course content is divided into five content modules that contain the content covered on each of the five exams. Within each module are (1) the exam date, (2) learning objectives for the module, (3) learning modules where the content is broken down into smaller segments, and (4) deep understanding question set. The content in the learning modules utilized the REACH model that was developed. The first module is the Introductory Module that describes the flipped classroom to the students. For example is Module D. In Module D, there were seven learning modules: (1) Organohalides, (2) Substitution and Elimination basics, (3) Substitution Reactions S_N2, (4) Substitution reactions S_N1, (5) Elimination Reactions E1, (6) Elimination Reactions E2, and (7) S_N1 vs S_N2 vs E1 vs E2. The contents of Learning Module 5: Elimination Reactions E1 are shown in Fig. 5.

Content

Read and Review – Notes

The notes are PowerPoint lecture slides and were provided by the textbook publisher. The provided slides are between 35 and 45 pages for each chapter. These

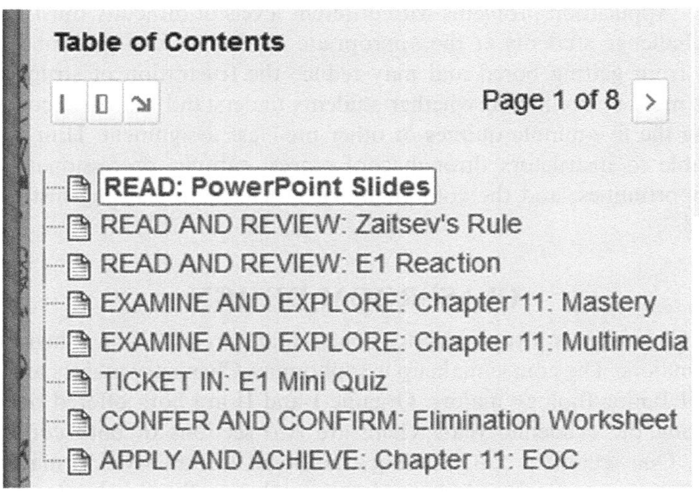

Fig. 5. Learning Module 5 - Table of Contents Blackboard Screenshot.

slides were divided into smaller files with topics aligned with each learning module.

Video instructional lectures: Kahn Academy provides free practice exercises and instructional videos for the concepts in Organic Chemistry. The instructional videos are whiteboard screencasts. This is particularly important for Organic Chemistry courses that require students to understand the flow of electrons in a reaction mechanism. The video lectures were embedded into Blackboard Learn from Kahn Academy.

Examine and Explore

The course uses the textbook's online homework system (OWLv2) by Cengage. OWLv2 is an interactive online homework platform that is textbook specific. There are three types of OWLv2 problem sets employed in this course: (1) Mastery Learning Activities, (2) Multimedia Activities, and (3) End-of-Chapter Questions. *OWLv2 Mastery Learning assignments* are concept-based questions. Students must answer a specific number of questions in the set correct before they can move on to the next topic. The questions are randomized for each attempt and students receive instant feedback and resource information. It allows students to practice at their own pace until they are able to master the concept. The mastery assignments have between 4 and 15 questions per chapter. They are due before the conclusion of the chapter and count as extra credit. *Multimedia OWL activities* are assigned for each chapter with the exception of Chapter 5 (Chiral Centers) and Chapter 12 (MS and IR) (not available in OWLv2). These activities assist with conceptual understanding through simulations, visualizations, and interactive examples. The multimedia activities are extra credit and are due before the conclusion of the chapter.

Apply and Achieve

End-of-Chapter OWL problems are the same problems that are at the end of each chapter in the textbook. These problems are required for each chapter in the module and are due the morning of the exam. There are between 21 and 38 questions in each chapter problem set.

Exams: Each semester has five 50-minute in-class exams. The lowest exam grade is dropped in the final average. The final exam in Organic Chemistry I is an instructor-designed two-hour exam. The final exam in Organic Chemistry II is the ACS standardized exam with 70 multiple-choice questions.

Confer and Confirm – Quizzes

After watching the video and/or reading the text, students take "Ticket In" five-question quizzes. The quizzes are designed by the instructor and administered through Blackboard Learn. They contain basic recall type questions. The students are permitted to take the quiz three times with the highest score counting toward their quiz average. For Organic I, there were 29 quizzes over the course of the semester; and for Organic II, there were 25 quizzes over the course of the semester.

In-class activities: Each class period is dedicated to working problem worksheets. Each student is required to work problems on the board during the class period. While working on problems, students are permitted to discuss the question with their boardmates. Other times, students are asked to work individually on the problems on the whiteboard. Once a student successfully has completed the question, they could take a picture of his or her board work to later transfer to the worksheet. This was implemented as time saver during class.

RESULTS AND DISCUSSION

Comparative Analysis of Students' Performance (Percentage of A, B, C, D, F, and W) Between Traditional & Flipped Classroom Teaching Methods

Organic Chemistry I

The Organic I flipped classroom had 49 students from Fall 2014 to Fall 2017. The lecture style semesters were Fall 2010–Fall 2013 with a total of 72 students. Both traditional and flipped classroom environments had a 2.5:1 ratio of female-to-male students. The GPA for the students enrolled in the flipped classroom was slightly higher than that of the traditional lecture (Table 1).

We compared the percentage of A, B, C, and the DFW rates for Fall 2010–Fall 2017 for Organic I. The Mann–Whitney U test is utilized to conduct the significance analysis. The Mann–Whitney U test is the nonparametric counterpart for the independent two-sample t-test. There are several reasons why the Mann–Whitney U test is a good fit for this study. To begin with, the sample sizes for each of the four levels are quite small (four sessions for each type of teaching method). For the lecture style teaching method, we have four semesters in total from Fall 2010 to Fall 2013. The percentage of A, B, C, and the DFW rates are computed for each semester. Thus, the sample size for each level is 6. The conditions for the active learning style are similar. We have four semesters from Fall 2014 to Fall 2017. In addition, the assumptions of normality and equality of variance required for the parametric tests are not met. Considering these various conditions, the quantitative nonparametric Mann–Whitney U test is a good fit for this study. The results from Table 2 indicate that none of the four performance levels are statistically different between the two teaching methods at the alpha level of 0.05.

The details of the comparisons between the two teaching methods regarding each of the four levels are available in Fig. 6. Even though not statistically different, students with the flipped teaching method show slightly better

Table 1. Student Population Organic I.

	Flipped	Lecture Style
Number of students	49	72
Gender M (%)	71	71
Gender F (%)	29	29
GPA	2.96	2.75

Table 2. Mann–Whitney U Test for Organic I.[a]

	Percentage A	B or Better	C or Better	DFW
Mann–Whitney U	2.000	2.000	2.000	2.000
Wilcoxon W	12.000	12.000	12.000	12.000
Z	−1.775	−1.732	−1.732	−1.732
Asymp. sig. (2-tailed)	0.076	0.083	0.083	0.083
Exact sig. [2*(1-tailed sig.)]	0.114[b]	0.114[b]	0.114[b]	0.114[b]

Notes: [a]Grouping variable: teaching method.
[b]Not corrected for ties.

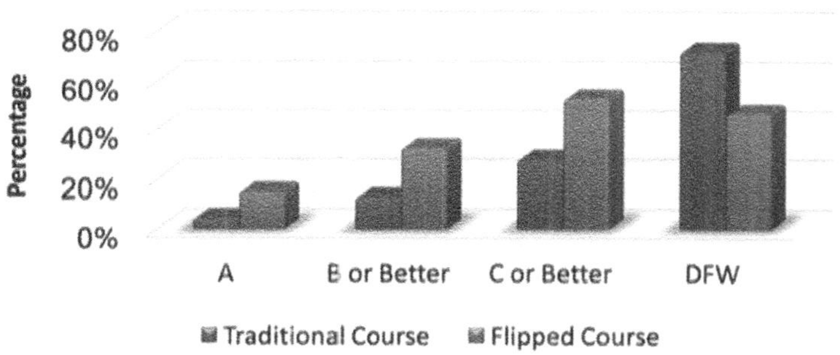

Fig. 6. Organic Chemistry I Final Grade Comparison.

performance, with a higher percentage of A, a higher percentage of B or better, a higher percentage of C or better, and a lower percentage of DFW. Cohen (1998) proposed the effect size r for the Mann–Whitney U test to determine the strength of the difference when no statistically significant difference was detected. The formula for the effect size is $r = \frac{Z}{\sqrt{N}}$. According to Cohen (1988), the r-value being equal to or greater than 0.1 indicates a small effect size; the r-value being equal to or greater than 0.3 indicates a medium effect size; and the r-value being equal to or greater than 0.5 indicates a large effect size. The sign of the r-value is of no significance. The effect sizes are computed for each of the four hypotheses. The results indicate that the active learning style has better performance over the four percentage levels, 100% of which show large effect sizes.

Organic Chemistry II

The Organic II flipped classroom had 34 students from Spring 2015 to Spring 2017. The lecture style semesters were Spring 2011–Spring 2013 with a total of 26 students. The overall GPA for all students was higher in Organic II than that in Organic I (Table 3).

The results from the Mann–Whitney U test at Table 4 indicate that none of the four performance levels are statistically different between the two teaching methods. Fig. 7 reveals that students' performance at each of the four levels are very close between students from both teaching methods.

Table 3. Student Population Organic II.

	Flipped	Lecture Style
Number of students	34	26
Gender M (%)	24	42
Gender F (%)	76	58
GPA	3.15	3.08

Table 4. Mann–Whitney U Test for Organic II.[a]

	Percentage A	B or Better	C or Better	DFW
Mann–Whitney U	4.500	3.000	3.000	3.000
Wilcoxon W	14.500	9.000	9.000	13.000
Z	−0.540	−1.061	−1.061	−1.061
Asymp. sig. (2-tailed)	0.589	0.289	0.289	0.289
Exact sig. [2*(1-tailed sig.)]	0.629[b]	0.400[b]	0.400[b]	0.400[b]

Notes: [a]Grouping variable: teaching method.
[b]Not corrected for ties.

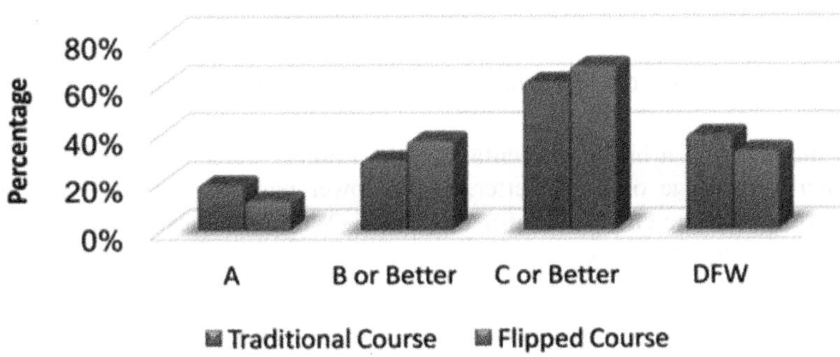

Fig. 7. Organic Chemistry II Final Grade Comparison.

Comparative Analysis of Students' Exam Results Between Traditional Method & Flipped Classroom Method

Organic Chemistry I

Students' performance on each of the six exams (Exam 1 through Exam 5, and the Final Exam) was analyzed. The questions on each of the six exams for both teaching styles covered the same content. A test bank was used to generate the questions for each exam.

The Cronbach's alpha reliability is conducted based on the six items to determine how closely related this set of items are as a group. The Cronbach's alpha is a measure of internal consistency, that is, how closely related a set of items are as a group. The Cronbach's alpha reliability for the six exams is 0.829, representing a good internal consistency. Thus, the six exams could be discussed as a whole. The combination of the six exams are recoded into a one new variable, Organic I. The independent two sample T-test was conducted based on the new variable Organic I between the two different teaching methods. The results from Table 5 indicate that the students' performance from Organic I is not significantly different between the two teaching methods. Other details indicate that the results through *Exam 1* to *Exam 5* are not significantly different, except *Exam 4*, with Sig value slightly less than level of significance value 0.05. However, the results from *Final* exam are statistically different between the two teaching methods.

Further details from Table 6 provide information about which teaching method leads to better performance. Overall, in Exam Average, for Organic I, students from the flipped teaching method have slightly better performance, even though not significant. In detail, students from the traditional teaching method have slightly higher scores for some of the exams, including *Exam 2, Exam 3, Exam 4*, and *Exam 5*, but have slightly lower scores for Exam 1. However, the final exam scores are significantly different between the students from the two different teaching methods, with the Sig value (0.018) being less than the alpha value (0.05). Interestingly, details from Table 6 indicate that students from the flipped classroom teaching method show a significantly higher mean score.

Fig. 8 shows that there was not much difference in performance on the 50-minute exams except for *Exam 4* with a marginal significance. However, the result for the *Final* exam is contradictory to that of the individual exams. Students in the flipped classroom method scored higher on the *Final* exam than students in the traditional teaching method.

Organic Chemistry II

For the Organic II course, students' performance on each of the five exams (Exam 1 through Exam 5) was analyzed. The final exam for Organic II was excluded because it was a standardized exam that covers topics from both semesters of Organic Chemistry. The Cronbach's alpha reliability for the five exams are 0.839, representing a good internal consistency. The combination of the five exams were recoded into one new variable, Organic II. The independent two-sample T-test was conducted based on the new variable Organic II between the two teaching methods. The results from Table 7 indicate that students' performance regarding Organic II is not statistically different. More details from Table 7 indicate that the students' performance regarding *Exam 5* is marginally statistically significant. Further details from Table 8 indicate that students from the flipped classroom teaching method show slightly higher mean scores to most of the exams. Fig. 9 presents a visual result that students from the flipped teaching methods perform better to most of the exams with regard to Organic II, which is different from the results in regard to the course Organic I, where students from the traditional teaching method perform better on most of the 50-minute exams.

Table 5. Independent Two-sample T-test for Organic I.

		Levene's Test for Equality of Variances		t-test for Equality of Means						
		F	Sig.	t	df	Sig. (2-tailed)	Mean Difference	Std. Error Difference	95% Confidence Interval of the Difference	
									Lower	Upper
Exam 1	EVA*	0.338	0.563	−0.270	73	0.788	−1.5665	5.8097	−13.1453	10.0123
	EVNA			−0.267	64.264	0.790	−1.5665	5.8699	−13.2921	10.1591
Exam 2	EVA	0.155	0.695	0.296	72	0.768	1.4197	4.8012	−8.1513	10.9906
	EVNA			0.294	63.552	0.770	1.4197	4.8262	−8.2231	11.0624
Exam 3	EVA	0.076	0.784	0.667	71	0.507	3.5623	5.3419	−7.0892	14.2139
	EVNA			0.661	60.629	0.511	3.5623	5.3878	−7.2125	14.3372
Exam 4	EVA	0.014	0.908	2.066	68	0.043	11.8972	5.7597	0.4039	23.3905
	EVNA			2.046	62.879	0.045	11.8972	5.8162	0.2740	23.5204
Exam 5	EVA	0.038	0.847	0.899	64	0.372	5.0333	5.5996	−6.1531	16.2197
	EVNA			0.900	62.138	0.372	5.0333	5.5931	−6.1467	16.2134
Final	EVA	0.337	0.564	−2.426	67	0.018	−25.8559	10.6573	−47.1278	−4.5839
	EVNA			−2.411	62.495	0.019	−25.8559	10.7256	−47.2927	−4.4190
Organic I	EVA	0.039	0.845	−0.503	73	0.617	−2.38945	4.75472	−11.86559	7.08668
	EVNA			−0.501	65.920	0.618	−2.38945	4.77407	−11.92141	7.14251

Notes: *EVA – equal variances assumed; EVNA – equal variances not assumed.

Table 6. Group Statistics for Organic I.

	Teaching Method	N	Mean	Std. Deviation	Std. Error Mean
Exam 1	Traditional	32	58.922	25.8682	4.5729
	Flipped	43	60.488	24.1334	3.6803
Exam 2	Traditional	3[illegible]	57.887	20.7520	3.7272
	Flipped	43	56.467	20.1053	3.0660
Exam 3	Traditional	30	62.260	23.0811	4.2140
	Flipped	43	58.698	22.0141	3.3571
Exam 4	Traditional	32	66.516	25.4433	4.4978
	Flipped	38	54.618	22.7318	3.6876
Exam 5	Traditional	30	58.267	22.4974	4.1074
	Flipped	36	53.233	22.7780	3.7963
Final	Traditional	3[illegible]	115.097	45.5297	8.1774
	Flipped	38	140.953	42.7839	6.9405
Organic I	Traditional	32	58.3732	20.68489	3.65661
	Flipped	43	60.7627	20.12717	3.06936

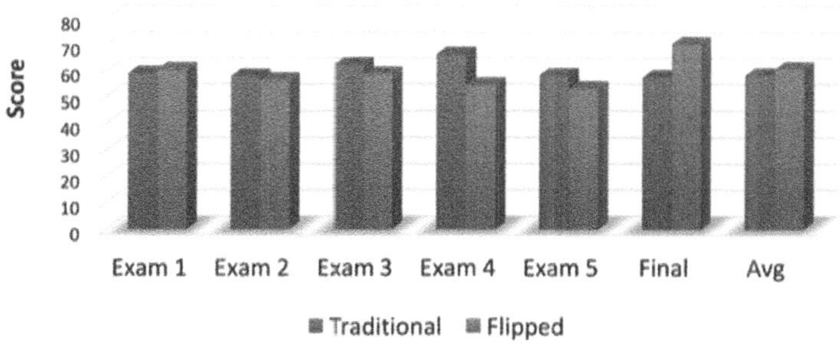

Fig. 8. Organic I Exam Grade Comparison.

Students' Activity Time of Flipped Classroom Teaching Method for Organic I and II

An important component of the flipped classroom is the pre-class preparation. Blackboard LMS has the capability to download the activity statistics for each class. It provides an overall summary of user activity that includes user activity for all areas of the course, as well as activity dates, times, and days of the week. The activity times for students enrolled in the course were analyzed.

In general, the activity times for Organic I are much higher than those for Organic II in 2015. However, the differences between Organic I and Organic II decrease gradually, while the activity times for Organic I are even lower than those for Organic II in 2017. The activity times peaked on Tuesday for Organic I, and the activity times peaked on Thursday for Organic II (Fig. 10). Further

Table 7. Independent Two-sample T-test for Organic II.

		Levene's Test for Equality of Variances		t-test for Equality of Means						
		F	Sig.	t	df	Sig. (2-tailed)	Mean Difference	Std. Error Difference	95% Confidence Interval of the Difference	
									Lower	Upper
Exam 1	EVA	1.299	0.260	−0.083	51	0.934	−0.4971	5.9733	−12.4890	11.4948
	EVNA			−0.082	45.361	0.935	−0.4971	6.0674	−12.7147	11.7205
Exam 2	EVA	1.211	0.276	−0.884	50	0.381	−6.5545	7.4114	−21.4407	8.3316
	EVNA			−0.911	49.268	0.366	−6.5545	7.1911	−21.0036	7.8945
Exam 3	EVA	0.664	0.419	0.799	52	0.428	5.4750	6.8537	−8.2780	19.2280
	EVNA			0.787	46.012	0.436	5.4750	6.9597	−8.5341	19.4841
Exam 4	EVA	0.006	0.939	−0.888	49	0.379	−4.8925	5.5066	−15.9583	6.1734
	EVNA			−0.893	46.114	0.377	−4.8925	5.4812	−15.9248	6.1398
Exam 5	EVA	4.632	0.037	−1.494	44	0.142	−10.6806	7.1481	−25.0866	3.7255
	EVNA			−1.392	28.269	0.175	−10.6806	7.6702	−26.3855	5.0244
Organic II	EVA	0.270	0.606	−0.725	52	0.471	−4.0013	5.5165	−15.0709	7.0683
	EVNA			−0.725	49.390	0.472	−4.0013	5.5177	−15.0873	7.0846

Notes: *EVA – equal variances assumed; EVNA – equal variances not assumed.

Table 8. Group Statistics for Organic II.

	Teaching Method	N	Mean	Std. Deviation	Std. Error Mean
Exam 1	Traditional	24	57.417	23.5255	4.8021
	Flipped	29	57.914	19.9705	3.7084
Exam 2	Traditional	22	58.795	23.3726	4.9831
	Flipped	30	65.350	28.3978	5.1847
Exam 3	Traditional	24	56.208	26.8789	5.4866
	Flipped	30	50.733	23.4531	4.2819
Exam 4	Traditional	22	62.045	19.1074	4.0737
	Flipped	29	66.938	19.7484	3.6672
Exam 5	Traditional	18	60.444	28.1743	6.6407
	Flipped	28	71.125	20.3104	3.8383
Organic II	Traditional	24	57.267	20.1646	4.1161
	Flipped	30	61.268	20.1264	3.6746

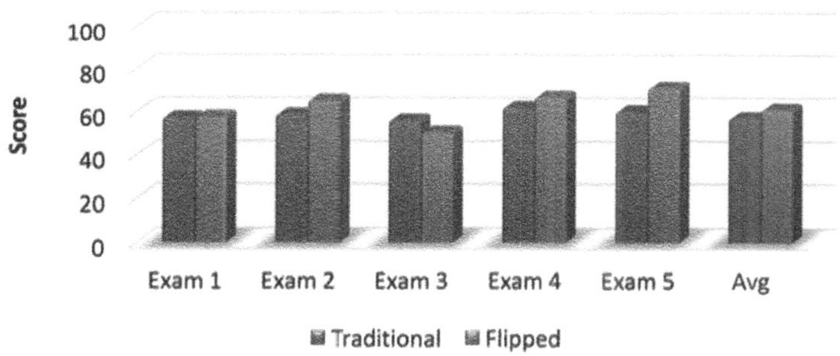

Fig. 9. Organic Chemistry II Final Grade Comparison.

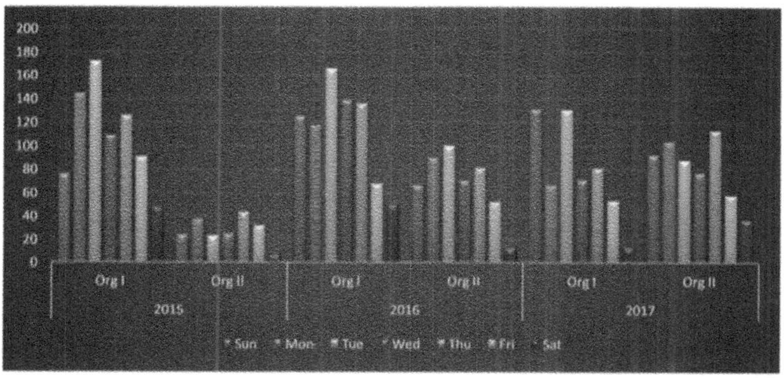

Fig. 10. Activity Time for Organic I & Organic II by Day.

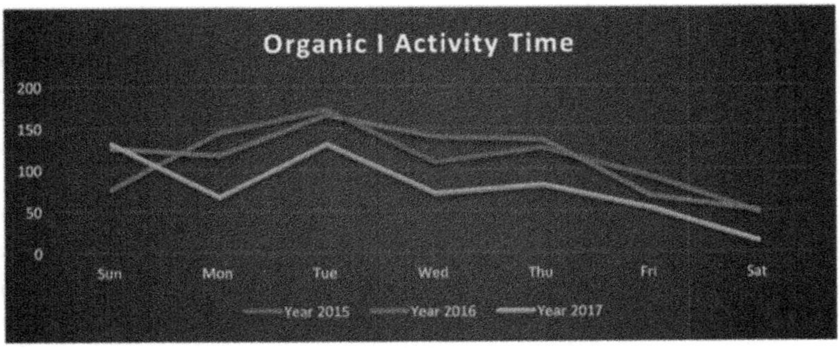

Fig. 11. Organic I Activity Time.

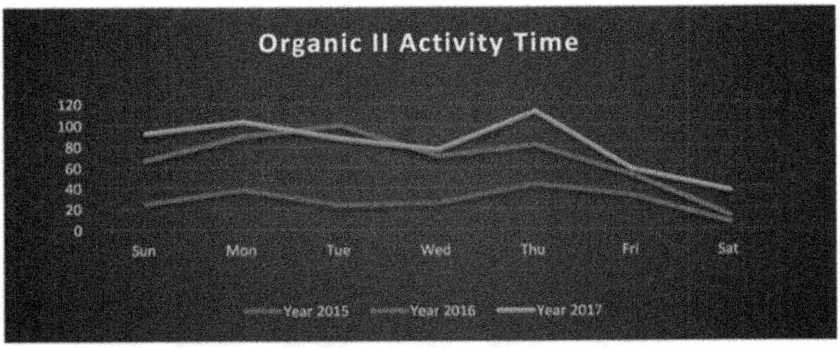

Fig. 12. Organic II Activity Time.

discussion about what happened on those days, such as whether there were major assignments or exams on the following day, is worth further exploration.

From Fig. 11 we can see that the activity time for Organic I decreased from 2015 to 2017. In the case for Organic II, an increase in activity time from 2015 to 2017 was observed (Fig. 12).

CONCLUSIONS AND FUTURE WORK

The focus of this project was how to transition effectively from a traditional style of teaching to a flipped classroom. The results were on student grade performance and DFW rate. To conclude, the flipped teaching method made a much bigger positive impact for the Organic I course rather than the Organic II course. Regarding Organic I, students from the flipped teaching method have a much higher percentage of A, a higher percentage of B or Better, and a higher percentage of C or Better. The DFW rate was lower for the flipped classroom. Interestingly, students from the traditional teaching method have slightly higher exam scores than those from flipped teaching method except for *Exam* 1. However, students from the flipped teaching method show significantly higher scores for the *Final* exam, which occupies a bigger portion in students' academic performance than the 50-minute exams. Further discussions are required as to why the flipped teaching method has the reverse impact on 50-minute exams and the final exam, even though the difference during the 50-minute exams were very small while the difference during the final exam is significantly different.

Regarding Organic II, the flipped teaching method led to a different result. Students from the flipped teaching method have a lower percentage of A, and a less significant decrease in DFW rate. When referring to the five exams, none of the five exams were significantly different between the two teaching methods. Students from the flipped teaching method have slightly higher scores through *Exam 1* to *Exam 5* except *Exam 3*. This is particularly interesting because the overall GPA of students enrolled in Organic II is higher than in Organic I. The data also demonstrated a lower Blackboard Learn activity time for the Organic II course. This indicates that these students are spending less time utilizing the resources such as videos, exercises, and pre-class quizzes. This reduced activity time may contribute to the results in the performance for Organic II. Another explanation is the different type of content in Organic I vs Organic II. Organic I content consists of resonance, electron flow, stereochemistry, and reaction transformations. These concepts are considered the fundamentals that are applied throughout Organic I and Organic II. The major difference is the number of synthetic reactions that the student must learn. In Organic I, students are introduced to 40 reactions compared to 108 in Organic II. Also, students are required to carry over the reactions learned in Organic I. As such, Organic II requires students to rely on memorization more than critical thinking. Active learning strategies help students with deeper learning such as critical thinking, which is used extensively for Organic I.

Future research in understanding the impact of this work must include surveys to gauge the students' perception of the flipped classroom. In addition, a survey specific to Organic II would include questions geared toward understanding why students are spending less time in the pre-class activities. This would allow the instructor to adjust the activities to meet the needs of the students.

REFLECTIONS AND RECOMMENDATIONS

My initial motive for this transition was the standard "*increased grades and decreased DFW rate*." However, I must admit that I have learned during the process that increased As and Bs and decreased DFW is not the only value gained in the implementation of a flipped classroom. After 12 years of dazed unengaged faces on my best days and the occasional snoring on my worst days, I have finally reached my academic nirvana (at least for two out of three class times a week). If you walk past my classroom, you will witness students engaged in scientific conversations and debates, self and classmate critiques, right and very wrong answers on the board, students teaching each other, students pushing through frustrations, students working on different problems, students taking walks to gain clarity [...] *strategically designed organized chaos*. I state with 100% certainty that the countless hours dedicated to the transition was worth the effort.

However, I would be remiss to exclude recommendations based on lessons learned during my journey. (1) The entire course content does not have to be developed prior to implementation of the flipped classroom. Faculty can start by designing and implementing in-class worksheets and active learning instruments. If there are other faculty teaching the same course, the worksheet design could be divided to reduce the workload. While pre-class assignments are a critical component, students can be directed to a suitable academic website (such as Kahn Academy) to watch videos and assign pre-class reading. (2) Regarding the implementation of board work, students have different levels of confidence and content knowledge at the beginning of the course. This leads to student resistance and anxiousness for some students. Over the course of several years, it is best to start with students completing problems as a group. One student from the group goes to the board to write the group's answer. This provides the student with support from their group members. As the course progresses, students work problems individually on the worksheets and then transfer the answers to the board. Usually by midterm, students are able to start to work the problem on the board. It is essential that students arrive at this level to assist the faculty in understanding the students' understanding of concepts and exactly where in the mechanism/problem the student is struggling and needs reinforcement. (3) Let the students know that you are trying something new together. Solicit their feedback on worksheets, videos quizzes etc. and listen carefully. I found that it invites the students into a critical part of their learning process and it provides us with insight into the mind of today's college student. I still practice this when I introduce a new activity or video to the class. The open flow of communication between the student and faculty is a valuable part of student learning.

ACKNOWLEDGMENTS

This research was supported in part by NIH NIGMS RISE – 5R25GM058904. The authors would like to thank Dr Richard Williams in the Department of Chemistry at Morgan State University.

REFERENCES

FLN. (2014). The Four Pillars of F-L-I-P™. Retrieved from https://flippedlearning.org/

Bergmann, J., & Sams, A. (2012). *Flip your classroom: Reach every student in every class every day*. Washington, DC: International Society for Technology in Education.

Cohen, J. (1988). *Statistical power analysis for the behavioral sciences*. Hilsdale. NJ: Lawrence Earlbaum Associates, 2.

Freeman, S., Eddy, S. L., McDonough, M., Smith, M. K., Okoroafor, N., Jordt, H., & Wenderoth, M. P. (2014). Active learning increases student performance in science, engineering, and mathematics. *Proceedings of the National Academy of Sciences*, *111*(23), 8410–8415.

Haak, D. C., HilleRisLambers, J., Pitre, E., & Freeman, S. (2011). Increased structure and active learning reduce the achievement gap in introductory biology. *Science*, *332*(6034), 1213–1216.

Hibbard, L., Sung, S., & Wells, B. (2015). Examining the effectiveness of a semi-self-paced flipped learning format in a college general chemistry sequence. *Journal of Chemical Education*, *93*(1), 24–30. doi:10.1021/acs.jchemed.5b00592

Jett, C. C. (2013). HBCUs propel African American male mathematics majors. *Journal of African American Studies*, *17*(2), 189–205.

Laws, P., Sokoloff, D., & Thornton, R. (1999). Promoting active learning using the results of physics education research. *UniServe Science News*, *13*, 14–19.

Lorenzo, M., Crouch, C. H., & Mazur, E. (2006). Reducing the gender gap in the physics classroom. *American Journal of Physics*, *74*(2), 118–122.

Minne, N. (1929). Molecular models in organic chemistry. *Journal of Chemical Education*, *6*(11), 1984.

Owens, E. W., Shelton, A. J., Bloom, C. M., & Cavil, J. K. (2012). The significance of HBCUs to the production of STEM graduates: Answering the call. *The Journal of Educational Foundations*, *26*(3/4), 33.

Science and Engineering Indicators Report. (2016). Washington, DC: National Science Foundation. Retrieved from https://www.nsf.gov/statistics/seind/.

Seery, M. K. (2015). Flipped learning in higher education chemistry: emerging trends and potential directions. *Chemistry Education Research and Practice*, *16*(4), 758–768.

Stull, A. T., Gainer, M., Padalkar, S., & Hegarty, M. (2016). Promoting representational competence with molecular models in organic chemistry. *Journal of Chemical Education*, *93*(6), 994–1001.

U.S. Department of Education. Institute of Education Sciences, N. C. f. E. S. Retrieved from https://nces.ed.gov/.

Wu, H. K., & Shah, P. (2004). Exploring visuospatial thinking in chemistry learning. *Science education*, *88*(3), 465–492.

INDEX